Relay Autotuning for Identification and Control

M. Chidambaram

Indian Institute of Technology, Chennai

Vivek Sathe

Dr. Babasaheb Ambedkar Technological University, Maharashtra

CAMBRIDGE
UNIVERSITY PRESS

CAMBRIDGE
UNIVERSITY PRESS

Cambridge House, 4381/4 Ansari Road, Daryaganj, Delhi 110002, India

Cambridge University Press is part of the University of Cambridge.

It furthers the University's mission by disseminating knowledge in the pursuit of education, learning and research at the highest international levels of excellence.

www.cambridge.org
Information on this title: www.cambridge.org/9781107058712

First published 2014

Printed in India by Shree Maitrey Printech Pvt. Ltd., Noida

A catalogue record for this publication is available from the British Library

Library of Congress Cataloging-in-Publication Data
Chidambaram, M.
Relay autotuning for identification and control / M. Chidambaram, Vivek Sathe.
pages cm
Summary: "Provides a simple method of designing P/PI controllers for series and parallel cascade control schemes for effective industrial operations"–Provided by publisher.
Includes bibliographical references and index.
ISBN 978-1-107-05871-2 (hardback)
1. Relay control systems. 2. Self-tuning controllers. 3. Electric relays–Automatic control.
I. Sathe, Vivek. II. Title.
TJ217.C4765 2014
629.8'36–dc23

ISBN 978-1-107-05871-2 Hardback

Contents

List of Figures

Figures in Appendices

List of Tables

Tables in Appendices

Acknowledgements

The authors owe a debt of gratitude to many friends, colleagues and students who helped us write this book. First of all, we gratefully acknowledge the contributions made by the graduate students – K. Sinivasan, V. Ramakrishnan, D.S. Ganesh, E. Sivakumar, D.S. Reddy, M.V. Sadasivarao – as part of their project works. We are also thankful for the support received from the Department of Chemical Engineering and the Continuing Education Centre, Indian Institute of Technology, Madras. The second author expresses his sincere appreciation to his son C. Srihari Prasath for his patience and understanding while writing the manuscript.

It would also be appropriate to express our regards to R. Mohan of Sigma Publishing who drew the figures of the book. We would be remiss if we did not thank Cambridge University Press and our editors involved in the process for their efficient professionalism and successfully bringing out the book.

Preface

To design proportional plus integral (PI)/PID controllers, the ultimate values of the controller gain (k_u) and frequency of oscillation (ω_u) should be known. The conventional Ziegler and Nichols continuous cycling method requires a large number of experiments to calculate these values. Åström and Hägglund (1984) suggested the use of ideal relay to generate closed loop oscillations. The ultimate gain and ultimate frequency can be found in a single-shot experiment. However, the method is still approximate because of the use of the principal harmonics approximation. Li et al. (1991) reported that for an open loop, stable first order plus time delay (FOPTD) system, an error of -18% to 27% is obtained in the calculation of k_u. Yu (1999) suggested a saturation feedback test to get better results for the ultimate gain and frequency. However, Yu (1999) did not report any result for large values of delay-to-time constant ratio.

An SOPTD model can incorporate higher order process dynamics better than an FOPTD model. The controller designed on the basis of the SOPTD model gives a better closed loop response than the one designed on an FOPTD model. It is better to have an SOPTD model with equal time constants since only three parameters are to be identified. Li et al. (1991) showed that the conventional analysis of the relay autotune method for an SOPTD model with equal time constants gives -11% to 27% error in the calculation of k_u. In identifying the transfer function models for unstable systems, it is necessary to estimate 5 parameters to describe such systems. There is no method available to estimate these parameters from a single relay test. Cascade control system is a multi-loop control scheme commonly used in the chemical process industries. It should be noted that there are two ways of relay tuning cascade control systems, namely, simultaneous tuning and sequential tuning procedures. However, there is a need to consider higher order harmonics in the analysis of relay tuning of such systems to obtain improved controllers settings.

The authors have organized the book in the following manner:

Chapter 1 gives an overview of the reported works on relay tuning of single loop control system, cascade controllers, estimation of model parameters and design of PI/PID controllers. Chapter 2 gives a method of considering higher order harmonics in the analysis of symmetric relay tuning method and analytical solutions for model parameters using a single asymmetrical relay test.

Chapter 3 gives a method of considering higher order harmonics in the analysis of series cascade control systems by the sequential tuning method. The method is also extended to parallel cascade control systems. Chapter 4 analyzes the simultaneous relay tuning of series and parallel cascade control systems.

Chapter 5 gives a simple method to design PID controllers for a series cascade control system. FOPTD models are assumed for both the inner and outer loop sub-systems. The method is based on matching the coefficients of the corresponding powers of s and s^2 in the numerator to α_1 and α_2 times that in the denominator of the closed loop transfer function model for a servo problem. The method is first applied to design a proportional (P) controller for the inner loop and then to design a proportional plus integral (PI) controller for the outer loop. The performance of the proposed controllers is evaluated for FOPTD models of the inner loop and the outer loop process transfer functions. This method is also extended to parallel cascade control systems.

Chapter 6 proposes two methods to improve the accuracy of the saturation relay method for FOPTD systems with a large ratio of time delay to time constant. Chapter 7 proposes a method to identify all the three parameters of a first order plus time delay (FOPTD) model using a single symmetrical relay test.

Chapter 8 gives a method to identify all the three parameters of an unstable FOPTD system using a single symmetric relay feedback test. It also proposes a method by incorporating higher order harmonics to explain the error in the calculation of k_u. Another method is proposed to estimate all the parameters of an unstable FOPTD system.

In Chapter 9, using a single symmetric relay feedback test, a method is proposed to identify all the three parameters of a stable second order plus time delay (SOPTD) model with equal time constants $[k_p \exp(-Ds)/(\tau s+1)^2]$. It provides a method to improve the accuracy in the k_u calculation by incorporating the higher order harmonics.

In Chapter 10, using a single asymmetric relay feedback test, a method is proposed to identify all the four parameters of a stable second order plus time delay (SOPTD) model $k_p \exp(-Ds)/[(\tau_1 s+1)\ (\tau_2 s+1)]$. In Chapter 11, a method is proposed to identify all the 5 parameters of an SOPTD system with a zero, $G(s) = k_p\ (\tau_1 s+1) \exp(-Ds)/[(\tau_2 s+1)(\tau_3 s-1)]$, using an asymmetric relay test.

In Chapter 12, the asymmetrical relay tuning method is extended to multivariable FOPTD systems. In Chapter 13, the asymmetrical relay tuning method is extended to multivariable SOPTD systems. Chapter 14 proposes a simple method of designing centralized multivariable PID controllers for multivariable systems. In Chapter 15, the use of detuned Ziegler–Nichols PID settings as initial guess values is suggested for the design of multivariable controllers by an optimization method using the genetic algorithm to reduce the number of iterations significantly. Chapter 16 summarizes the results and conclusions of the work.

There are three appendices.

Appendix A gives the *Simulink* block diagrams used in the present work.

Appendix B gives the *Matlab* programs used in Chapter 3, Chapter 5 and Chapter 6.

Appendix C gives an improved relay tuning method for integrating plus FOPTD systems.

1 Introduction

Identification of transfer function models of a system is required for an improved tuning of controllers. Several methods have been reported in the literature for identification of transfer function models with two, three and four parameters (pure delay system, first order plus time delay (FOPTD), second order plus time delay (SOPTD), etc.) using relay feedback approach. In this section, the basics of conventional relay feedback method and modifications in the original autotuning method are reviewed for single-input single-output systems. Excellent reviews on relay tuning methods are given by Yu (1999, 2006), Hang et al. (2002) and Wang et al. (2003). Methods of designing PI/PID controllers based on the transfer function models are also briefly reviewed.

1.1 Relay Feedback Method

Åström and Hägglund (1984) suggested the use of an ideal (on–off) relay (Fig. 1.1) to generate a sustained oscillation in the closed loop. For positive gain process, on–off relay is defined by $u = u_{max}$ if $e \geq 0$, and $u = u_{min}$, if $e < 0$. For negative gain processes, on–off relay is defined by $u = u_{min}$, if $e \geq 0$, and $u = u_{max}$ if $e < 0$. Amplitude (a) and period of oscillation (p_u) are noted from the sustained oscillation. This is a closed loop method for identification of transfer function models. The method is based on the observation that when an open loop output lags the input by π radians, the closed loop system may oscillate (Fig. 1.2) with a period P_u. The ultimate gain (K_u) and ultimate frequency (ω_u) can be calculated from the oscillatory response (Aström and Hägglund, 1984) from

$$k_u = 4h / (\pi \text{a}) \tag{1.1}$$

$$\omega_u = 2\pi / P_u \tag{1.2}$$

where 'h' is magnitude of the relay. Advantages of this method are: it is simple, time efficient and a closed loop method.

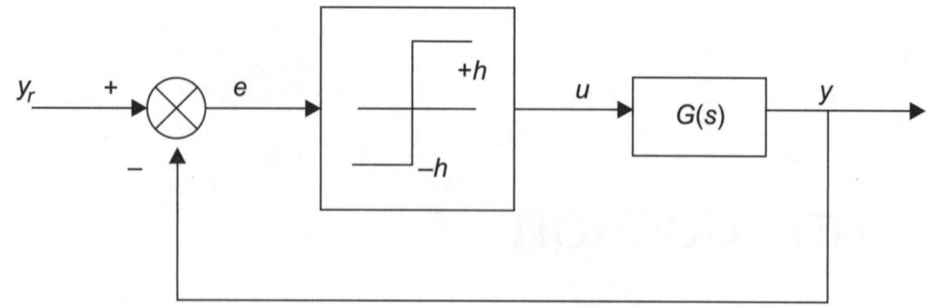

Fig. 1.1 *Block diagram for symmetric relay feedback system*

Let input to the relay be given by a sin(ωt). Then output of the relay will be a square wave, as shown in Fig. 1.2. We can approximate the square wave by a Fourier series. In general, any function can be expressed as

$$u(t) = A_0 + \sum_{n=1}^{\infty} B_n \sin(n\omega t) + \sum_{n=1}^{\infty} A_n \cos(n\omega t) \tag{1.3}$$

$$B_n = (1/\pi) \int_0^{2\pi} u(t) \sin(n\omega t) dt; \quad A_n = (1/\pi) \int_0^{2\pi} u(t) \cos(n\omega t) dt \tag{1.4}$$

Since the symmetric square wave is an odd function, $A_n = 0$ ($n = 0,1,2,\ldots$), $B_n = 0$ ($n = 0,2,4,\ldots$). Eq.(1.4) becomes

$$B_n = (2/\pi) \int_0^{\pi} h \sin(\omega t) dt \tag{1.5}$$

$$= 4h/(n\pi) \tag{1.6}$$

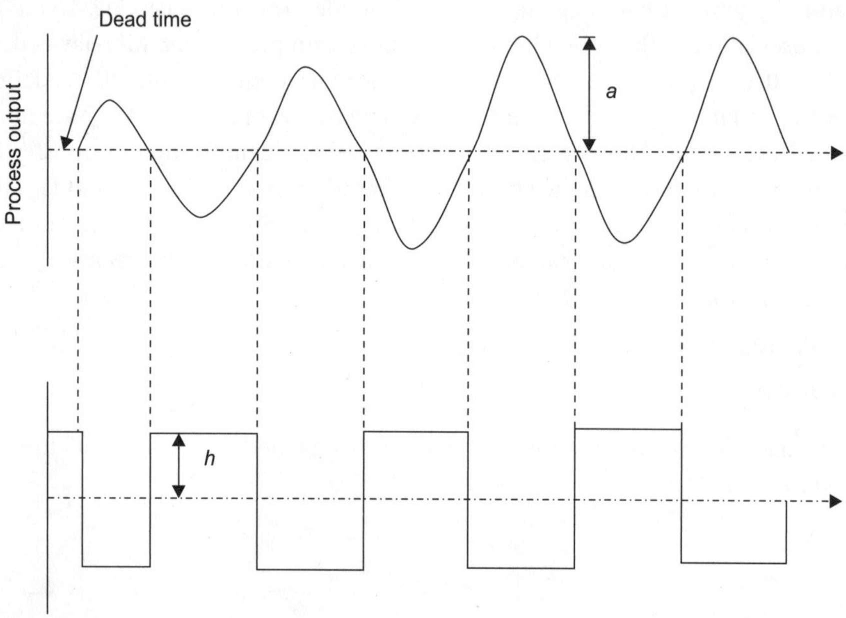

Fig. 1.2 *Response of a relay feedback system*

Hence, $u(t)$ from Eq.(1.3) can be written as

$$u(t) = (4h/\pi)\ [\sin(\omega t) + (1/3)\sin(3\omega t) + (1/5)\sin(5\omega t) + \cdots] \tag{1.7}$$

In other words, response of the relay to a sinusoidal input is a periodic signal that contains a fundamental component as well as higher order harmonics. The coefficients of higher order harmonic functions are smaller than those of the fundamental one and the higher harmonics continue to decrease as the order of the harmonics increases. Assume further that the process dynamics is of low pass characteristics and that contribution from the first harmonics dominates in the output. Error signal has the following amplitude

$$a = (4h / \pi)\ |G(j\omega t)| \tag{1.8}$$

Condition for the sustained oscillation is given by phase angle criterion

$$< G(j\omega t) = -\pi \tag{1.9}$$

and

$$k_u = 1 / |G(j\omega_u)| \tag{1.10}$$

$$= 4h / (\pi a) \tag{1.11}$$

where k_u can be regarded as equivalent gain of the relay for transmission of the sinusoidal signal with amplitude a. Frequency of the limit cycle is thus automatically adjusted to the frequency ω_u, at which the open loop process has a phase lag of π radians. Physically, k_u is the ultimate gain that brings the system to stability boundary under the proportional controller action. An experiment with the relay feedback will thus give the period and amplitude of loop transfer function of the process at frequency at which the process lag is π radians. Notice that an input signal whose energy content is concentrated at ω_u is generated automatically in the relay feedback experiment. In Fig. 1.1, it is assumed that the process has a positive steady-state gain and, accordingly, the relay is defined as $u = u_{max}$ if $e \geq 0$ and otherwise if $e < 0$. If the process has a negative steady-state gain, then the relay is defined as $u = u_{max}$ if $e \leq 0$ and $u = u_{min}$ if $e > 0$.

1.2 Identification by Symmetric Relay Feedback Method

Luyben (1987) used the autotuning method to identify FOPTD and SOPTD models when the steady-state gain is known *a priori*. From the initial response, value of time delay (D) is found out. Using values of k_u and ω_u in the phase angle and amplitude criteria for first order plus time delay model, values of time constant (τ) and process gain (k_p) are estimated:

$$-\tan^{-1}(\tau\,\omega_u) - (D\,\omega_u) = -\pi \tag{1.12}$$

$$k_p\,k_u / (1+\tau^2\omega_u^{\,2})^{0.5} = 1 \tag{1.13}$$

The method proposed by Luyben (1987) for obtaining simple transfer function models from a single autotune test requires the steady-state gain to be known *a priori*. This limits its usefulness for identification.

Li et al. (1991) presented a modified procedure that did not require the knowledge of steady-state gain. The method used two autotune tests. First is a normal test; second is run with an additional dead time so that the phase angle is shifted by about 45° and a smaller ultimate frequency is obtained. Then the least squares method is used to determine unknown parameters: two time constants and steady-state gain. The method is demonstrated on several simulated processes and successfully tested on an experimental heat-exchanger process. Autotune testing of both continuous and sampled-data systems is considered.

Majhi and Atherton (1999) proposed an improved method for calculating the transfer function model for the processes by relay tuning. In this method, output response is aligned with input response by shifting to left in order to note the time delay. Sets of analytical expressions are derived from a single symmetric/asymmetric relay feedback test. Using the derived expressions, exact parameters of the open loop stable and unstable FOPTD and SOPTD transfer function models are obtained.

Thyagarajan and Yu (2003) conducted relay feedback tests on processes with different orders and a wide range of dead time to time constant ratios. On the basis of the shape of response from the relay feedback test, these processes are broadly classified into three major categories. Procedures are given to find parameters for the corresponding model structures and then different tuning rules are employed to find appropriate PI controller settings. The procedure is tested against linear systems with and without noise. Several simulation results are given to show the effect of including the shape factor. Improvement is achieved from the conventional relay feedback test and no additional testing is required.

1.3 Identification using Asymmetrical Relay

Huang et al. (1996a, b) proposed a new method that uses data from a relay feedback loop to derive the SOPTD transfer function models. The relay feedback test is followed by an open loop test. Through one simple run of the proposed test, all the parameters in an SOPTD model can be identified with sufficient accuracy. The resulting SOPTD models have a close step response behavior to that of the process. The proposed method also works well in presence of a constant unknown disturbance. When SOPTD processes have a large dead time, the simulation results show that the proposed method works well with slight modifications. The disadvantage of this method is that it requires two tests – one asymmetric relay test followed by an open loop test. Also, the method cannot be applied to open loop unstable processes.

Shen et al. (1996a) proposed the input-biased relay feedback test (Fig. 1.3) for system identification. In addition to the critical point (k_u and ω_u), the steady-state gain can also be found out in a single relay test. Function analysis of the biased relay feedback is described and extensions to system identification are proposed. Potential problems associated with biased relay are given and methods to avoid them are suggested. The Autotune Variation (ATV) method is proposed for the identification of parametric transfer functions. Two linear examples are used to test the effectiveness of the proposed method. Results show that biased relay gives a satisfactory performance.

Shen et al. (1996b) proposed a saturation relay feedback test (Fig. 1.4) system to improve the accuracy of the estimates of ultimate gain and ultimate frequency. The study shows that too small a slope in the saturation relay may fail to generate a limit cycle and, subsequently, lead to a failed experiment. A procedure to overcome the trade-off between the accuracy and failure is proposed. The results show that a significant improvement can be achieved. The method first uses a conventional relay feedback and, from the test, slope of the saturation relay is calculated. Then the saturation relay test is carried out.

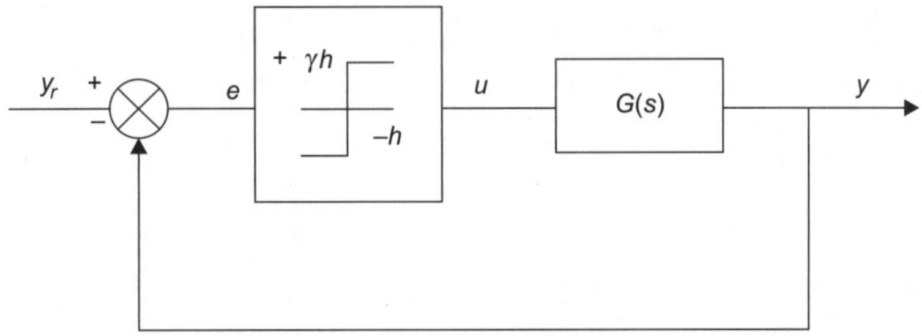

Fig. 1.3 *Block diagram for asymmetric relay feedback system*

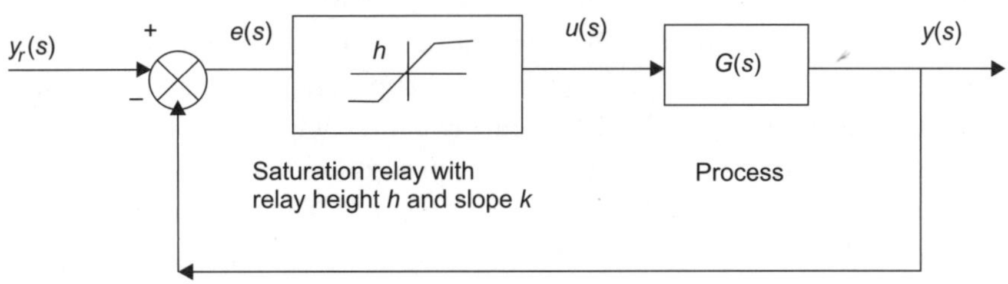

Fig. 1.4 *Block diagram for saturation relay feedback system*

Methods to improve the autotune identification methods under load disturbances are reported by Hang et al. (1993), Shen et al. (1996c), Park et al. (1997), Wang and Hwang (2003) and Leva (2005).

1.4 Identification of Unstable Processes

Kavdia and Chidambaram (1996) applied the relay tuning method for tuning controllers online for unstable processes. Response of the closed loop system with the controller settings of the identified model is compared with that of the actual system's controller parameters.

Huang and Chen (1999) proposed an autotuning procedure for PID controllers for a more general class of unstable processes that have second order dynamics. The second order dynamics are represented by models that have both stable and unstable poles together with dead time. A biased relay feedback is used to generate a constant limit cycle for identifying the model. On finishing the identification, simple tuning rules are provided to tune the parameters of PID controllers. Two unstable non-linear processes are also used to illustrate this proposed autotuning method.

Thyagarajan and Yu (2003) proposed a method to identify the approximate UFOPTD model using the shape of the response. The time constant (τ) is calculated using the equation

$$2\exp(t_{a/2}/\tau) - \exp(t_a/\tau) = 1 \tag{1.14}$$

Here t_a is the time required to reach the peak amplitude, $t_{a/2}$ is the time required to reach one half of the peak amplitude. Once τ is known, the other two model parameters (D and k_p) are obtained using the ultimate properties (k_u and ω_u) as

$$D = [-\pi + \tan^{-1}(\tau\omega_u)]/\omega_u \tag{1.15}$$

$$k_p = [1 + (\tau\omega_u)^2]^{0.5}/k_u \tag{1.16}$$

1.5 Autotuning of Cascade Control System

Cascade control system (Fig. 1.5) is a multi-loop control scheme, commonly used in chemical process industries. In cascade control systems, there are two loops: secondary (inner) loop and primary (outer) loop. The effect of disturbances entering the secondary loop is effectively controlled by the cascade control scheme. Manual tuning of cascade controllers is a tedious and time-consuming task. In view of the widespread application of relay tuning, it would be useful to automate the tuning procedure. There are two ways of tuning cascade control systems, namely, simultaneous tuning and sequential tuning procedures. Hang et al. (1994) proposed a sequential relay autotuning of cascade control loops (Fig. 1.6). The sequential tuning procedure involves two steps: The inner loop is tuned before tuning the outer loop. Using a relay for the inner loop and keeping the outer loop open, the inner loop is then tuned. The inner loop controller is designed and set and then the outer loop is kept under relay feedback in order to tune the outer loop controller.

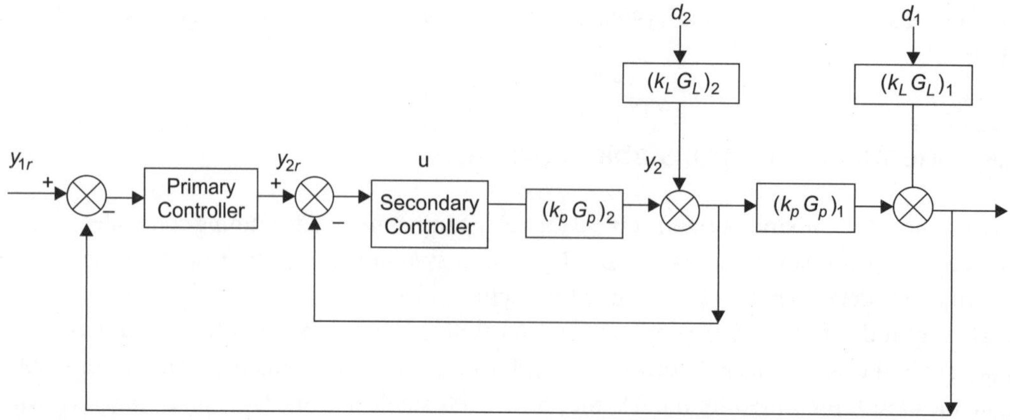

Fig. 1.5 *Series cascade control scheme*

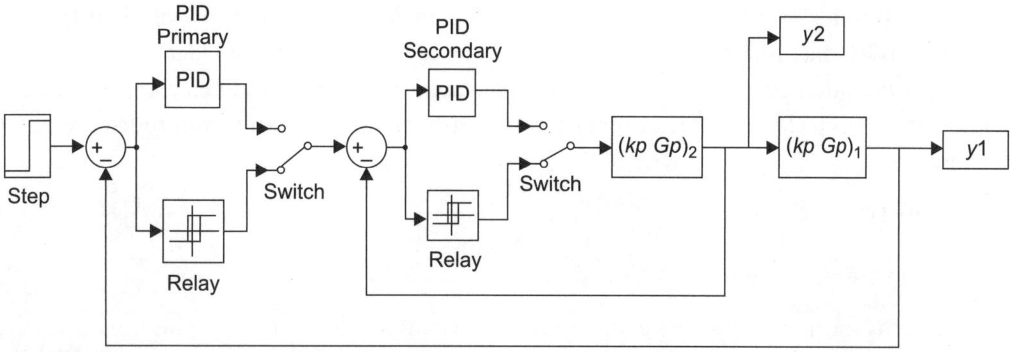

Fig. 1.6 *Relay feedback tuning of cascade control system*

Tan et al. (2000) proposed a new simultaneous online automatic tuning method for cascade control using a relay feedback approach. Departing from the traditional approach towards tuning of cascade control systems, where the secondary and primary loops are tuned in strict sequence, the proposed approach is to carry out the entire tuning process in one experiment. For ease of practical applications, the entire procedure of controller design may be automated and carried out online. A direct controller tuning approach to tune the controllers is proposed here. Robustness analysis of the new cascade control design is further carried out by drawing on existing results for SISO feedback systems. Simulation results are provided to illustrate the applicability and effectiveness of both the online autotuning approach and the new cascade control design.

Saraf et al. (2003) proposed a simultaneous autotuning method for cascade control of unstable CSTR processes. The sequential tuning method is not applicable for unstable systems. The simultaneous autotuning method is also applicable for stable systems. The method involves replacing both the controllers by the relays and carrying out the autotuning procedures to obtain limit cycles. The output of the primary auto relay loop is the input of the secondary auto relay loop. The secondary loop is forced to oscillate at the frequency of the primary loop with two relays inserted in the feedback loop because of the condition $\omega_{c,\,pri} < \omega_{c,\,sec}$. Thus the ultimate gain and ultimate period are first obtained for the primary loop. The primary controller parameters can also be obtained. Then the secondary loop executes relay feedback with the primary controller in place in order to obtain its own ultimate period, and the secondary controller parameters are obtained.

1.6 Relay Tuning of Multivariable System

Luyben (1987) used independent relay feedback method for identification of transfer function matrix of multivariable systems. The steady-state gains of each element of transfer function matrix is obtained from the step response test. Then each loop is subjected to relay feedback, keeping other loops open. From the limit cycle data, the ultimate gain and the ultimate frequencies are noted. The delays are also noted from the initial part of the responses. Then, from the amplitude and phase angle criteria, each element in the same row of plant transfer model is modeled as first order or second order system with dead time. Same procedure is used to get the other rows of transfer function matrix. Process gain and delay are assumed to be known. For each element of transfer function matrix, phase angle and amplitude criteria are given by

$$-\tan^{-1}(\tau\omega_c) - D\omega_c = -\pi \tag{1.17}$$

$$k_p k_{c,u} / (1+\tau^2\omega_c^2)^{0.5} = 1 \tag{1.18}$$

The equations for finding process gain and time constant for the first order plus time delay are given as

$$k_p = \left(1+\tau^2\omega_c^2\right)^{0.5} / k_{c,u} \tag{1.19}$$

$$\tau = \left(1/\omega_c\right)\left[\tan\left(\pi - D\omega_c\right)\right] \tag{1.20}$$

Since, this method is only a partially closed loop test, the system is more susceptible to disturbances and also is not suitable for unstable systems. Also the time delays are found from initial part of the response, which may not be accurate for systems with higher order dynamics.

Friman and Waller (1994) studied autotuning by relay feedback with an emphasis on PID control of TITO systems. They studied relay identification and autotuning in combination with integrator-plus-dead-time and gain-delay models for various systems (SISO and MIMO). In relay feedback for MIMO identification and control, relay is connected over one element in transfer function matrix and all elements in the same column can be identified from measured outputs. For gain plus delay model, the delay and gain are obtained as

$$D = 0.5 P_u \tag{1.21a}$$

$$k_p = (y_{max} - y_{min})/(2h) \tag{1.21b}$$

While, for integrator-plus-dead-time model, model parameters are obtained as

$$D = 0.25 P_u \tag{1.22a}$$

$$k_p = 2(y_{max} - y_{min})/(P_u h) \tag{1.22b}$$

After identifying the transfer function matrix, systematic controller design is carried out by model-based techniques. The approach is illustrated through a number of simulated SISO examples and two experimental 2 × 2 systems (a mixing tank and a distillation column). Other references in the MIMO systems are given by Semino and Scali (1996; 1998) and Shen and Yu (1994).

1.7 PI/PID Controller Design

For many of the control problems, a satisfactory performance is obtained using PI/PID controllers. Excellent reviews on the design of PID controllers are available (Ang et al., 2005;

Aström and Hägglund, 1995, 2001; Cominos and Munro, 2002; Datta et al., 2000; Johnson and Moradi, 2005; O'Dwyer, 2003; Sung and Lee, 1996; Tan et al., 1999; Visioli, 2005). There are special journal issues brought out on PID control (Aström et al., 2001; Isaksson and Hägglund, 2002). For controller design purposes, many processes are adequately represented by first order plus time delay models. The methods available for design of PID controllers can be broadly classified into:

(i) methods based on stability analysis (Aström and Hägglund, 2004; Chidambaram, 1998; Cohen and Coon, 1953; Ziegler and Nichols, 1942), (ii) methods based on gain and phase margin method (Wang et al. 1999a, 1999b; Wang and Shao, 1999; Wang and Cai, 2002), (iii) synthesis method (Chen and Seborg, 2002; Smith and Corripio, 2001), (iv) pole placement method (Clement and Chidambaram, 1997a), (v) IMC method (Abbas, 1997; Chien and Fruehauf, 1991; Lee et al., 1998a; Rivera et al., 1986), (vi) equating coefficient method (Padmasree and Chidambaram, 2005), (vii) optimization method (He et al., 2000; Pedret et al., 2002; Syrcos and Kookos, 2005; Toscano, 2005; Visioli, 2001) and (viii) other methods.

Foley et al. (2005) compared the performance of PID controller settings proposed by Rivera et al. (1986), Skogestad (2003), Wang and Shao (2000) and Chen and Seborg (2002). The correct choice of tuning method depends on the process control objective as well as the plant dead time to time constant ratio. WS method is recommended when the response to set point changes is of greatest concern. For purely regulatory applications, CS method gives the best overall results for lag-dominant processes ($D/\tau < 0.35$). IMC-improved PI (Rivera et al., 1986) method is better suited for the regulation of delay dominant systems ($D/\tau > 0.35$). When set point following and disturbance rejection are of roughly equal importance (as for slave loops in a cascade structure), IMC method proposed by Skogestad (2003) should be used if D/τ is small and Rivera et al. (1986) method for large values of D/τ.

Methods of designing PI/PID controllers for stable systems with inverse response systems are given by stability analysis (Ziegler and Nichols method, 1942), IMC method (Scali and Rachid, 1998; Skoestad, 2003; Wang et al., 2001), gain and phase margin method (Luyben, 2001), synthesis method (Padmasree and Chidambaram, 2005) and equating coefficient method (Padmasree and Chidambaram, 2005).

The design of PID controllers for unstable FOPTD model has attracted attention recently (Chidambaram, 1997; Padmasree and Chidambaram, 2005). The performance specifications that are usually obtained for stable FOPTD model may not be obtained for unstable FOPTD systems (Stein, 2003). The methods of designing PID controllers for unstable FOPTD systems are given by (i) modified Ziegler–Nichols method (DePaor and O'Malley, 1989; Venkatashankar and Chidambaram, 1994; Ho and Xu, 1998), (ii) IMC method (Rotstein and Lewin, 1991; Lee et al., 1998a; Marchetti et al., 2001), (iii) pole placement method (Clement and Chidambaram, 1997b), (iv) optimization method (Manoj and Chidambaram, 2001; Visioli, 2001), (v) synthesis method (Chandrasekhar et al., 2002), (vi) two degrees of freedom method (Jacob and Chidambaram, 1996; Huang and Chen, 1997; 1999), (vii) equating coefficient method (Padmasree et al., 2004). In many of these methods, one or two adjustable parameters are used to calculate the PID settings.

Padmasree et al. (2004) compared the performances of PID controller design proposed for FOPTD systems by Huang and Chen (1999), Rotstein and Lewin (1991), Visioli (2001) and their proposed method. The performance of Padmasree et al. (2004) is the best.

The important formulae used for controller tuning are as follows:

(1) Ziegler–Nichols closed loop tuning method (Ziegler and Nichols, 1942)

P-control: $k_c = 0.5\, k_u$ (1.23a)

PI-control: $k_c = 0.45\, k_u$; $\tau_I = p_u/1.2$ (1.23b)

PID-control: $k_c = 0.6\, k_u$; $\tau_I = p_u/2$; $\tau_D = p_u/8$ (1.23c)

where, k_u – ultimate gain; p_u – period of oscillations

(2) IMC tuning equations (Luyben, 2001)

$$\lambda = \max\,(1.7\, D, 0.2\, \tau) \quad (1.24a)$$

$$k_c\, k_p = (2\, \tau + D)/\, 2\lambda \quad (1.24b)$$

$$\tau_I = \tau + (D/2) \quad (1.24c)$$

where λ is the tuning parameter

(3) Simple IMC PID settings (Skogestad, 2003)

Series PID form: $c(s) = k_c\, [(\tau_I s + 1)\, /\, \tau_I s]\, [\tau_D s + 1]$

Process	$y(s)/u(s)$	k_c	τ_I	τ_D
First	$k_p \exp(-\theta s)/(\tau s+1)$	$\tau_1/[k_p(\tau_c+\theta)]$	min $\{\tau_1, 4(\tau_c+\theta)\}$	–
Second	$k_p \exp(-\theta s)/[(\tau_1 s+1)(\tau_2 s+1)]$	$\tau_1/[k_p(\tau_c+\theta)]$	min $\{\tau_1, 4(\tau_c+\theta)\}$	τ_2

Where τ_c- tuning parameter (for fast response and robustness $\tau_c = \theta$)

(4) Ideal form of PID controller (Skogestad, 2003)

Parallel form: $c(s) = k_c\, [1 + (1/\, \tau_I s) + (\tau_D s)]$ (1.25)

$k_c = k_c[1 + (\tau_D\, /\tau_I)]$; $\tau_I s = \tau_I\, [1 + (\tau_D\, /\tau_I)]$; $\tau_D = \{\tau_D\, /[1 + (\tau_D\, /\tau_I)]\}$ (1.26)

Process	k_c	τ_I	τ_D
$\tau_1 \leq 8\theta$	$0.5\, (\tau_1+\tau_2)/(k_p\, \theta)$	$\tau_1+\tau_2$	$\tau_2/[1+(\tau_2/\tau_1)]$

Contd.

Contd.

$\tau_1 \geq 8\theta$	$[0.5\ \tau_1\ (8\theta+\tau_2)]/(8\ k_p\ \theta^2)$	$8\theta+\tau_2$	$\tau_2/[1+(\tau_2/8\theta)]$

(5) Padmasree et al. (2004) tuning formulae for FOPTD unstable system

Parameter	Equation $[\varepsilon = (\theta/\tau)]$	Range
$k_c\ k_p$	$1.2824\ \varepsilon^{-0.8325}$	$0.01 \leq \varepsilon \leq 1.2$
τ_I/τ	$5.573\varepsilon - 0.0063$	$0.01 \leq \varepsilon \leq 0.5$
	$0.483\ \exp(3.3739\varepsilon)$	$0.5 \leq \varepsilon \leq 1.2$
τ_D/τ	$0.507\varepsilon + 0.0028$	$0.01 \leq \varepsilon \leq 1.2$

(6) Rao and Chidambaram (2007) proposed a synthesis method of designing a controller for the system

$$G_c(s) = k_p(1 - ps)\exp(-\tau_D s)/[a_1 s^2 + a_2 s + 1] \tag{1.27}$$

Here $p = \pm\ \tau_1$. The controller G_c is a PID controller with additional lead and lag terms:

$$G_c(s) = k_c\left(1 + \frac{1}{\tau_I s} + \tau_D s\right)\frac{(\alpha s + 1)}{(\beta s + 1)} \tag{1.28}$$

where $k_c = n_1/K_p h$, $\tau_I = n_1$, $\tau_D = n_2/n_1$, $\beta = x_1/a_2$ and $\alpha = 0.5\theta$

$$n_2 = (y_3 z_1 - y_1 z_2)/(y_2 y_3 - y_1 y_4);\ n_1 = (y_2 z_2 - z_1 y_4)/(y_2 y_3 - y_1 y_4);\ \beta = x_1 / a_2 \tag{1.29}$$

where

$$y_1 = -a_2^2 + 0.5\theta p a_2;\ y_2 = -0.5\theta p a_1 - a_2 p - 0.5\theta a_2$$

$$y_3 = -a_1 a_2 - a_2 p - 0.5\theta a_2;\ y_4 = -0.5\theta p + a_2$$

$$z_1 = a_2\lambda^3 + 1.5\theta a_2 \lambda^2 - 0.5\theta a_1 \lambda^3 - 3\lambda a_2^2 - \theta a_2^2 - p a_2^2$$

$$z_2 = 3a_2\lambda^2 + 1.5\theta a_2 \lambda - 0.5 p\theta a_2 - 0.5\theta\lambda^3 - 3a_1 a_2 \lambda - a_1 a_2 \theta - a_1 a_2 p \tag{1.30}$$

Rao and Chidambaram (2007) derived design equations for PID controllers with lead/lag filter for systems described by integrator plus first order models or by higher order models. From the literature survey, it can be observed that the analysis of relay tuning incorporating higher order harmonics for series and parallel cascade controller systems is not yet addressed. The simple tuning method proposed by Padmasree et al. (2004) can be extended to cascade control systems. Identification of transfer function models of unstable FOPTD systems and SOPTD systems needs attention.

2 Improved Autotune Identification Methods

This chapter reviews the methods proposed by Srinivasan and Chidambaram (2003; 2004) to accurately estimate the model parameters of a first order plus time delay (FOPTD) transfer function model using (1) the conventional relay autotune method and (2) asymmetric relay autotune method. Usually, the value of delay is assumed or noted from the initial portion of the response of the system. Whenever identifying a higher order dynamics system by an FOPTD model, this method wrongly identifies the time constant as negative (Li et al., 1991) due to an error in identifying the time delay, which is due to an error in the model structures. Using conventional relay autotune method, an additional equation is formulated to calculate accurately the parameters of the FOPTD model. Even when the actual system is FOPTD and the time delay to time constant ratio is larger, higher order harmonics cannot be neglected in the output response. Hence, there is a need to consider the higher order harmonics of the relay oscillations to get improved accurate values for the controller ultimate gain. For the asymmetric relay tuning method, analytical solutions are given for the evaluation of the model parameters.

2.1 Introduction

Luyben (1987) used the relay feedback method to identify the model parameters (k_p, τ and D) of an FOPTD model. Using the controller ultimate gain and period of oscillation, two equations are formulated using the amplitude criterion and phase angle criterion. The value of time delay is noted from the initial portion of the relay output. The values of k_p and τ are then calculated from the two derived equations. For identifying the FOPTD model for a system whose dynamics are of really higher order, the method of Luyben gives a negative time constant (Li et al., 1991). The error lies in the difference in the model structures.

Further, when the ratio of time delay to time constant is larger, Li et al. (1991) reported that the model identified by relay autotune method gives as high as 27% error in the value of k_u. For smaller D/τ, Li et al. (1991) found by simulation that the relay autotune method gives –18% error in the value of k_u. The autotune method uses only principal harmonics for the calculation of model parameters. Li et al. (1991) and Leva (2005) proposed the use of two relay feedback tests (one conventional relay test and the second with added known delay). In these methods, all the parameters of the model are identified. Shen et al. (1996a) used a biased relay for getting the model parameters using a single relay test. Huang et al. (1996a, b) proposed an asymmetrical relay method. Yu (1999) gave an excellent survey of relay feedback method. Park et al. (1997) proposed modifications of the method under load disturbances. Consideration of higher order dynamics is given by Chiang et al. (1992), Scali et al. (1999), Kaya and Atherton (1999a; 1999b), Luyben (2001) and Atherton (2000). Scali et al. (1999) proposed the use of several relay tuning tests to be carried out. The Luyben method (2001) is not effective for unstable systems. In the methods proposed by Atherton (2000) and Kaya and Atherton (1999a; 1999b), computational complexity is involved.

In the first part of this chapter, (i) a modification in analyzing the data of the conventional relay method is proposed by formulating an additional equation so as to estimate a more accurate value of the process gain k_p, and hence more accurate FOPTD model parameters and (ii) a method to consider higher order harmonics in analyzing the conventional relay testing, for getting improved model parameters, is also proposed. In the second part of this chapter, analytical solutions are given for the model parameters for the asymmetric relay tuning method.

2.2 Estimation of Time Delay

Padmasree and Chidambaram (2001) proposed a method for formulating an equation for time delay of the FOPTD model when analyzing the closed loop response of a PID controlled system. In the present work, their method in formulating the additional equation is applied here for analyzing the relay feedback oscillation.

Let $G(s)$ represent the process transfer function of the system

$$G(s) = y(s) / u(s) \tag{2.1}$$

From the definition of Laplace transform, we get

$$y(s) = \int y(t)\exp(-st)dt \tag{2.2}$$

Let us evaluate the above integral (from $t = 0$ to ∞) for $s_1 = 8/t_s$, where t_s is the time at which few (say 3) repeated cycles of oscillation appear in the output. The reason for taking $s_1 = 8/t_s$ is that, for $t > t_s$, because of a very small value of the term [exp($-st$)], all further contribution by the subsequent terms is negligible while evaluating the integral value. Now we have the numerical value for $y(s_1)$. Similarly, from the $u(t)$ versus time data, $u(s_1)$ is calculated. From the assumed FOPTD model as well, we can formulate the equation for $y(s_1)$ as

$$y(s_1) = u(s_1)k_p \exp(-Ds_1) / (\tau s_1 + 1) \tag{2.3}$$

Thus we are able to formulate the above additional equation using $y(s_1)$ and $u(s_1)$. Note that we must record the transient data of the relay until two or three sustained oscillations are obtained in the process output. From the amplitude criterion and phase angle criterion, we get the following equations for the parameters of the FOPTD model as

$$\tau = v / \omega_u \tag{2.4}$$

$$D = [\pi - \tan^{-1}(v)] / \omega_u \tag{2.5}$$

Using Eqs (2.4) and (2.5) in Eq.(2.3) we get

$$k_p \exp[(-s_1 / \omega_u)\{\pi - \tan^{-1}(v)\}]/[(vs_1 / \omega_u)] + 1] = y(s_1) / u(s_1) \tag{2.6}$$

where $v = [(k_u k_p)^2 - 1]^{0.5}$ and k_u is the controller ultimate gain.

Using the values of $y(s_1)$ and $u(s_1)$ in the above equations and solving the resulting non-linear algebraic equation, we get the value of k_p. Using the value of k_p in Eqs (2.4) and (2.5), we get the values of the FOPTD model parameters τ and D. Let us consider an example given by Li et al. (1991)

$$y(s) / u(s) = \exp(-2s) / [(10s + 1)](s + 1)] \tag{2.7}$$

Using the relay feedback method (with relay height as ±1), we get the oscillations as shown in Fig. 2.1. The steady oscillations show a pure sinusoidal waveform, therefore, assumption of a relay method to neglect higher order harmonics is valid for this example. By the Luyben (1987) method, the delay is noted as 2 from the initial response, and from the oscillations, it is noted that $\omega_u = 0.597$ and $a = 0.194$. Using the relation $k_u = 4h/(\pi a)$, we get $k_u = 6.553$.

Using the amplitude criterion and phase angle criterion for an FOPTD system, Li et al. (1991) reported the values of k_p as −0.501 and τ as −5.03. Since negative values for τ and k_p are obtained, they suggested that FOPTD model is not valid for this system and fitted a second order plus time delay (SOPTD) model using a second relay test with a known additional delay time.

By the proposed method of analyzing the conventional relay method, we note, from Fig. 2.1, the value of t_s as 68.9 and hence using Eq.(2.2) we get $y(s_1) = -0.124$ for the calculated value of $u(s_1) = -0.0407$. Using Eqs (2.6), (2.4) and (2.5), we get $k_p = 1.08$, $\tau = 11.725$ and $D = 2.87$. The proposed method is able to identify an FOPTD model. Figure 2.2 shows the open loop response of the actual system and the identified FOPTD model. A good match is obtained. Figure 2.2 shows that the response of the identified FOPTD model by the present method is better than that of SOPTD model proposed by Li et al. (1991). Thus by properly interpreting the conventional relay oscillations, we can estimate the model parameters of an FOPTD model.

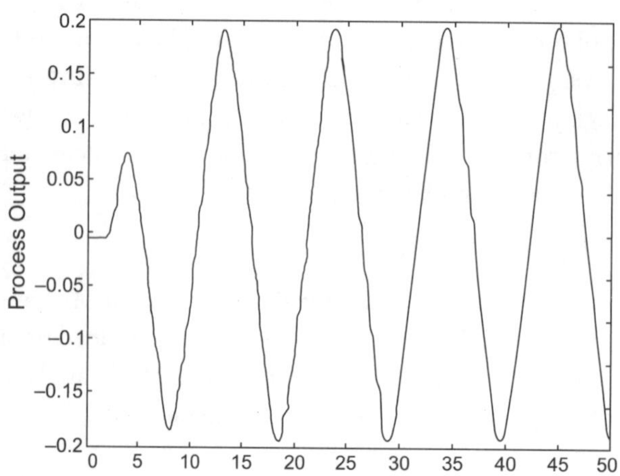

Fig. 2.1 *Relay feedback oscillations for the system exp(−2s)/[(10s + 1)(s + 1)] relay height = 1*

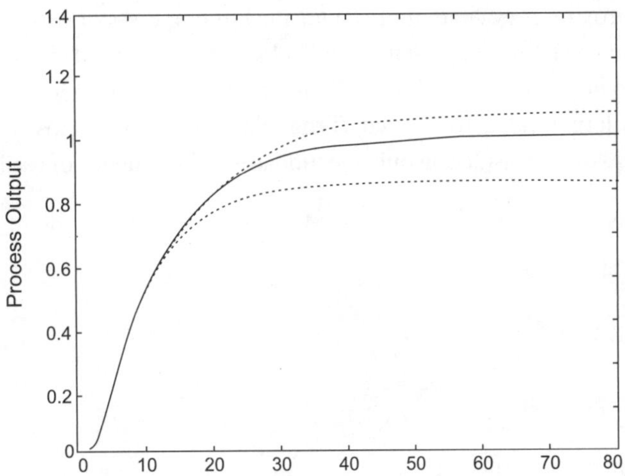

Fig. 2.2 *The open loop comparisons of the original system and the identified model*
Legend: middle line–actual system; upper line–FOPTD model by the present method; lower line–SOPTD model by Li et al. method

2.3 Method for Considering Higher Order Harmonics

From the Fourier series analysis, it can be easily shown (Åström and Hägglund, 1984) that a relay consists of many sinusoidal waves of odd multiples of fundamental frequency 'ω' and with the amplitude $4h/(n\pi)$ (n = 1,3,5,…). The input to the process thus consists of many sine waves. For an FOPTD system, output wave is also a sinusoidal wave with different amplitude and different frequency. The oscillations what we observe, i.e., $y(t)$ is the addition of several sine waves

$$y(t) = [a_1 \sin(\omega_u t + \phi_1)] + (1/3)a_3 \sin(3\omega_u t + \phi_3) + (1/5)a_5 \sin(5\omega_u t + \phi_5) + \cdots] \tag{2.8}$$

where

$$a_1 = 1/[1+(\tau\omega_u)^2];\ a_3 = 1/[1+(3\tau\omega_u)^2];\ a_5 = 1/[1+(5\tau\omega_u)^2],\ \ldots \text{ etc.} \tag{2.9}$$

$$\phi_1 = -D\omega_u - \tan^{-1}(\tau\omega_u) = -\pi \tag{2.10a}$$

$$\phi_3 = -3D\omega_u - \tan^{-1}(3\tau\omega_u);\ \phi_j = -jD\omega_u - \tan^{-1}(j\tau\omega_u) \tag{2.10b}$$

$$y(t) = a_1\left[\sin\left(\omega_u t + \phi_1\right)\right] + (1/3)b_3 \sin\left(3\omega_u t + \phi_3\right) + (1/5)b_5 \sin\left(5\omega_u t + \phi_5\right) + \cdots] \tag{2.11}$$

where $b_3 = \{[1+(\tau\omega_u)^2]/[1+(3\tau\omega_u)^2]\}^{0.5};\ b_5 = \{[1+(\tau\omega_u)^2]/[1+(5\tau\omega_u)^2]\}^{0.5}$; etc., (2.12)

If $\tau\omega_u$ is assumed to be very large (∞), then $y(t)$ will consist of only the fundamental harmonics. In Eq.(2.8), the terms containing $a_3 \sin(3\omega_u t + \phi_3)$, $a_5 \sin(5\omega_u t + \phi_5)$, etc. will be neglected. As stated earlier, this assumption gives a large error in the calculation of k_u and hence in the estimated model parameters of the FOPTD model. In what follows, we will consider higher order dynamics for the calculation of the model parameters.

Let us derive approximate evaluation of $y(t)$ for the limiting cases of smaller $\tau\omega_u$ and separately for larger $\tau\omega_u$. Whenever the relay oscillation is close to a rectangular wave form, the results for smaller $\tau\omega_u$ is to be used. If the relay oscillation is closer to a triangular waveform, then the following result for larger $\tau\omega_u$ is to be used. If the relay oscillation is close to a sine wave form, then the standard equation considering only the fundamental frequency of oscillation can be used.

Case 1: When $\tau\omega_u$ is smaller

$$\phi_3 = -3D\omega_u - \tan^{-1}(3\tau\omega_u) \quad (2.13a)$$

$$= -3D\omega_u - 3\tan^{-1}(\tau\omega_u) = 3\phi_1 = -3\pi \quad (2.13b)$$

Similarly, $\phi_5 = -5\pi;\ \phi_N = -N\pi$ (2.13c)

Hence, Eq.(2.11) can be written as

$$y(t) = a_1[-\sin(\omega_u t) - (1/3)b_3\sin(3\omega_u t) - (1/5)b_5\sin(5\omega_u t) - \cdots] \quad (2.14)$$

For smaller value of $\tau\omega_u$, $\tau\omega_u$ can be neglected when compared to 1 and hence the values of b_1, b_3, … b_N can each be approximated to 1. Hence, Eq.(2.14) becomes

$$y(t) = a_1\left[-\sin(\omega_u t) - (1/3)\sin(3\omega_u t) - (1/5)\sin(5\omega_u t) - \cdots\right] \quad (2.15)$$

Here the value of 'a_1' is to be calculated. The value of 'a_1' is not the amplitude observed from the output oscillation.

Li et al. (1991) assumed that the system is a low pass filter so that higher order harmonics of the input sine waves are filtered and the system oscillation contains only the fundamental frequency. The value of k_u is obtained as (Li et al., 1991)

$$k_u = 4h/(a_0\pi) \quad (2.16)$$

where a_0 is observed output amplitude. In Eq.(2.15), when only the first term is considered then $a = a_0$. Since $y(t)$ contains many sine waves and cannot be represented by a single sine wave of amplitude of a_0, the evaluated k_u deviates from the actual value. Li et al. (1991) reported that an error of –18% to 27% is noted in the estimated value of k_u by using Eq.(2.16).

This problem can be overcome as follows: From the output oscillations, it is possible to calculate $y(t)$ at any time 't'. 'ω_u' is the frequency of the observed oscillations. From Eq.(2.15), we get

$$a_1 = y(t)/\Sigma[\sin(i\omega_u t)/i] \quad (2.17)$$

Let us consider the time (t^*) at which

$$\omega_u t^* = 0.5\pi \quad (2.18)$$

Then Eq.(2.15) becomes

$$y(t^*) = a\left[1 - (1/3) + (1/5) - (1/7) + (1/9) - \cdots\right] \quad (2.19)$$

where a = modulus of a_1. Let the number of terms to be considered in the above equation be denoted as N. Using the limiting value for the summation term (0.25π), we get the following from Eq.(2.19)

$$a = 1.273\, y(t^*) \tag{2.20}$$

The time at which the value of $y(t^*)$ is to be noted from the response is given by

$$t^* = 0.5\pi / \omega_u \tag{2.21}$$

From the relay oscillation test, we can note down the value of ω_u, and t^* can be calculated. The value of $y(t^*)$ is noted from the process output. The value of k_u is given by

$$k_u = 4h/(a\pi) \tag{2.22}$$

The method is tested on FOPTD systems with various values of D/τ (refer to Table 2.1, for model parameters) and the results of k_u by the present method is given in Table 2.2a along with the results of Li et al. (1991). From the identified values of k_u and the ultimate frequency of oscillation, the parameters of an FOPTD are evaluated. Table 2.1 gives the values of the identified model parameters compared with the actual model parameters for different case studies of larger delay to time constant.

If the system allows all the frequencies of oscillations (in that case, the output oscillation is also a rectangular waveform) then the limiting case of $\pi/4$ can be used in the above analysis for the summation term in order to estimate the model parameters. However, if the process filters out beyond certain frequencies, then the output will very much deviate from the rectangular waveform. In that case, we have to use an appropriate value of N in order to get accurate values of the FOPTD model parameters.

Equation (2.19) shows that the sum of terms in the bracket varies from 1 to 0.8 for $N = 1$ to ∞. Hence, an error of 0.27 times of the actual value is obtained when $N = 1$ is used instead of $N = \infty$. The number of terms to be considered in Eq.(2.19) is any one of 1,3,5,7,9,… and ∞. When the number of terms in Eq.(2.19) is ∞, then Eq.(2.20) is considered for 'a'. Li et al. (1991) used $N = 1$(i.e., $a = a_o$). The present study shows that the value of $N = 5$ gives a better result of the calculated k_u and of the model parameters of FOPTD. Figure 2.3 shows the system oscillations (both the dynamics and steady-state) of relay test for the system $\exp(-4s)/(\tau s + 1)$ with $\tau = 1$ and 0.4 (i.e., $D/\tau = 4$ and 10). If the system's steady-state oscillation resembles that of a pure sinusoidal wave (as shown in Fig. 2.1), then a value of $N = 1$ can be used. If the oscillation deviates from a pure sine wave and depends on the extent of deviation, $N = 3$ or 5 or 7 can be considered. For example, for Fig. 2.3a, a value of $N = 3$ or 5 can be considered. When the oscillation is near to that of a rectangular waveform then a value of $N = 9$ can be considered (example for the condition of Fig. 2.3b). An empirical relation between D/τ and N is fitted as

$$N = -0.0395(D/\tau)^2 + 1.4746\,(D/\tau) - 1.9266 \tag{2.23}$$

The nearest integer value of N from the above equation is to be used.

In the later section we will show, by a simulation study on a higher order system, the improved estimate of k_u and hence of the identified model parameters of an FOPTD system. Now let us consider the limiting case of larger $\tau\omega_u$ for calculation of k_u.

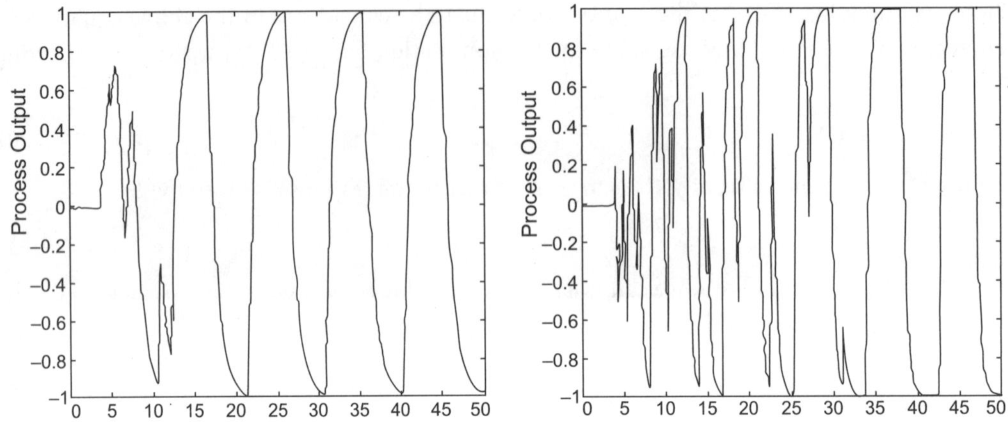

Fig. 2.3 *Relay feedback oscillations for the system (a) exp(−4s)/(s+1); (b) exp(−4s)/(0.4s+1) relay height = 1*

Case 2: Limiting case of larger $\tau\omega_u$

$$\phi_1 = -D\omega_u - \tan^{-1}(\tau\omega_u) = -\pi \tag{2.24a}$$

$$= -D\omega_u - (\pi/2) = -\pi \tag{2.24b}$$

hence, $D\omega_u = 0.5\pi$ (2.25)

$$\phi_3 = -3D\omega_u - \tan^{-1}(3\tau\omega_u) \tag{2.26a}$$

$$= -3(D\omega_u) - \pi/2 = -4\pi/2 \tag{2.26b}$$

Similarly, $\phi_5 = -6\pi/2; \ldots; \phi_N = -(N+1)\pi/2$ (2.27)

Hence, Eq.(2.11) can be written as

$$y(t) = a_1[-\sin(\omega_u t) - (1/3)\, b_3 \sin(3\omega_u t) - (1/5)\, b_5 \sin(5\omega_u t) + \cdots] \tag{2.28}$$

where $b_3 = \{[1+(\tau\omega_u)^2]/[1+(3\tau\omega_u)^2]\}^{0.5}$; $b_5 = \{[1+(\tau\omega_u)^2]/[1+(5\tau\omega_u)^2]\}^{0.5}$

For larger value of $\tau\omega$, 1 can be neglected when compared with the value of $\tau\omega_u$ and hence the values of $b_1, b_3, \ldots$ can be approximated as

$$b_3 = 1/3;\ b_5 = 1/5;\ b_j = 1/j, \text{ etc.} \tag{2.29}$$

Hence, Eq.(2.28) becomes

$$y(t) = a_1[-\sin(\omega_u t) + (1/9)\sin(3\omega_u t) - (1/25)\sin(5\omega_u t) + \cdots] \tag{2.30}$$

Here, the value of 'a_1' is to be calculated. Value of 'a_1' is not the amplitude observed from the output oscillation.

Li et al. (1991) reported that an error of −19% is noted in the estimated value of k_u by using Eq.(2.22) for larger values of $\tau\omega_u$. This problem can be overcome as follows: From the output oscillations, it is possible to calculate $y(t)$ at any time 't'. 'ω_u' is the frequency of observed oscillations. From Eq.(2.30), we get

$$a = y(t) / \Sigma\, [\sin(i\omega_u\, t)/i^2] \tag{2.31}$$

Let us consider the time (t^*) at which

$$\omega_u\, t^* = 0.5\pi \tag{2.32}$$

Then Eq.(2.30) becomes

$$y(t^*) = a\,[1 + (1/9) + (1/25) + (1/49) + (1/81) + \cdots] \tag{2.33}$$

Let the number of terms to be considered in the above equation be denoted as N. Using the limiting value for the summation term ($0.125\pi^2$), we get from Eq.(2.33)

$$a = 0.810\, y(t^*) \tag{2.34}$$

The time at which the value of $y(t^*)$ is to be noted from the response is given by

$$t^* = 0.5\, \pi / \omega_u \tag{2.35}$$

Table 2.2b compares the values of k_u from Eqs (2.22) and (2.33) for different number of terms to be considered in Eq(2.33) for the lower values of D/τ. The proposed method of considering the higher order harmonics is able to explain the reported maximum error of −18% and +27% as shown in Tables 2.2b and 2.2c respectively.

Table 2.1 *Comparison of estimated k_u and identified FOPTD model parameters for the present method (when $\omega t^* = \pi/2$) and Li et al. (1991) method*

Simulated model parameters				**Ultimate gains**			**Frequency from oscillation**	**Identified model parameters**			
No	k_p	τ	D	k_u (ana)	k_u (Li)	k_u (Pro)	ω (osc)	k_p (Li)	k_p (Pro)	τ (Li)	τ (Pro)
1	1	0.2	4	1.01	1.27	1.00	0.76	0.79	1.00	0.14	0.14
2	1	0.4	4	1.04	1.27	1.068	0.74	0.80	0.98	0.28	0.42
3	1	0.5	4	1.06	1.27	1.077	0.72	0.82	0.97	0.37	0.45
4	1	1.0	4	1.19	1.30	1.18	0.67	0.86	0.96	0.74	0.81
5	1	1.0	5	1.13	1.28	1.13	0.55	0.83	0.96	0.73	0.78

Note: The time delay values for the present method are 4.133, 3.83, 3.93, 3.94, 4.95 respectively for Nos 1 to 5

Table 2.2a *Effect of including higher order harmonics on k_u (Eqs 2.19 and 2.22)*

D/τ	$N=1$ k_u	$N=3$ k_u	$N=5$ k_u	$N=7$ k_u	$N=9$ k_u	$N=\infty$ k_u	k_u exact
4	1.30	1.220	1.175	1.155	1.145	1.106	1.19
5	1.28	1.172	1.128	1.109	1.098	1.061	1.13
8	1.27	1.118	1.077	1.059	1.050	1.013	1.06
10	1.27	1.109	1.068	1.051	1.040	1.005	1.04
40	1.27	0.85	1.103	0.922	1.063	1.00	1.00

Table 2.2b *Details of calculations for Table 2.2a*

D/τ	a_0 observed amplitude	ω_u frequency	t^* $(=0.5\pi/\omega_u)$	$y(t^*)$	a calculated amplitude
4	0.9817	0.67	2.34	0.904	1.089
5	0.9933	0.69	2.28	0.942	1.127
8	0.9997	0.72	2.17	0.987	1.182
10	1.0000	0.74	2.14	0.995	1.225
40	1.0000	0.77	2.04	1.000	1.230

Table 2.2c *Effect of including higher order harmonics on k_u (use of Eqs 2.33 and 2.22)*

D/τ	$N=1$ k_u	$N=3$ k_u	$N=5$ k_u	$N=7$ k_u	$N=9$ k_u	$N=\infty$ k_u	k_u exact
0.2	7.023	8.73	8.98	9.09	9.15	9.36	8.50
0.08	16.64	19.78	20.34	20.58	20.72	21.21	20.28
0.04	32.39	38.06	39.15	39.61	39.88	40.82	39.98

Table 2.2d *Details of calculations for Table 2.2c*

D/τ	a_0 observed amplitude	ω_u frequency	t^* $(=0.5\pi/\omega_u)$	$y(t^*)$	a calculated amplitude*
0.2	0.1813	0.4274	3.67	0.1678	0.1458
0.08	0.0765	0.408	3.85	0.0741	0.0644
0.04	0.0391	0.4002	3.925	0.0385	0.0334

Note: *using first three terms in the series

2.4 Simulation Study

Let us consider an example with a higher order dynamics:

$$y(s)/u(s) = 1\exp(-4s)/(0.5s+1)^3 \tag{2.36}$$

Using the relay feedback method (with relay height as ±1) we get an oscillation shown in Fig. 2.4. From Fig. 2.4, we noted the output, ω_u as 0.5889 and 'a' as 0.9969. It shows that the oscillations contain a higher order harmonics (since the steady oscillation is not a pure sinusoidal wave form). From Fig. 2.4, we note t_s = 56 and hence s_1 = 8/56. The values of $y(s_1)$ and $u(s_1)$ are calculated. Five terms (N = 5) are used in evaluating the sum term of Eq.(2.19). The calculated value of 'a' is 1.165. The identified value of k_u = 1.093. Hence, the present method gives the model parameters of FOPTD as k_p = 0.9714, D = 4.7534 and τ = 0.605. Li et al. (1991) reported the identified value of k_u as 1.28 and that of the model parameters as k_p = 1.0841, D = 4.04 (noted from the initial response) and τ = 1.6362. They assumed that all higher order harmonics are filtered out by the system (Fig. 2.4 shows that it is not so). The open loop response of the actual system and that of the identified FOPTD models are compared in Fig. 2.5. Response by the present model is better than that of Li et al. (1991). A PID controller is designed based on k_u and P_u by using Ziegler–Nichols continuous cycling tuning method. The PID settings by the present identified model are: k_c = 0.707; τ_I = 5.34 and τ_D = 1.334 and by the Li et al. (1991) model as k_c = 0.768; τ_I = 5.34 and τ_D = 1.335. Performances of the PID controllers are evaluated based on the original transfer function model of the process. The performance of the PID controller based on the present model gives an improved performance as shown in Fig. 2.6.

PID controllers are also designed based on the FOPTD model parameters by using the Ziegler–Nichols open loop tuning formula. Since the ratio of delay to time constant is more than 1, the appropriate modified Ziegler–Nichols tuning formulae (Chidambaram, 1998) are used: $k_c = 0.6\,[1 + (\pi\tau/D)^2]^{0.5}$, $\tau_I = D$ and $\tau_D = 0.25D$. The PID settings for the present identified model are k_c = 0.6462, τ_I = 4.75 and τ_D = 1.18 whereas for the model identified by Li et al. (1991) are k_c = 0.971, τ_I = 4.04 and τ_D = 1.01.

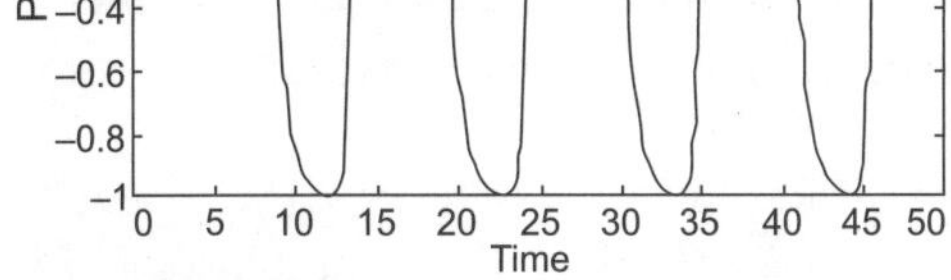

Fig. 2.4 *Relay feedback oscillations for the system exp(−4s)/(0.5s+1)^3 relay height = 1*

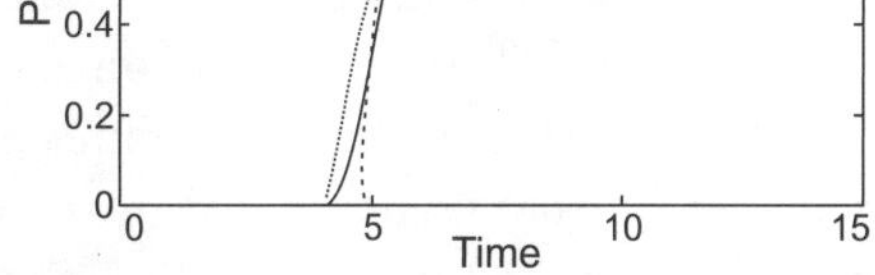

Fig. 2.5 *Open loop comparisons of actual system and the identified FOPTD models Legend: solid–actual system exp(−4s)/(0.5s+1)^3; long dash–FOPTD model by present method; dash dot–FOPTD model by Li et al. method*

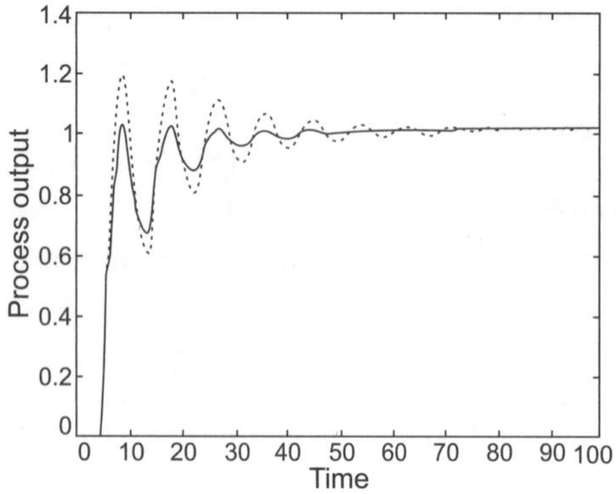

Fig. 2.6 *The closed loop response of the system exp(–4s)/(0.5s+1)³ (controller design by the closed Ziegler–Nichols method)*
Legend: solid–controller settings based on the actual system; dash–controller settings based on the identified model

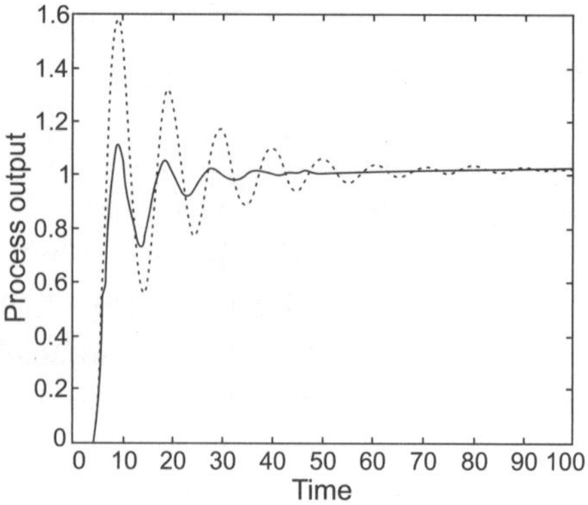

Fig. 2.7 *The closed loop performance of the system exp(–4s)/(0.5s+1)³ (controller design by the open Ziegler–Nichols method)*
Legend: same as in Fig. 2.6

Figure 2.7 shows the performance comparison of the closed loop system evaluated on the original higher order system using PID settings based on identified model. The performance of the present method gives a better response than that identified by the Luyben (1987) method.

The effect of measurement noise is studied by adding a random noise (with mean = 0 and standard deviation = 0.5%) to the process output, and the corrupted signal is used for relay feedback. The present method gives $s_1 = 0.1344$, $\omega_u = 0.72$, $y_{s1} = 0.2168$, $u_{s1} = 0.3882$ and hence the identified model parameters of the FOPTD are obtained as $k_p = 1.005$, $D = 3.82$ and $\tau = 0.5$. These values are closer to the actual model parameters of $k_p = 1$, $D = 4$ and $\tau = 0.5$.

2.5 Modified Asymmetrical Relay Method for Improved System Identification

In this section, the asymmetrical relay feedback method is modified to get improved parameter estimates of an FOPTD transfer function model. The method uses a single relay test. Analytical solution is provided for the model parameters.

Shen et al. (1996a) used a biased relay (Refer to Fig. 2.8) to obtain the model parameters using a single relay test. In this method, the process gain (k_p) is calculated from $u(t)$ and $y(t)$ data as

$$k_p = \int_0^{2\pi} e(t)\,d(\omega t) \Big/ \int_0^{2\pi} u(t)\,d(\omega t) \tag{2.37}$$

Here $e(t) = y_r - y(t)$, where y_r is the set point and for relay test $y_r = 0$. By substituting the values of k_p, ω_u and calculated k_u [from the relation $k_u = 4h/(\pi a)$] in the amplitude criterion and the phase angle criterion, the model parameters τ and τ_d are calculated analytically from

$$k_u = (1/k_p)\ [1 + \tau^2\omega_u^{\ 2}]^{0.5} \tag{2.38}$$

$$\pi - \tan^{-1}(\tau\omega_u) - \tau_d\omega_u = 0 \tag{2.39}$$

Shen et al. (1996a) showed that the method gives improved model parameters with a single test. In their method, though a good estimation of k_p is obtained, there is an error in calculating k_u because of using the relation $k_u = 4h/(\pi a)$. They also used the observed frequency of oscillation for the ultimate frequency of oscillation. Shen et al. (1996a) gave simulation results only for stable systems. Huang et al. (1996a, b) used an asymmetrical relay test to get output response for three or four cycles followed by turning off the relay and recording the output response till the output is constant (the later part of the test is an open loop test). There is no need to calculate k_u in this method. Equations are given for calculating the FOPTD model parameters. It should be noted that the relay used by Shen et al. (1996a) is shifting the (ideal) relay horizontally whereas that of Huang et al. (1996a, b) is shifting the ideal relay vertically (Fig. 2.8). The Huang et al. (1996a, b) method is not applicable for identifying an unstable FOPTD model since the method requires an open loop response as well. In the present work, using only a single asymmetrical relay test (without turning off the relay), a method is proposed for formulating the necessary equations so as to get better estimates of the FOPTD model parameters than that of Shen et al. (1996a) and Huang et al. (1996a, b).

2.6 Method

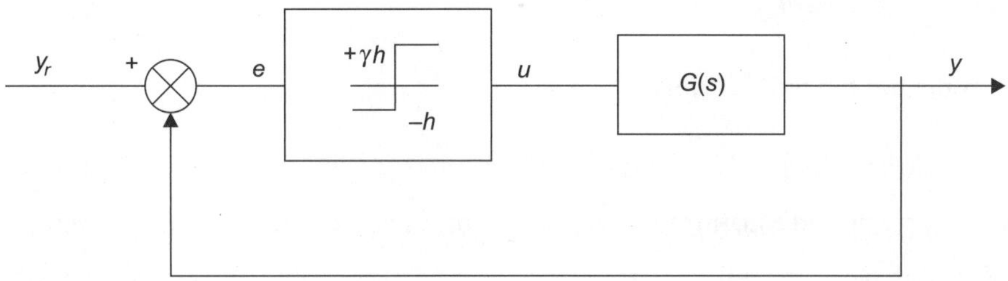

Fig. 2.8 *Block diagram for asymmetric relay feedback system*

In the present method, we consider an asymmetrical relay as shown in Fig. 2.8 where for $e(t) \geq 0$, $u = \gamma h$ and when $e(t) < 0$, $u = -h$, where γ is the displacement in the relay height (h). $\gamma = 1$ gives a symmetrical relay and $\gamma > 1$ (or <1) gives an asymmetrical relay. With this relay, the responses in y and u are recorded. Let us denote $G(s)$ as the transfer function model to be identified. Since $y(t)$ and $u(t)$ are piece-wise continuous and periodic functions, we can write the expression for the Laplace transform of $y(t)$ as (Kreyszig, 1996)

$$L[y(t)] = y(s) = [1/(1-\exp\{-Ps\})]\int_0^P y(t)\exp(-st)\,dt \tag{2.40}$$

A similar expression can be written for $u(s)$ as well. Hence, we can write

$$G(s) = y(s)/u(s) = [\int_0^P y(t)\exp(-st)dt]/[\int_0^P u(t)\exp(-st)dt] \qquad (2.41)$$

$G(j\omega)$ can be written by substituting $s = j\omega$ in the above equation, where $P = 2\pi/\omega$ and ω is the frequency of oscillation observed in the output response. By using the Euler's equation for $\exp(-j\omega t)$ Eq.(2.41) can be written as

$$G(j\omega) = (c_1 - jd_1)/(c_2 - jd_2) \qquad (2.42)$$

where $c_1 = \int_0^P u(t)\cos(\omega t)dt;\ d_1 = \int_0^P u(t)\sin(\omega t)dt \qquad (2.43a)$

$$c_2 = \int_0^P y(t)\cos(\omega t)dt;\ d_2 = \int_0^P y(t)\sin(\omega t)dt \qquad (2.43b)$$

Equation (2.43a) to (2.43b) can be evaluated numerically using Simpson's rule. Equation (2.42) can be written as

$$G(j\omega) = a + jb \qquad (2.44)$$

where $a = (c_1 c_2 + d_1 d_2)/(c_2^2 + d_2^2) \qquad (2.45a)$

$$b = (d_2 c_1 - d_1 c_2)/(c_2^2 + d_2^2) \qquad (2.45b)$$

Let us identify the parameters of an FOPTD model:

$$G(s) = k_p \exp(-\tau_d s)/(\tau s + 1) \qquad (2.46)$$

For this model, we can write $G(j\omega)$ as

$$a + jb = k_p[\cos(\tau_d\omega) - j\sin(\tau_d\omega)]\ /\ [1 + j\tau\omega] \qquad (2.47)$$

On cross-multiplying and equating the real part and the imaginary part separately to zero, we get

$$a - b\tau\omega - k_p\cos(\tau_d\omega) = 0 \qquad (2.48)$$

$$a\tau\omega + b + k_p\sin(\tau_d\omega) = 0 \qquad (2.49)$$

Substituting the value of k_p obtained from Eq.(2.37), and substituting ω_u for ω, an analytical solution of above two equations gives the values for τ and τ_d.

The modifications proposed in the present work are use of Eq.(2.37) for k_p and use of Eq.(2.42) to formulate the two equations [Eqs (2.48) and (2.49)] for the model parameters. Whereas the equations [Eqs (2.38) and (2.39)] used by Shen et al. (1996a) assumed that all the higher harmonics die out and the frequency of oscillation corresponds to that of the ultimate frequency. As stated earlier, Shen et al. (1996a) calculated k_u value from $4h/(\pi a)$. Hence the use of amplitude and phase angle criteria is made by them. Li et al. (1991), Shen et al. (1996a) and Huang et al. (1996a, b) considered only stable transfer function. In the present method, we do not use the equation $k_u = 4h/(\pi a)$ and also do not need ω_u. We will show the improvement of the proposed method in the following case studies. The general guidelines reported in literature for selecting the value of γ is 2 to 4 for stable systems and 1.05 to 1.1 for unstable systems. In the present work $\gamma = 2$ is used for stable systems and $\gamma = 1.1$ is used for unstable systems.

2.7 Simulation Results

2.7.1 Case study 1: Consider the transfer function given by Huang et al. (1996a, b):

$$G(s) = 2\exp(-0.1s)/[(2s + 1)(s + 1)]$$

The asymmetrical relay (with $h = 1$ and $\gamma = 2$) is used and the oscillations in the output $y(t)$ and $u(t)$ are noted. The use of Eqs (2.37), (2.48) and (2.49) gives the model parameters as $k_p = 2.19$, $\tau = 7.92$ and $\tau_d = 0.443$, whereas the method of Huang et al. (1996a, b) gives $k_p = 2.0$, $\tau = 2.67$ and $\tau_d = 0.42$ respectively. Application of the Li et al. (1991) method gives a negative value for the time constant ($k_p = 13.5$, $\tau = -0.1$, $\tau_d = 0.1$). From the phase angle criterion for the actual system, we get by numerical solution $\omega_u = 3.841$ and $k_u = 15.28$. Using these values and that of k_p in the phase angle and amplitude criteria equations for an FOPTD, we get the model parameters: $k_p = 2$, $\tau = 8$ and $\tau_d = 0.42$. The present method gives model parameters closer to these values. Using the identified model parameters, PID controllers are designed by Ziegler–Nichols ultimate cycling method. The closed loop responses are evaluated on the actual system. The responses are shown in Fig. 2.9. Table 2.3 shows the controller settings and the ISE values. Table 2.3 and Fig. 2.9 show that the present method gives a closed response very close to the performance based on the controller designed for the actual system.

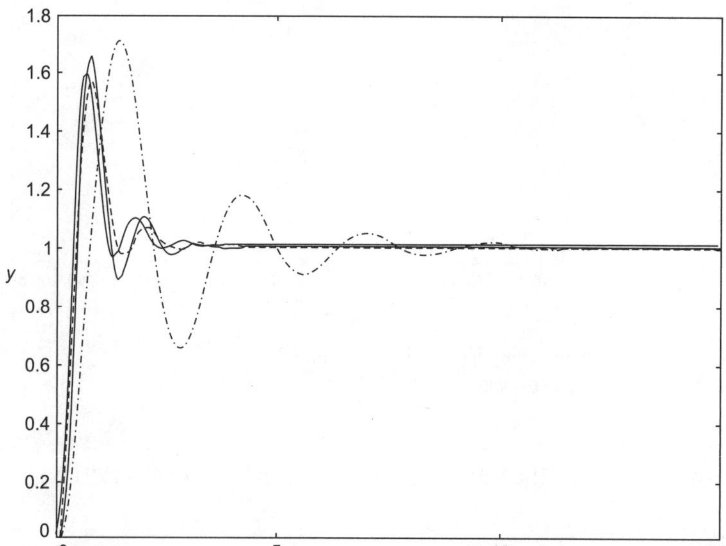

Fig. 2.9 *Closed loop response comparisons for case study 1*
Legend: solid–actual; dash–proposed method; chain–Huang et al. method; dot–Shen et al. method

2.7.2 Case study 2: Consider the transfer function

$$G(s) = \exp(-2s)/(10s + 1)$$

The present method gives the model parameters as $k_p = 1.03$, $\tau = 10.3$ and $\tau_d = 2.3$, whereas the ideal relay method of Li et al. (1991) gives $k_p = 0.988$, $\tau = 8.02$ and $\tau_d = 2$, respectively. In the Li et al. (1991) method, the value of delay is noted from the initial response as 2. They suggested carrying out an open loop step test to get the parameters of the delay time. A significant improvement is obtained by the present method. The use of Shen et al. (1996a) method gives the model parameters as $k_p = 0.999$, $\tau = 8.118$ and $\tau_d = 2.005$. Figure 2.10 and Table 2.3 show that the present method gives ISE values of the closed loop system similar to that of the actual system. The effect of measurement noise is studied by adding a random noise (with 0.5% standard deviation) and the corrupted signal is used for feedback. The present method gives the model parameters as $k_p = 0.9336$, $\tau_d = 2.03$ and $\tau = 9.3638$. These values are close to the values obtained without any measurement noise. This indicates that the present method is robust for measurement noise.

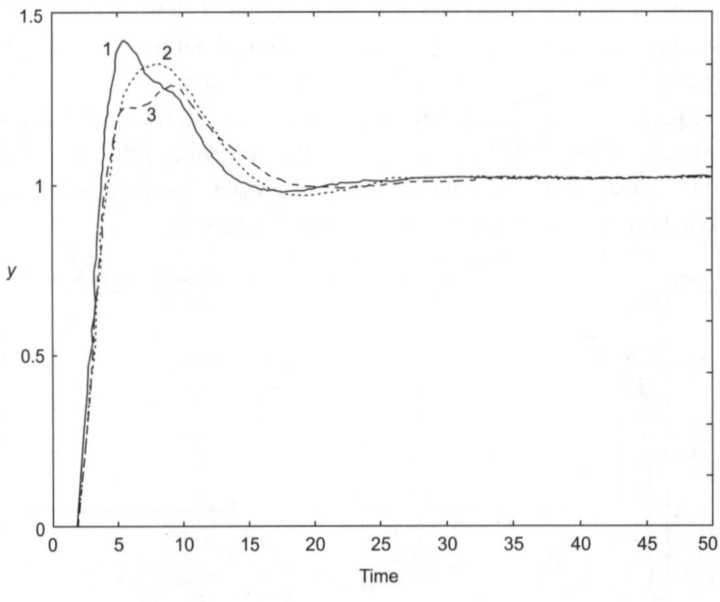

Fig. 2.10 *Closed loop response comparisons for case study 2*
Legend: 1–actual; 2–present method; 3–Li et al. method

2.7.3 Case study 3: Consider the transfer function given by Li et al. (1991):

$$\exp(-2s)/(10s^2+11s+1)$$

Application of the ideal relay method by Li et al. (1991) gives the model parameters of FOPTD as $k_p = -0.501$, $\tau_d = 2$ and $\tau = -5.030$. We get negative values for time constant and gain. According to Li et al. (1991), in such a case, the FOPTD model is not valid and thus recommend to fit an SOPTD model. Instead, the proposed method is able to identify an FOPTD model (with k_p = 1.017, τ_d = 2.872 and τ = 11.71). Applying the Shen et al. (1996a) method gives the model parameters as k_p = 0.998, τ = 10.82 and τ_d = 2.0. PID controllers are designed for the identified models and also for the actual system. Figure 2.11 and Table 2.3 show that the present method gives the closed loop response close to that of the actual system.

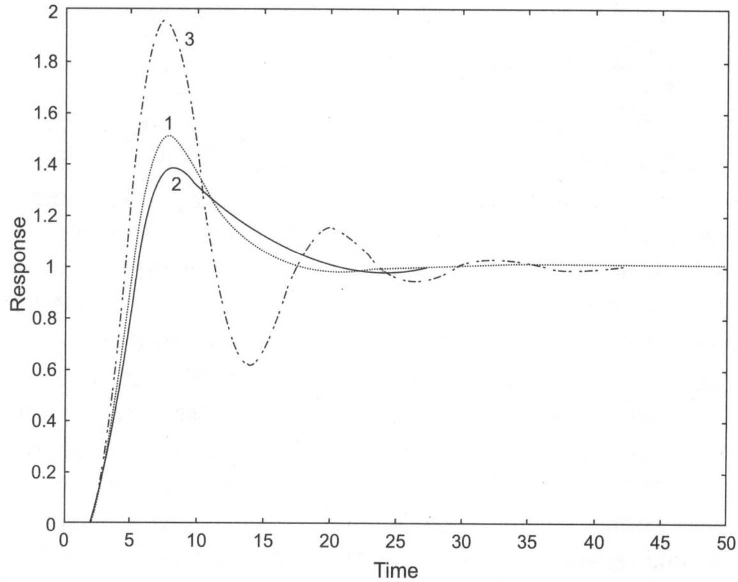

Fig. 2.11 *Closed loop response comparisons for case study 3*
Legend: 1–actual system; 2–present method; 3–Huang et al. method

Table 2.3 *Closed loop performance comparisons of different methods of identification with that of the actual system*

Case study	Method	$(k_c k_p)_{max}$	ω_u	k_c	τ_I	τ_D	ISE*
1	Actual	30.7	3.841	18.44	0.8174	0.204	0.4928
	Proposed	28.11	3.547	16.86	0.886	0.222	0.4694
	Huang et al.	10.03	3.738	6.01	0.84	0.21	0.9508
	Shen et al.	31.88	3.673	18.75	0.836	0.218	0.4684
	Li et al.	(gives negative time constant, not applicable)					
2	Actual	7.91	0.785	4.748	4.0	1.000	3.3826
	Proposed	7.10	0.683	4.261	4.6	1.150	3.2380
	Li et al.	6.38	0.785	3.725	4.0	1.000	3.5146
	Shen et al.	6.43	0.783	3.861	4.01	1.000	3.5037
3	Actual	7.07	0.6	4.242	5.23	1.308	4.5172
	Proposed	6.48	0.5466	3.887	5.744	1.436	4.2917

Contd.

Table 2.3 *Contd.*

	Shen et al.	8.55	0.785	5.13	4.0	1.000	6.7261
	Li et al.	(gives negative time constant, not applicable)					

Note: *ISE values evaluated on the actual system using the above PID settings

2.7.4 Case study 4: Let us consider an unstable FOPTD system as:

$k_p \exp(-\tau_d s)/(\tau s - 1)$ with $k_p = 1$, $\tau_d = 0.5$ and $\tau = 1$

The asymmetrical relay test is conducted (with $h = 1.0$, for $e \geq 0$, $u = h$, for $e < 0$, $u = -\gamma h$, with $\gamma = 1.1$). By the proposed method we get $k_p = 1.005$, $\tau = 1.0094$ and $\tau_d = 0.61$. The model parameters obtained by the method of Li et al. (1991) are $k_p = 0.94$ and $\tau = 0.77$ and $\tau_d = 0.5$. It should be noted that the value of τ_d is assumed in the Li et al. (1991) method. The Shen et al. (1996a) method gives $k_p = 1.005$, $\tau = 0.7953$ and $\tau_d = 0.5082$. These values are similar to that obtained by Li et al. (1991). It should be noted that an appropriate equation $[\tan^{-1}(\tau\omega_u) - \tau_d\omega_u = 0]$ for an unstable system is used instead of Eq.(2.39). As stated earlier, the Huang et al. (1996a, b) method is not applicable for an unstable systems. On the basis of the FOPTD model, PID settings are calculated based on ISE minimization method (Jhunjhunwala and Chidambaram, 2001). A PID controller is also designed on the actual system parameters. The PID controller settings and the closed loop ISE values are given in Table 2.4. Figure 2.12 shows the closed loop responses. Table 2.4 and Fig. 2.12 show that the model identified by the proposed method gives a response closer to that of the actual system.

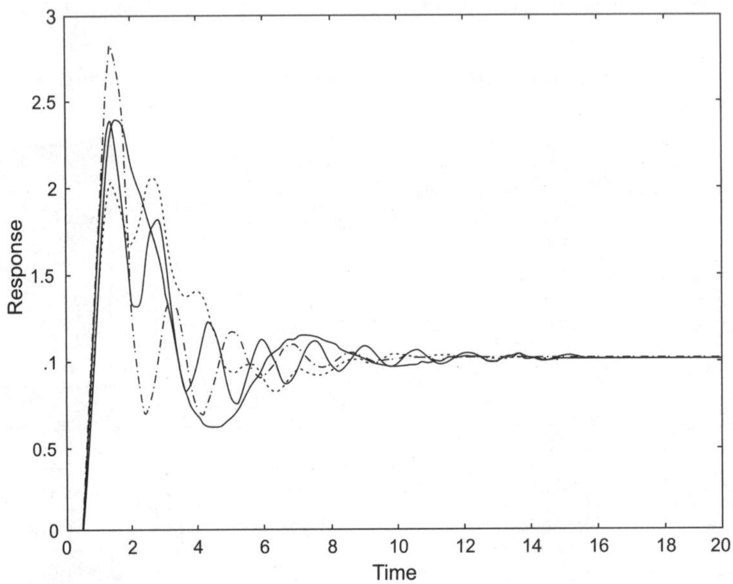

Fig. 2.12 *Closed loop servo response of the unstable FOPTD system*
Legend: solid–actual system; dash–present work; chain–Shen et al. method; dot–Li et al. method

Table 2.4 *PID controller settings and ISE values for case study 4 (unstable system)*

S.No	Settings	Actual system	Present	Li et al.	Shen et al.
1	k_c	2.381	2.033	2.533	1.971
2	τ_I	2.379	2.951	1.832	1.815
3	τ_D	0.290	0.348	0.223	0.293
4	ISE*	2.357	2.617	2.933	3.182

Note: *ISE values calculated on the actual system with the above PID settings.
PID settings by the method of Jhunjhunwala and Chidambaram (2001)

2.8 Conclusions

A method is suggested to derive an additional equation so that the process gain can be estimated using the conventional relay autotune method. This method avoids getting a negative time constant of an FOPTD model. For systems showing higher order harmonics in the response, a modification of the calculation for the parameters of the FOPTD model using the conventional relay feedback method is also proposed. This method does not assume the complete filtering of higher order harmonics, is also simple, gives an improved value for the controller ultimate gain, gives more accurate results (of k_u and of the identified FOPTD model parameters) than that proposed by Luyben (1987) and Li et al. (1991) and gives improved open loop as well as closed loop performances.

The proposed modification in the asymmetrical relay test gives improved values of the parameters of the FOPTD model. Analytical solutions are given for the evaluation of the model parameters. The method needs only one relay test to identify the parameters of an FOPTD model. The method is also applicable for identifying an unstable FOPTD model. Simulation results are given for stable and unstable transfer function models. The method gives better closed loop response than that of Huang et al. (1996a, b) and Shen et al. (1996a).

3 Cascade Controllers Tuning by Relay Autotune Method

In this chapter, the method discussed in Chapter 2 to analyse the conventional on–off relay oscillations for a single loop feedback controller is extended to the relay tuning of (i) series cascade controllers and (ii) parallel cascade controllers. Using the ultimate gain and ultimate cross-over frequency of the two loops, the inner loop (PI) and outer loop (PID) controllers are designed by the Ziegler–Nichols (1942) tuning method. The performances of the controllers are compared with the results based on conventional relay analysis. The improved method for analyzing biased autotune method proposed for single feedback controller as discussed in Chapter 2 is also extended to relay autotune of series and parallel cascade control systems.

3.1 Introduction

Cascade control scheme utilizes two control loops: secondary or inner loop embedded within a primary or outer loop. The disturbances entering the inner loop are reduced or eliminated before their effect is felt on the outer loop output variable (y_1). Owing to the presence of two control loops, two controllers need to be tuned. Hang et al. (1994) proposed a relay autotuning of cascade control loops. They used conventional on–off relay testing and using the value of k_u from $4h/(\pi a)$ and the Ziegler–Nichols (1942) tuning formulae, the controllers are tuned. With the inner loop under PI control action, the relay test is repeated for the outer loop (refer to Fig. 3.1).

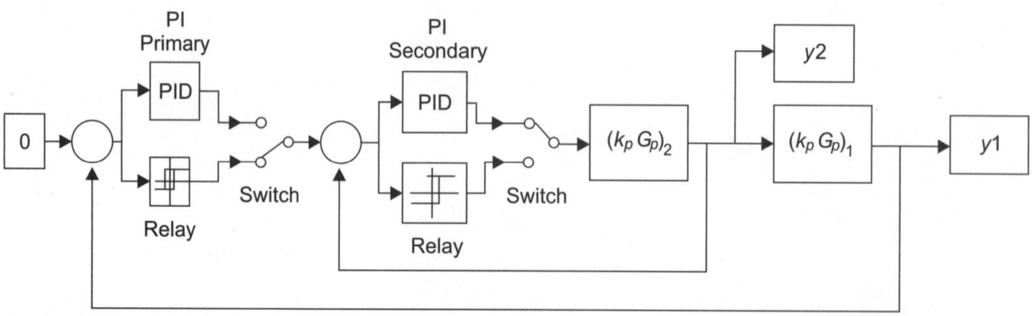

Fig. 3.1 *Relay feedback tuning of cascade control system*

In the present work, the methods of Srinivasan and Chidambaram (2003; 2004) are applied to tune cascade controllers. The improved performance of the proposed controller is compared with that of the method proposed by Hang et al. (1994). As suggested by Luyben and Luyben (1997), PI controller for the inner loop and PID controller for the outer loop are used.

3.2 Method 1

The series cascade control scheme is considered here. First, the conventional on–off relay is considered. The relay is used in the inner loop and the outer loop is kept under manual mode. The relay oscillations are noted. For simulation study, the process models assumed are $(k_pG_p)_2 = \exp(-0.1s)/(0.025s + 1)$ and $(k_pG_p)_1 = \exp(-0.1s)/(0.025s + 1)$. For single loop systems, Li et al. (1991) reported that only for large (D/τ) ratio, the conventional relay analysis gives a significant error in calculating the value of k_u. Using a symmetric relay height of ±1, the oscillation in the inner loop output variable y_2 is noted (refer to Fig. 3.2). The amplitude and frequency of oscillations are noted as 0.9814 and 27.0128 respectively. Using the relation $k_u = 4h/(\pi a_0)$, the value of k_u is obtained as 1.2974. On the basis of the transfer function model, the exact value of k_u is calculated as 1.18. Thus significant error is obtained in k_u by the conventional relay analysis. Using the results of relay testing, the PI settings are calculated using the Ziegler–Nichols continuous cycling tuning method as $k_{c2} = 0.5838$ and $\tau_{I2} = 0.1938$. Using this PI setting in the inner loop instead of the relay and introducing a relay in the outer loop, the resulting oscillation in the outer loop gives an amplitude of 0.695 and frequency of 13.3685 (refer to Fig. 3.3). Hence the value of $k_u = 1.832$ is obtained using $4h/\pi a_2$.

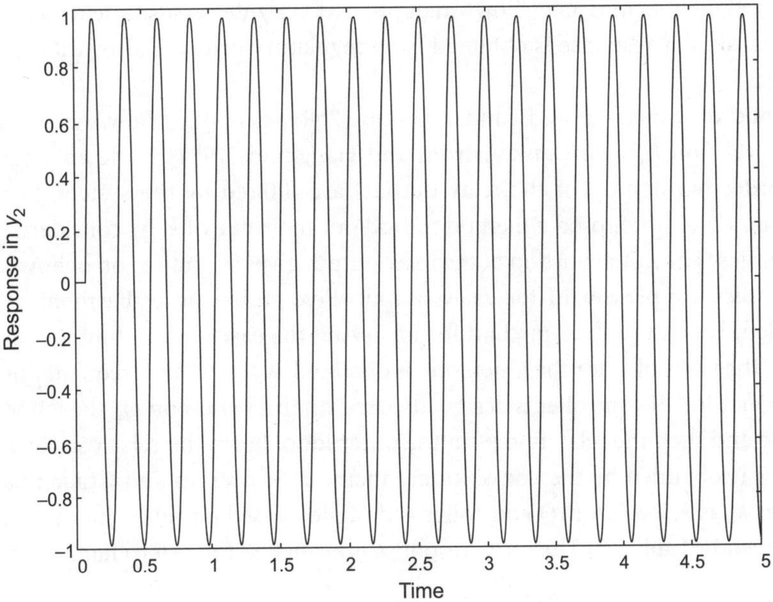

Fig. 3.2 *Response in y_2 using symmetric relay (with relay height ±1) in inner loop*

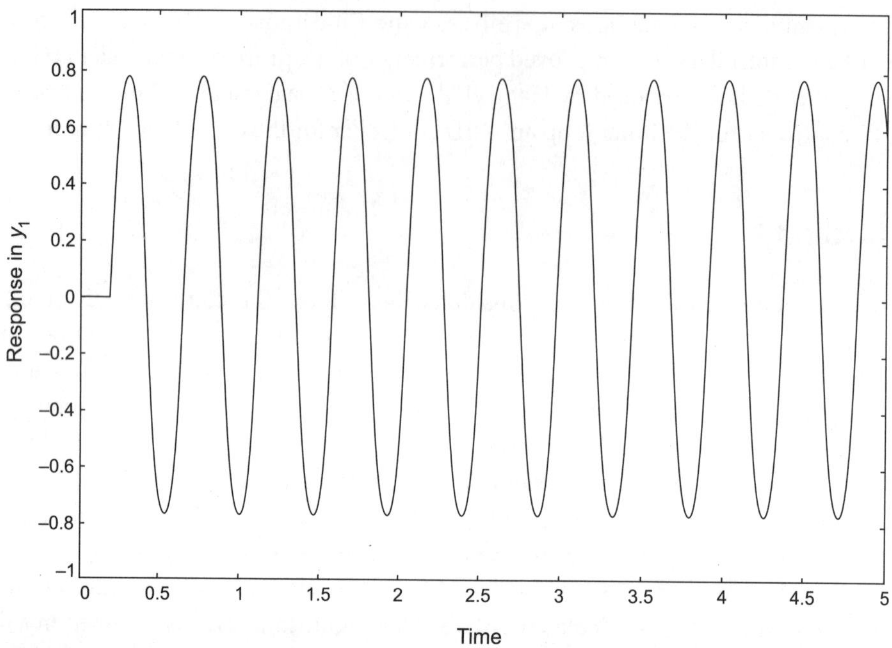

Fig. 3.3 *Response in y_1 for symmetric relay in outer loop (with inner loop on PI settings)*

On the basis of the relay test results, the outer loop PID controller is designed by the Ziegler–Nichols method as $k_{c1} = 1.0992$, $\tau_{I1} = 0.235$ and $\tau_{D1} = 0.0587$. The closed loop servo response is evaluated for a unit step change in the set point and the response is shown in Fig. 3.4a. A highly oscillatory response is obtained. Figure 3.4b shows manipulated variable versus time data corresponding to Fig. 3.4a. A similar response is obtained for a regulatory problem also [with $(k_L G_L)_2 = 1$] as shown in Fig. 3.5.

The method of Srinivasan and Chidambaram (2004) is applied now. As stated earlier, in deriving the relation $k_u = 4h/(\pi a_0)$, Aström and Hägglund (1984) made an assumption that all higher order harmonics (of the relay output) are filtered by the system. Srinivasan and Chidambaram (2003) proposed a method to find out the value of k_u by considering the higher order harmonics. The initial portion of the relay output gives an indication of how many higher order harmonics are present in the relay output. Five higher order harmonics ($N = 5$) are recommended. Let us use their method for analyzing the cascade autotuning. For the system under study, the value of k_u for the inner loop is obtained as 1.1794 and frequency of oscillations as 27.0128. Once the PI controller is designed, based on this value and ω_u, the inner loop is kept in a PI mode and then the relay is kept in the outer loop. From the relay oscillation, the value of $k_u = 1.832$ is obtained by the conventional method. Srinivasan and Chidambaram (2004) method gives k_u as 1.5637. A PID controller is designed based on this value. The details of the results are given in Table 3.1. The servo response is evaluated for a step change in the set point.

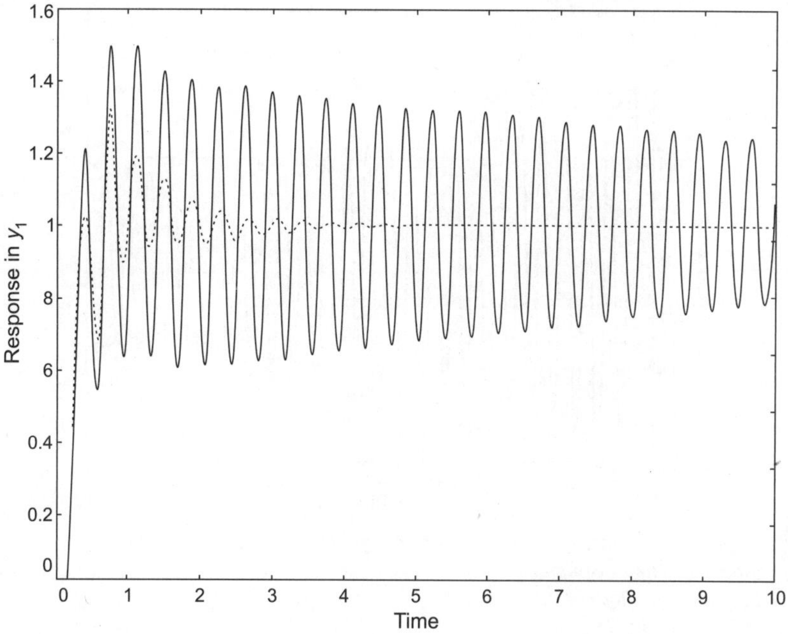

Fig. 3.4a *Servo response in y_1 using PID controller for outer loop and PI for inner loop for the first example Legend: outer oscillatory response–Hang et al. method; inner solid–method 1; inner dash–method 2*

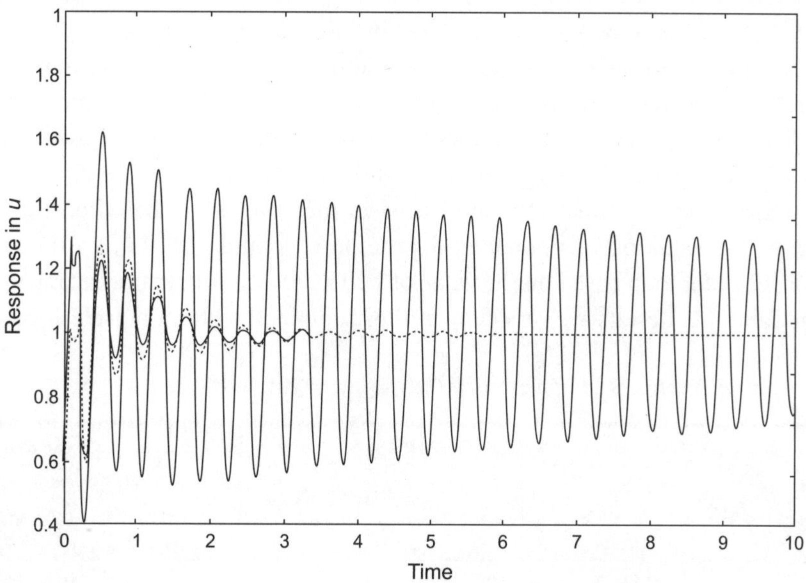

Fig. 3.4b *Response in u corresponding to Fig. 3.4a. PID controller for outer loop and PI for inner loop (for first example)*
Legend: as in Figure 3.4a

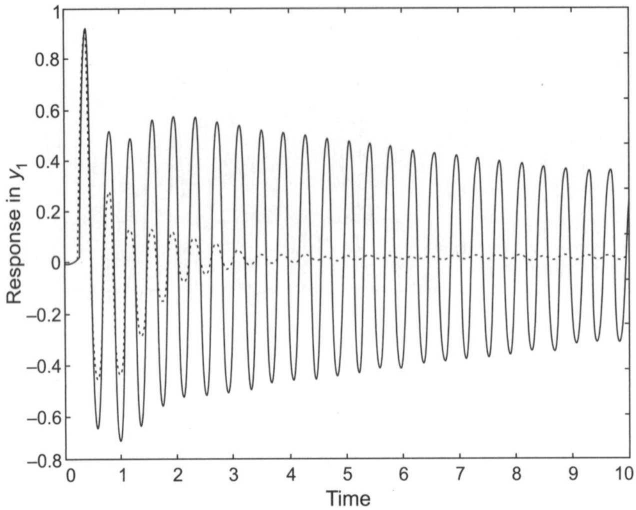

Fig. 3.5 *Regulatory response in y_1 for a disturbance in inner loop. PID controller for outer loop and PI for inner loop (first example)*
Legend: as in Figure 3.4a

The response in y_1 is shown in Fig. 3.4a. The response by the conventional analysis using single harmonics gives a highly oscillatory response, whereas the method of Srinivasan and Chidambaram (2004) gives an excellent response. Similar performance is obtained for a regulatory problem $[(k_L G_L)_2 = 1]$ as shown in Fig. 3.5. The ISE values are given in Table 3.2.

The effect of measurement noise is studied by adding a random noise (Gaussian distribution with zero mean and standard deviation of 0.5%) in inner loop and the corrupted signal is used for the feedback. The present method gives $k_u = 1.1737$, and frequency of oscillations (ω) as 26.8512. A PI controller is designed by the Ziegler–Nichols method. The inner loop is kept under PI controller and symmetric relay is used in the outer loop. Based on the oscillation obtained in y_1, the value of k_u is obtained for the outer loop as 1.5915. A PID controller is designed for the outer loop based on the value of k_u. The results for symmetric relay with noise are given in Table 3.1. Results are very close to those obtained without noise.

Table 3.1 *Controller settings comparison for $(D/\tau)_{\text{inner loop}} = 4.0$, $(D/\tau)_{\text{outer loop}} = 4.0$*

Loop	Controller settings	Symmetric relay			Asymmetric relay	Asymmetric relay with noise ($\sigma = 0.5\%$)
		$N = 1$	$N = 5$	$N = 5$(with noise, $\sigma = 0.5\%$)		
Inner	k_{c2}	0.5838	0.5307	0.5282	0.5306	0.5309
	τ_{I2}	0.1938	0.1938	0.1950	0.1917	0.2
Outer	k_{c1}	1.0992	0.9382	0.9549	0.9780	1.07
	τ_{I1}	0.2350	0.2350	0.2360	0.24	0.2324
	τ_{D1}	0.0587	0.0587	0.059	0.06	0.0581

Table 3.2 *Performance comparison of proposed methods and the Hang et al. method for* $(D/\tau)_{\text{inner loop}} = 4.0$, $(D/\tau)_{\text{outer loop}} = 4.0$

Comparison parameters	Symmetric relay		Asymmetric relay
	$N = 1$	$N = 5$	
ISE	0.7989	0.2765	0.2808
Overshoot	0.5135	0.2022	0.2494
Settling time	55	7	7

3.3 Method 2

In this section, the biased relay autotune method is applied with a relay height of +2 and −1. The method proposed by Srinivasan and Chidambaram (2003) (refer to chapter 2) is extended to series cascade systems. The value of k_u for the inner loop based on the assumed FOPTD model ($k_P = 0.8722$, $\tau = 0.0088$, $D = 0.1073$), obtained by the relay method, is 1.1792. A PI controller is designed by the Ziegler–Nichols method. The inner loop is kept under a PI controller and asymmetric relay is then used in the outer loop. Based on the oscillation obtained in y_1 and hence based on the identified FOPTD model ($k_P = 1.0684$, $\tau = 0.11$, $D = 0.162$), the value of k_u is obtained for the outer loop as 1.63. A PID controller is designed for the outer loop based on the value of k_u obtained. The results of the relay test are given in Table 3.1. The servo response in y_1 for a unit step change in the set point is shown in Fig. 3.4. The performance is as good as the performance of the proposed symmetric relay method.

The effect of measurement noise is studied by adding a random noise (Gaussian distribution with zero mean and standard deviation of 0.5%) in the inner loop and a corrupted signal is used for feedback. Results for the asymmetric relay with noise are given in Table 3.1. The results are close to the results obtained without any noise.

In the modified symmetrical relay method proposed by Srinivasan and Chidambaram (2004), the value of order (N) of higher order harmonics is to be selected. Whereas in the asymmetric method of Srinivasan and Chidambaram (2003), no such value of N is required. In the asymmetric method, the model is to be identified and then the controller settings are calculated. In the symmetrical method, the controller settings are calculated on the basis of the ultimate values obtained from the relay test. Figure 3.5 shows the regulatory response for a step change in the inner loop disturbance. The proposed two methods give an improved performance than that by the conventional relay analysis (refer to Table 3.2).

In literature, for a single loop asymmetric relay tuning method, the value of $\gamma = 2$ is used. Therefore, the relay height of +2 and −1 is used in the present study as well. However, simulation studies are carried out with different relay heights of $\gamma = 2$, 2.5, 3 and 4. The results are summarized in Table 3.3. It is observed that the resulting PI controller settings are not changed significantly.

Table 3.3 *Effect of change in relay height on PI settings using asymmetric relay testing*

Asymmetric relay		PI controller		
h	γ	k_u	k_c	τ_I
0.5	4.0	1.1859	0.5336	0.2050
1.0	2.0	1.1972	0.5387	0.1975
1.0	2.5	1.1902	0.5356	0.2000
1.0	3.0	1.1837	0.5327	0.2008

A second example is also considered. The model assumed for the inner loop is $(k_pG_p)_2 = \exp(-0.1s)/(0.01s + 1)$ and for the outer loop is $(k_pG_p)_1 = \exp(-0.1s)/(0.1s + 1)$. This example considers a lower value of (D/τ) in the outer loop. Using a symmetric relay with a height of ±1, the oscillation in the inner loop output variable y_2 is recorded. The amplitude and frequency of oscillations are noted. By the conventional method of analyzing the relay testing, PI settings are calculated by the Ziegler–Nichols tuning method. Using this PI setting for the inner loop, and symmetric relay in the outer loop, the oscillations in the outer loop are noted with an amplitude of 0.44 and frequency of oscillation (ω_u) as 11.6788. Hence, by the conventional method of relay analysis, k_u is calculated as 2.8937. Based on k_u and ω_u, the outer loop PID controller is designed and the values of controller settings are given in Table 3.4.

The methods 1 and 2 are also applied. For method 2, the value of $\gamma = 2$ is used. The results are shown in Table 3.4. The closed loop servo response is evaluated for a unit step change in the set point. The response in y_1 is shown in Fig. 3.6a. Figure 3.6b shows u versus time data corresponding to Fig. 3.6a. Regulatory response is evaluated for a unit disturbance in the inner loop. The response in y_1 is shown in Fig. 3.6c. Since the error is small, integral of the absolute value of the error is calculated for performance comparisons and the results are shown in Table 3.4. Methods 1 and 2 show good performances.

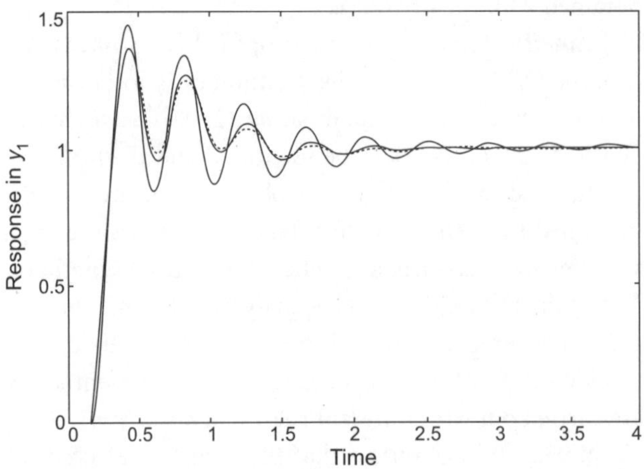

Fig. 3.6a *Servo response in y_1 using PID controller for outer loop and PI for inner loop for the second example*
Legend: as in Figure 3.4a

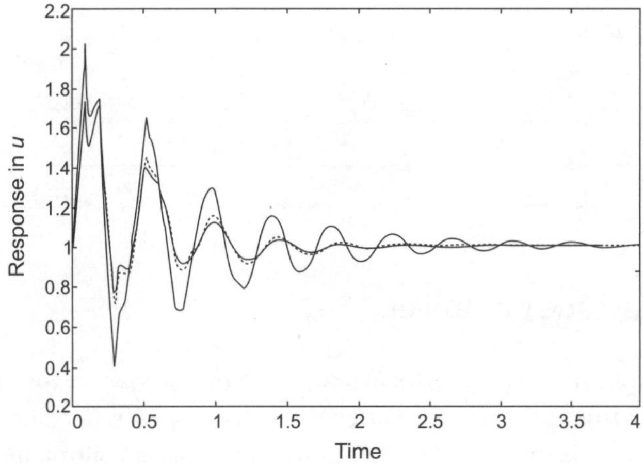

Fig. 3.6b *Response in u corresponding to Fig. 3.6a. PID controller for outer loop and PI for inner loop (second example)*
Legend: as in Figure 3.4a

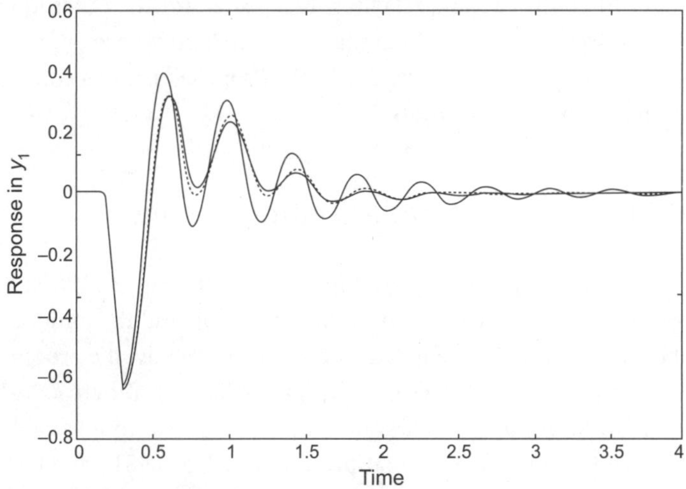

Fig. 3.6c *Regulatory response in y_1 for a disturbance in inner loop. PID controller for outer loop and PI for inner loop (for second example)*
Legend: as in Figure 3.4a

Table 3.4 *Controller settings comparison for the second example*

Loop	Controller settings	Symmetric relay		Asymmetric relay
		$N = 1$	$N = 9$	
Inner	k_{c2}	0.5730	0.4681	0.4694
	τ_{I2}	0.1783	0.1783	0.1828

Contd.

Table 3.4 *Contd.*

Outer	k_{c1}	1.74	1.82	1.86
	τ_{I1}	0.269	0.2750	0.2806
	τ_{D1}	0.0673	0.0688	0.0702
IAE		0.5009	0.411	0.419

3.4 Parallel Cascade Controllers

A parallel cascade control scheme utilizes two control loops: secondary (or 'slave') or inner loop receives its set point from primary (or 'master') or outer loop. A schematic diagram of parallel cascade control scheme is shown in Fig. 3.7. In parallel cascade control, manipulated variable affects both variables directly (Luyben and Luyben, 1997). Because of the presence of two control loops, two controllers are to be tuned. Hang et al. (1994) proposed a relay autotuning of series cascade control loops. They used conventional on–off relay testing and, using the value of k_u from $4h/(\pi a_0)$ and Ziegler–Nichols (1942) tuning formulae, the controllers are tuned. With the inner loop under PI control action, relay test is repeated for the outer loop. In the present work, the methods of Srinivasan and Chidambaram (2003; 2004) are applied to tune parallel cascade controllers. The improved performance of the proposed parallel controller is compared with that of the conventional relay analysis.

3.5 Method 1 for Parallel Cascade Control Systems

Parallel cascade control scheme is considered in this section (Fig. 3.7); first the conventional on–off relay is considered. The relay is used in the inner loop and the outer loop is kept under manual mode. The relay oscillations are noted. For simulation study, the process modes assumed are $(k_pG_p)_2 = k_{p2}\exp(-4s)/(s+1)$ and $(k_pG_p)_1 = k_{p1}\exp(-6s)/(s+1)$. Where $k_{p1} = 1.0$ and $k_{p2} = 1.0$. Using a symmetric relay height of 1, oscillation in the inner loop output variable y_2 is noted (Fig. 3.8). The amplitude and frequency of oscillations are noted as 0.9815 and 0.6713 respectively. Using the relation $k_u = 4h/(\pi a_0)$, the value of k_u is obtained as 1.2972. Based on the transfer function model, exact value of k_u is calculated as 1.18. Thus, significant error is obtained in k_u by the conventional relay analysis. Using the results of relay testing, PI settings are calculated using the Ziegler and Nichols (1942) continuous cycling tuning method as $k_{c2} = 0.5737$ and $\tau_{I2} = 7.8$.

Using these PI settings in the inner loop (instead of the relay) and introducing a relay in the outer loop, the oscillations in the outer loop is noted with an amplitude of 1.1204 and a frequency of 0.4839 (Fig. 3.9). The value of $k_u = 1.1364$ is obtained from $4h/(\pi a_0)$. Based on the relay test results, the outer loop PID controller is designed by using the Ziegler–Nichols (1942) method as $k_{c1} = 0.6818$, $\tau_{I1} = 6.4915$ and $\tau_{D1} = 1.6229$. Closed loop servo response is evaluated for a unit step change in the set point and the response is shown in Fig. 3.10. An oscillatory response is obtained. A similar response is obtained for a regulatory problem [with $(k_LG_L)_2 = 1$], as shown in Fig. 3.11.

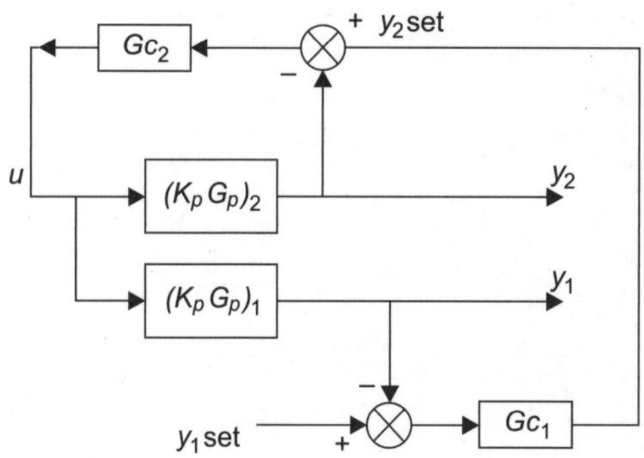

Fig. 3.7 *Parallel cascade control scheme*

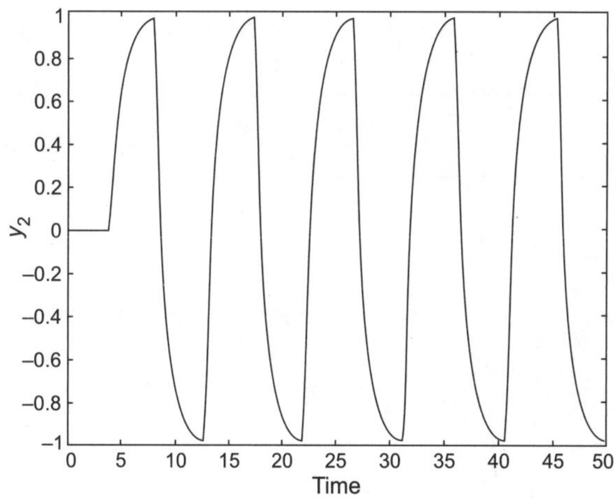

Fig. 3.8 *Response in y_2 using symmetric relay (with relay height ±1) in inner loop*

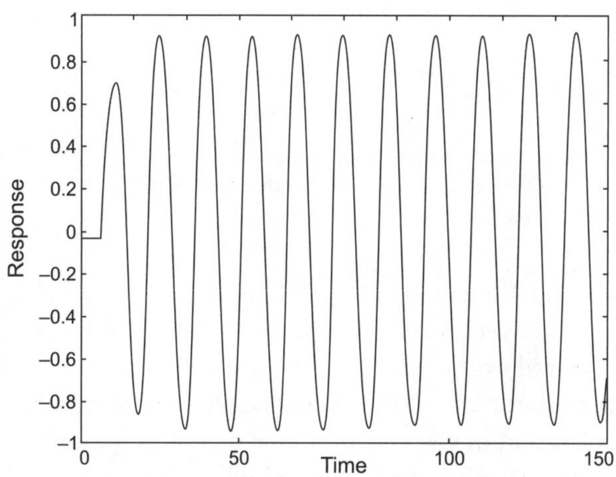

Fig. 3.9 *Response in y_1 for symmetric relay in outer loop (with inner loop on PI settings)*

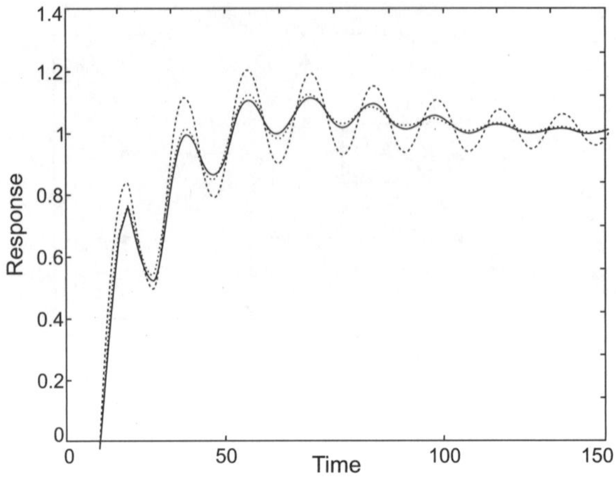

Fig. 3.10 *Servo response in y_1 using the Ziegler–Nichols settings. PID controller for outer loop and PI for inner loop*
Legend: outer oscillatory response–conventional analysis; inner solid–method 1; inner dash–method 2

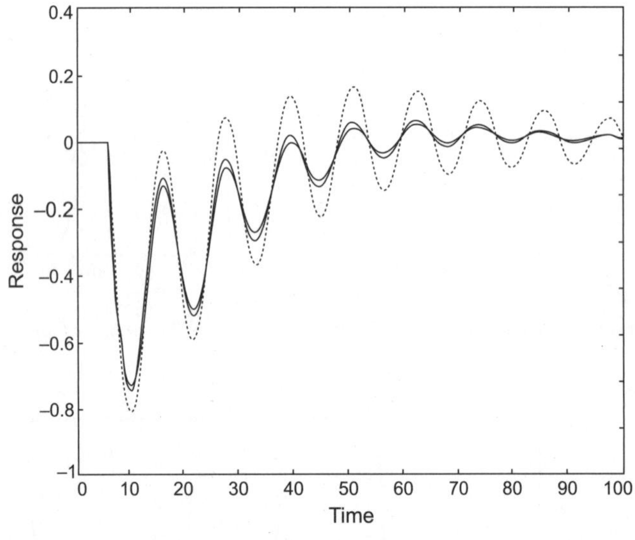

Fig. 3.11 *Regulatory response in y_1 for a disturbance in the inner loop using the Ziegler–Nichols settings. PID controller for the outer loop and PI for the inner loop*
Legend: as in Fig. 3.10a

Method of Srinivasan and Chidambaram (2004) is applied now. They suggested that the assumption made in the conventional relay oscillation is that all higher order harmonics (of the relay output) are filtered by the system. They gave a method to find out the value of k_u by considering the higher order harmonics (refer Chapter 2 for summary of the method). The initial portion of the relay output gives an indication of how many higher order harmonics are present in the relay output. A value of 5 higher order harmonics ($N = 5$) is recommended. Let us use their method for analyzing parallel cascade autotuning.

For the system under study, value of k_u in the inner loop is obtained as 1.1764 and frequency of oscillations $\omega_u = 0.6713$. Once the PI controller is designed on the basis of these values, the inner loop is kept under PI and then the relay is kept in the outer loop. From the relay oscillation, the value of $k_u = 1.1364$ is obtained by the conventional method. Srinivasan and Chidambaram (2004) method gives k_u as 1.0946 and $\omega_u = 0.4785$. A PID controller is designed based on this value. Details of the results are given in Table 3.5. The servo response is evaluated for a step change in the set point. The response in y_1 is shown in Fig. 3.9. The response by the conventional analysis using single harmonics gives an oscillatory response; whereas using the present method an excellent response is obtained. Similar performance is obtained for a regulatory problem $[(k_L G_L)_2 = 1]$ as shown in Fig. 3.11. The IAE values for servo and regulatory performance are given in Table 3.6.

Table 3.5 *Controller settings comparison using the Ziegler–Nichols settings for parallel cascade control system*

Loop	Controller settings	Symmetric relay		Asymmetric relay
		N = 1	N = 5	
Inner	k_{c2}	0.5837	0.5294	0.54
	τ_{I2}	7.8000	7.8000	7.9
		N = 1	N = 9	
Outer	k_{c1}	0.6818	0.6568	0.664
	τ_{I1}	6.4915	6.5660	6.6097
	τ_{D1}	1.6229	1.6415	1.6524

Table 3.6 *Performance comparison of proposed methods and conventional method for* $(D/\tau)_{\text{inner loop}} = 4.0$, $(D/\tau)_{\text{outer loop}} = 6.0$

Comparison parameters	Symmetric relay				Asymmetric relay	
	Conventional		Proposed method-1		Proposed method-2	
Overshoot	0.214		0.1196		0.1314	
Settling time	98		68		74	
	Servo	Regulatory	Servo	Regulatory	Servo	Regulatory
IAE	16.9819	15.1818	15.013	11.6161	15.0683	11.8893

3.6 Method 2 for Parallel Cascade Control Systems

In this section, the biased relay autotune is applied with a relay height of +2 and –1. In literature, for a single loop asymmetric relay tuning method, the value of $\gamma = 2$ is used. Therefore, the relay height of +2 and –1 is used in the present study as well. However, simulations studies are carried out with different relay heights of $\gamma = 2, 2.5, 3$ and 4. The results are summarized in Table 3.7. It is observed that the resulting PI controller settings are not changed significantly.

Table 3.7 *Effect of change in relay height using asymmetric relay testing on PI settings for parallel cascade control system*

Asymmetric relay		PI controller		
h	γ	k_u	k_c	τ_I
0.5	4.0	1.1859	0.5336	0.2050
1.0	2.0	1.1972	0.5387	0.1975
1.0	2.5	1.1902	0.5356	0.2000
1.0	3.0	1.1837	0.5327	0.2008

The method discussed in the preceding chapter is extended here to parallel cascade systems. Value of k_u for the inner loop based on the identified FOPTD model ($k_P = 0.9903$, $\tau = 0.9731$, $D = 4.0125$) obtained by the relay method is 1.20 and frequency of oscillation $\omega = 0.6621$. The PI controller is designed by the Ziegler–Nichols (1942) method. The inner loop is kept under PI and the asymmetric relay is then used in the outer loop. Based on the oscillation obtained in y_1 and hence based on the identified FOPTD model ($k_P = 1.0002$, $\tau = 0.9981$, $D = 6.0029$), the value of k_u is obtained for the outer loop as 1.1066 with $\omega = 0.4753$. The PID controller is designed for the outer loop based on the values of k_u and ω_u. Results of the relay test are given in Table 3.5. The servo response in y_1 for a unit step change in the set point is shown in Fig. 3.10. The performance is as good as the proposed method 1. In the proposed method 1, the value of order (N) of higher order harmonics is to be selected. Whereas in the asymmetric method by Srinivasan and Chidambaram (2003), no such value of N is required. In the asymmetric method, the model is to be identified and then the controller settings are calculated. In the symmetrical method, the controller settings are calculated on the basis of the ultimate values obtained from the relay test. Figure 3.11 shows the regulatory response for a step change in the inner loop disturbance.

3.7 Conclusions

The modified analysis of relay autotuning method proposed for single feedback system and modified analysis of asymmetric autotuning method discussed in the previous chapter are extended to tune series and parallel cascade controllers. Both the methods effectively take care of higher order harmonics. The performances of the PI–PID controllers are compared with the conventional relay analysis (principal harmonic analysis). The methods give a superior performance than that of the conventional analysis.

4 Simultaneous Relay Autotuning of Cascade Controllers

The present chapter is concerned with simultaneous relay autotuning of cascade controllers. Traditionally, autotuning procedures for cascade controllers are applied using a sequential, one-loop-at-a-time method. Saraf et al. (2003) proposed a simultaneous method of the relay tuning of series cascade control system. In this method, they assumed the principal harmonic analysis of relay oscillations. This results in error in calculation of the ultimate gains. The higher order harmonics are to be considered in the analysis of simultaneous relay tuning of series and parallel cascade control systems. Using improved ultimate gains, the inner loop (PI) and outer loop (PID) controllers are designed by the Ziegler–Nichols (Z-N) tuning method. The improved performance of the control is compared with the principal harmonic analysis reported by Saraf et al. (2003)

4.1 Introduction

Hang et al. (1994) proposed a sequential relay autotuning of series cascade control of stable systems. In this method, the outer loop is open. The on–off relay is used in the inner loop and the value of k_u is obtained from $4h/(\pi a_0)$. The controllers in the inner loop are tuned using the Ziegler–Nichols tuning formulae. With the inner loop under PI control action, the relay test is repeated for the outer loop.

Simultaneous relay tuning method is required whenever the system is to be kept under a closed loop in order to reduce the effect of disturbances or when the open loop system is unstable. As stated earlier, Saraf et al. (2003) assumed the principal harmonic analysis of relay oscillations. The method results in an error in the calculation of the ultimate gains. In the present work, the method of Srinivasan and Chidambaram (2004) that considers higher order harmonics is applied to simultaneous relay tuning of cascade control systems. Using the improved ultimate gain, the inner loop (PI) and outer loop (PID) controllers are designed by using the Ziegler–Nichols continuous cycling tuning method. The improved performance of the proposed method is compared with that of the Saraf et al. (2003) (considered as conventional method throughout the text).

4.2 Method for Series Cascade Control Systems

Series cascade control scheme is first considered here. For a simulation study, the process model is assumed as $y_1(s)/u(s) = k_{p2}\exp(-0.1s)/(0.025s + 1)$ and $y_2(s)/y_1(s) = k_{p1}\exp(-0.1s)/(0.025s + 1)$, where $k_{p1} = 1.0$ and $k_{p2} = 1.0$. The conventional on–off relay is used for both the loops simultaneously (Fig. 4.1). Using a symmetric relay height of ±1, the oscillation in the outer loop output variable y_1 is noted (Fig. 4.2). The amplitude and frequency of oscillations are noted as 0.9983 and 13.09 respectively. Using the relation $k_u = 4h/(\pi a_0)$, the value of $k_{u,1}$ is obtained as 1.2754. Using the results of relay testing, the PID settings are obtained by Ziegler–Nichols tuning formulae (continuous cycling method) as $k_{c,1} = 0.7652$, $\tau_{I,1} = 0.24$ and $\tau_{D,1} = 0.06$

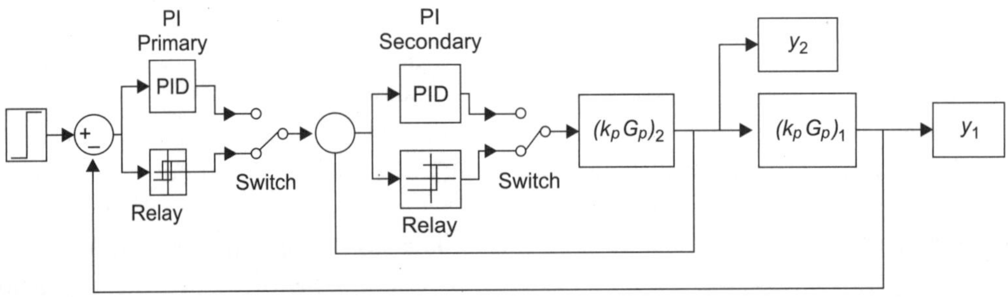

Fig. 4.1 *Relay feedback tuning of cascade control system*

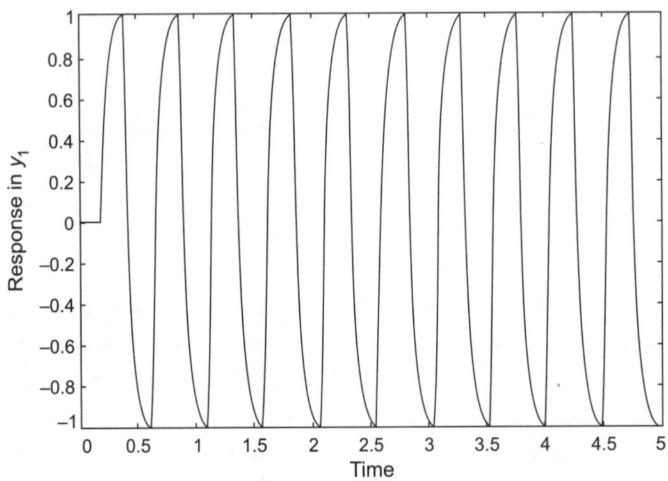

Fig. 4.2 *Response in y_1 using symmetric relay with relay height ±1 in both the outer and inner loops*

With PID settings in the outer loop (instead of the relay) and a relay in the inner loop, oscillations in the inner loop is noted with the amplitude of 0.9966 and frequency of 19.635 (Fig. 4.3). The value of $k_{u,2} = 1.2776$ is obtained from $4h/(\pi a_0)$. Based on the relay test results, the inner loop PI controller is designed using the Ziegler–Nichols continuous cycling method as $k_{c,2} = 0.575$ and $\tau_{I,2} = 0.267$.

The method of Srinivasan and Chidambaram (2004) is applied now. As stated earlier, they gave a method to find out the value of k_u by considering the higher order harmonics. The initial portion of the process output gives an indication of how many higher order harmonics are present in the relay output. A value of 5 higher order harmonics ($N = 5$) is recommended. Let us use their method for analyzing the series cascade autotuning.

For the system under study, the conventional on–off relay is used for both the loops simultaneously. Using a symmetric relay height of ±1, oscillations in the outer loop output variable y_1 is noted (Fig. 4.2). At $N = 5$, Srinivasan and Chidambaram (2004) method gives corrected $k_{u,1}$ as 1.0881 and $\omega_{u,1} = 13.09$. Refer to Table 4.1 for detailed calculations. A PID controller by the *Z-N* continuous cycling method, is designed based on these values (Table 4.2).

Now, the outer loop is kept under calculated PID controller settings and the relay is kept in the inner loop. Srinivasan and Chidambaram (2004) method gives $k_{u,2}$ as 1.09 and $\omega_{u,2} = 20.944$ at $N = 9$. A PI controller is designed based on this value. The details of the results are given in Table 4.1.

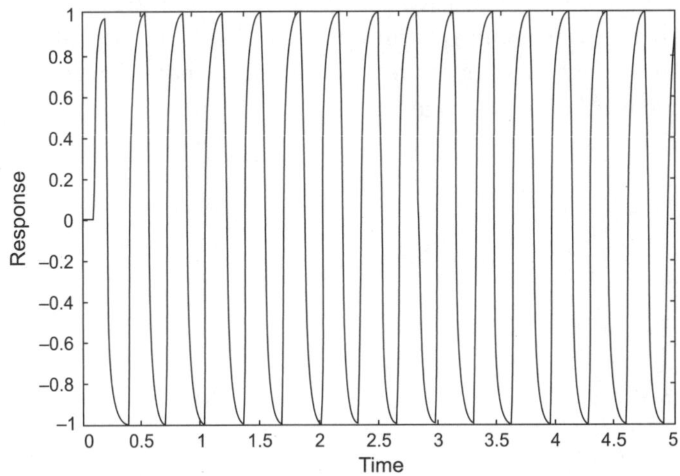

Fig. 4.3 *Response in y_2 for symmetric relay in inner loop and PID in outer loop*

Table 4.1 *Calculation details for proposed method (series cascade control system)*

Outer loop								
P_u	ω_u	t^*	$y(t^*)$	a_0	a_c	$k_{u,1}$		
						$N = 1$	$N = 3$	$N = 5$
0.48	13.09	0.12	0.977	0.9983	1.17	1.2754	1.1295	1.0881
Inner loop								
P_u	ω_u	t^*	$y(t^*)$	a_0	a_c	$k_{u,2}$		
						$N = 1$	$N = 5$	$N = 9$
0.3	20.944	0.075	0.9494	0.9949	1.1676	1.2798	1.1197	1.09

Note: a_0 – observed amplitude, a_c – calculated amplitude and $t^* = 0.5\pi/\omega_u$

Table 4.2 *Controller settings comparison using the Ziegler–Nichols settings (series cascade control systems)*

		Saraf et al. method	Proposed method
Number of harmonics considered		$N = 1$	$N = 5$
Outer loop	$k_{c,1}$	0.7652	0.6529
	$\tau_{I,1}$	0.24	0.24
		0.06	0.06
Number of harmonics considered		$N = 1$	$N = 9$
Inner loop	$k_{c,2}$	0.575	0.49
	$\tau_{I,2}$	0.267	0.25

Note: PI controller: $k_c = 0.45\ k_u$, $\tau_I = p_u/1.2$
PID controller: $k_c = 0.6\ k_u$, $\tau_I = p_u/2$, $\tau_D = p_u/8$

The regulatory response is evaluated for a step change in a disturbance entering the inner loop $[(k_L\,G_L)_2 = 1]$. The response in y_1 is shown in Fig. 4.4. It is observed that Srinivasan and Chidambaram (2004) method gives an improved performance. Similar performance is also obtained for a servo problem as shown in Fig. 4.5.

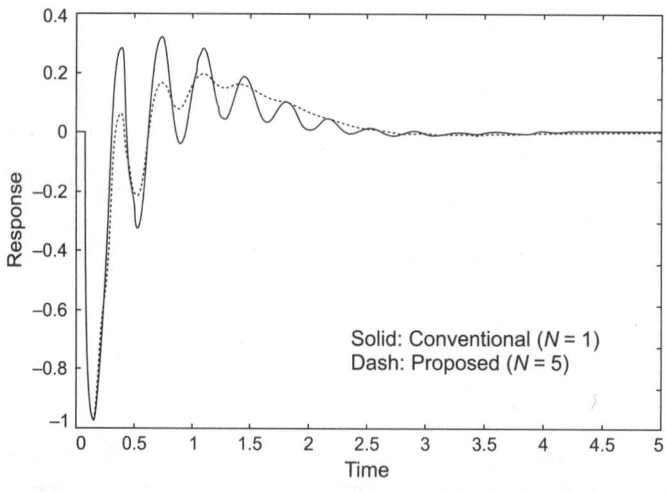

Fig. 4.4 *Regulatory response in y_1 for a disturbance entering in the inner loop using PID controller for outer loop and PI for inner loop*
Legend: solid line–Saraf at al. method; dash–proposed method

Figure 4.6 shows a response in u corresponding to Fig. 4.5. It is observed that consideration of higher order harmonics gives improved performance. The effect of measurement noise is studied by adding a random noise with Gaussian distribution with zero mean and a standard deviation of 0.2% in the inner loop and corrupted signal is used for the feed back. The present method gives $k_{u1} = 1.087$ and $\omega_u = 13.09$. A PID controller is designed by the Ziegler–Nichols method. The outer loop is kept under the PID controller and the relay is used in the inner loop (refer to Fig. 4.7). Based on the oscillations obtained in y_2, the value of k_u is obtained for the inner loop as 1.118. A PI controller is designed for the inner loop based on the value of k_u. The results for the relay testing with noise are given in Table 4.3. The results are close to the results obtained without any noise. Figure 4.8 shows a response for a unit step change in the set point. The method gives improved results.

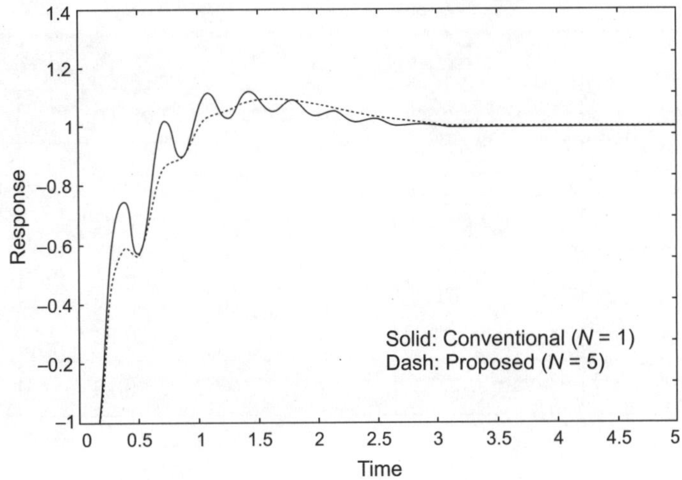

Fig. 4.5 *Servo response in y_1 using PID controller for the outer loop and PI for the inner loop*
Legend: as in Fig. 4.4

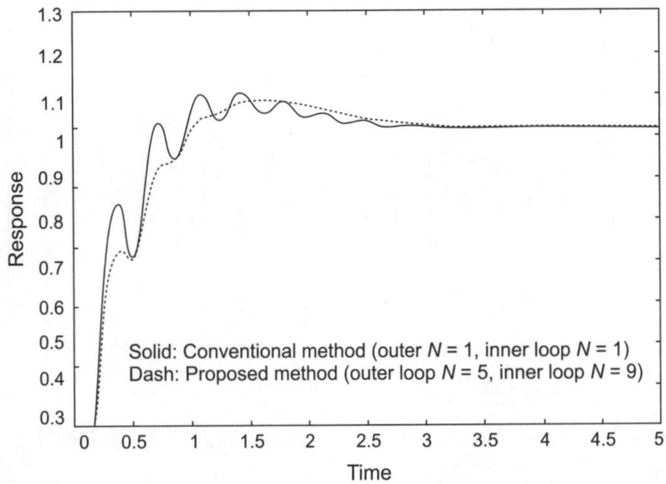

Fig. 4.6 *Response in u corresponding to Fig. 4.5 using PID controller for the outer loop and PI for the inner loop*
Legend: as in Fig. 4.4

The effect of load is studied by adding a load of 0.02 in the inner loop. Transfer function for the load is assumed as that of the process transfer function. The load affected output signal is used for the feedback relay and as the system output for k_u and p_u identification. Effect of the load is reported in Table 4.3. Figure 4.9 shows a response for a unit step change in the set point with 0.02 load in the inner loop. The results are close to those obtained without any load.

Table 4.3 *Controller settings for simultaneous relay series cascade control system*

Parameter		Conventional method		Proposed method	
		Noise 0.2%	Load 0.02	Noise 0.2%	Load 0.02
Outer loop	$k_{c,1}$	0.77	0.75	0.65	0.64
	$\tau_{I,1}$	0.24	0.18	0.24	0.18
	τ_D	0.06	0.045	0.06	0.045
Inner loop	$k_{c,2}$	0.58	0.565	0.52	0.5
	$\tau_{I,2}$	0.27	0.25	0.25	0.24

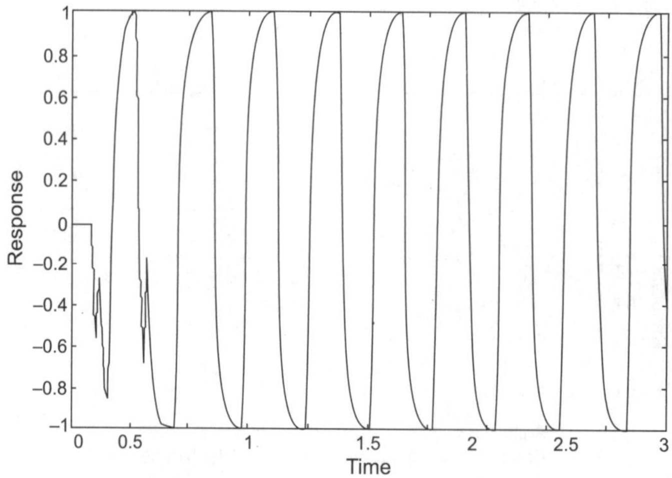

Fig. 4.7 *Response in y_2 for symmetric relay in the inner loop and PID in the outer loop in presence of measurement noise in the inner loop*

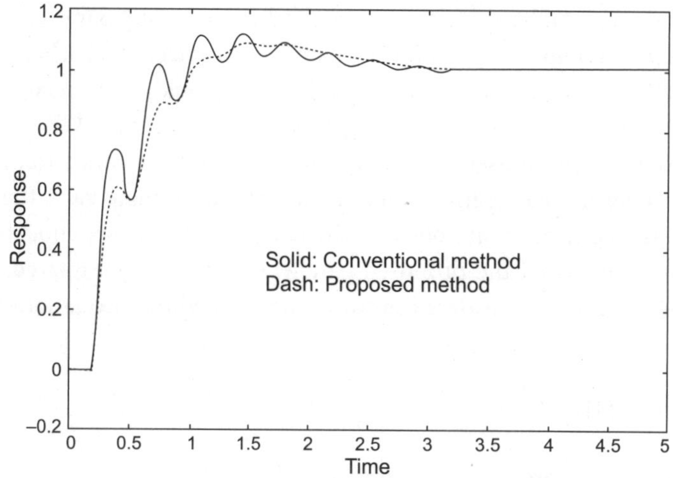

Fig. 4.8 *Servo response in y_1 using PID controller for the outer loop and PI for the inner loop with noise in the inner loop*
Legend: as in Fig. 4.4

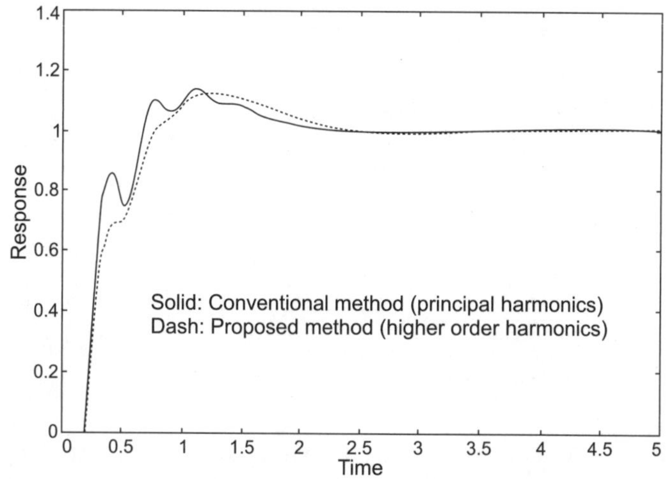

Fig. 4.9 *Servo response in y_1 using PID controller for the outer loop and PI for the inner loop with load in the inner loop*
Legend: as in Fig. 4.4

4.3 Method for Parallel Cascade Control Systems

A parallel cascade control scheme utilizes two control loops: secondary (or slave or inner loop) and primary (or master or outer loop). The secondary loop receives its set point from the primary loop. Figure 4.10 shows a schematic diagram of the parallel cascade control scheme. In a parallel cascade control, manipulated variable affects both variables directly. There are two controllers to be tuned.

In the present work, simultaneous tuning method proposed by Saraf et al. (2003) for the series cascade control system is applied for the parallel cascade control scheme. In the method of Saraf et al. (2003) they assumed the principal harmonic analysis of relay oscillations. This results in an error in the calculation of ultimate gains. In the present work, method of Srinivasan and Chidambaram (2004) proposed for a single loop system is extended to get more accurate values of the controller ultimate gains. Using the improved ultimate gain, the inner loop (PI) and outer loop (PID) controllers are designed by using the Ziegler–Nichols tuning formulae. The improved performance of the proposed parallel controller is compared with that of the analysis of Saraf et al. (2003) (considered as conventional method throughout the text).

4.4 Simulation Study

For a simulation study, the process model is assumed as

$$y_1(s)/u(s) = k_{p2}\exp(-0.1s)/(0.025s + 1) \tag{4.1}$$

$$y_2(s)/u(s) = k_{p1}\exp(-0.1s)/(0.025s + 1). \tag{4.2}$$

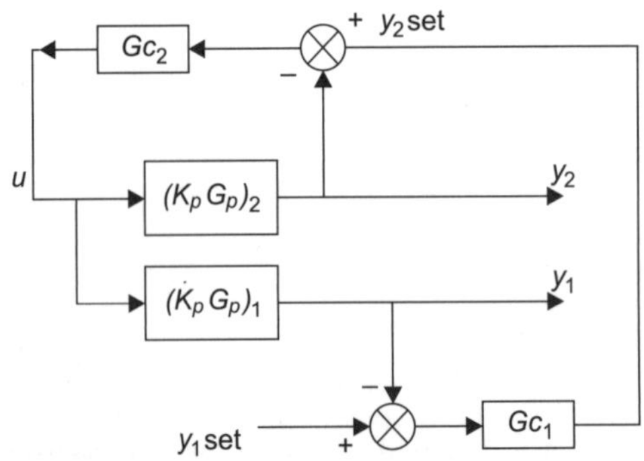

Fig. 4.10 *Parallel cascade control scheme*

Where $k_{p1} = 1.0$ and $k_{p2} = 1.0$. The conventional on–off relay is used for both the loops simultaneously. Output of the primary loop is input of the secondary auto relay loop. The secondary loop is forced to oscillate at the frequency of the primary loop because of the condition $\omega_{cu}^{\text{primary}} \leq \omega_{cu}^{\text{secondary}}$. Thus, ultimate gain and ultimate frequency are first obtained for the primary loop. Hence, the primary controller parameters can be obtained. Then, the secondary loop executes relay feedback with a primary controller in place in order to obtain its own ultimate period of oscillations, and the secondary controller parameters are obtained. Using a symmetric relay height of ±1, oscillations in the outer loop output variable y_1 is noted (Fig. 4.11). Amplitude and frequency of oscillations are noted as 0.9798 and 27.3182 respectively. Using the relation $k_u = 4h/(\pi a_0)$, value of $k_{u,1}$ is obtained as 1.30. Using the results of the relay testing, PID settings are obtained by the Ziegler and Nichols tuning formulae (continuous cycling method) as $k_{c,1} = 0.78$, $\tau_{I,1} = 0.115$ and $\tau_{D,1} = 0.0228$.

With this PID setting in the outer loop (instead of the relay) and a relay in the inner loop, oscillations in the inner loop is noted with an amplitude of 0.9687 and frequency of 29.9199 (Fig. 4.12). The value of $k_{u,2}$ = 1.3144 is obtained from $4h/(\pi a_0)$. Based on the relay test results, the inner loop PI controller is designed using the Ziegler–Nichols method as $k_{c,2}$ = 0.5915 and $\tau_{I,1}$ = 0.175. The method of Srinivasan and Chidambaram (2004) is applied now. As stated earlier, they suggested a method to find out the corrected value of $k_{u,}$ by considering the higher order harmonics. The initial portion of the process output gives an indication of how many higher order harmonics are present in the relay output. A value of 5 higher order harmonics (N = 5) is recommended. Let us use their method for analyzing the parallel cascade autotuning.

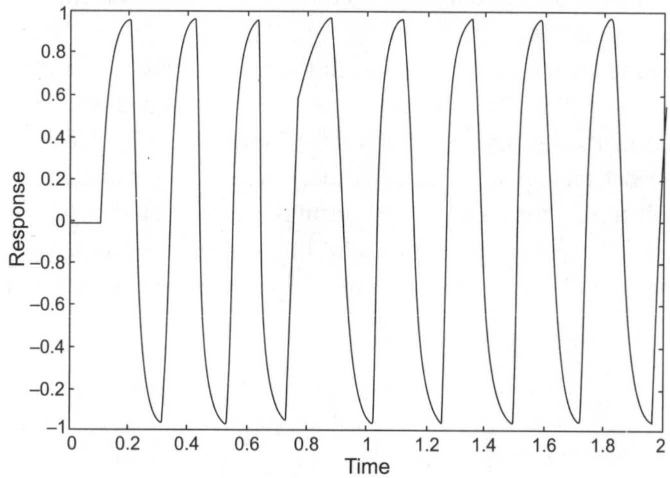

Fig. 4.11 *Response in y_1 using symmetric relay with relay height ±1 (parallel cascade control system)*

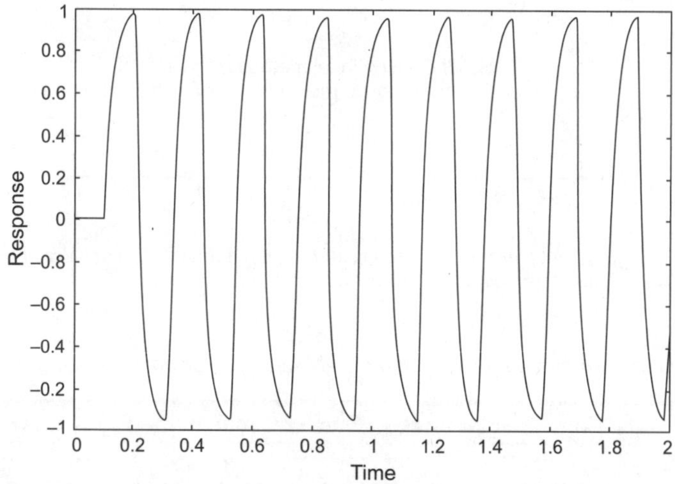

Fig. 4.12 *Response in y_2 using symmetric relay with relay height ±1 (parallel cascade control system)*

4.5 Higher Order Harmonics in both the Loops

In the present section the higher order harmonics are considered for both the loops. The conventional on–off relay is used for both the loops simultaneously. Using a symmetric relay height of ±1, oscillations in the outer loop output variable y_1 is noted. Amplitude and frequency of oscillations are noted as 0.9798 and 27.3182 respectively. Using Srinivasan and Chidambaram 2004 method for the outer loop, the exact value of $k_{u,1}$ is calculated as 1.18. Thus, significant error is obtained in $k_{u,1}$ by the conventional relay analysis. Using Srinivasan and Chidambaram (2004) method and considering higher order harmonics (N = 5) value of $k_{u,1}$ as 1.18 and $\omega_{u,1}$ = 27.3182 are obtained. PID settings are calculated using the Ziegler–Nichols tuning formulae as $k_{c,1}$ = 0.7, $\tau_{I,1}$ = 0.115 and $\tau_{D,1}$ = 0.0288. Using these PID settings in the outer loop and a relay in the inner loop, oscillations in the inner loop is noted with an amplitude of 0.9697 and frequency of 29.92. The value of $k_{u,2}$ = 1.313 is obtained from $4h/(\pi a_0)$. Srinivasan and Chidambaram method is applied now for the inner loop. Considering the higher order harmonics (N = 9 since there is an initial dynamics), value of $k_{u,2}$ is calculated as 1.18. By using the Ziegler–Nichols tuning formulae, the PI settings are calculated as $k_{c,2}$ = 0.53, $\tau_{I,2}$ = 0.175. Table 4.4 shows the details of the calculations for the controller settings. The servo response is evaluated for a unit step change in the set point. The response in y_1 is shown in Fig. 4.13.

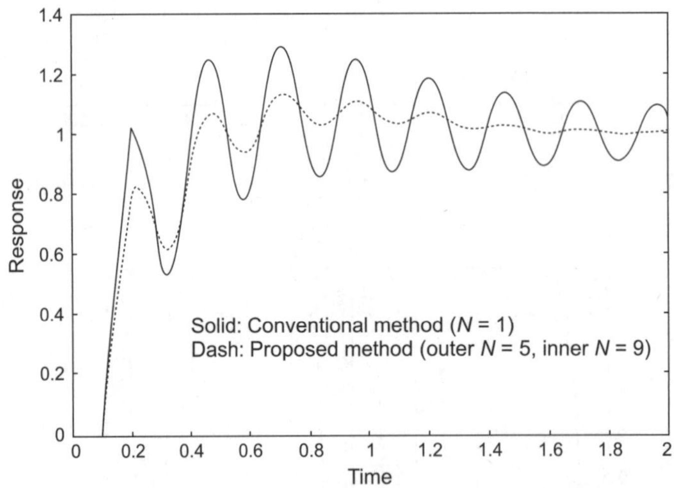

Fig. 4.13 *Servo response in y_1 for unit step change in the set point with PI controller in the inner loop and the outer loop on PID settings*

Table 4.4 *Controller settings comparison using Z–N settings (for parallel cascade control systems)*

Loop	Controller settings	Conventional method	S-C method $N = 5$
Outer loop	$k_{c,1}$	0.78	0.70
	$\tau_{I,1}$	0.115	0.115
	$\tau_{D,1}$	0.0288	0.0288

Contd.

Table 4.4 *Contd.*

Inner loop	$k_{c,2}$	0.5915	0.53
	$\tau_{I,2}$	0.175	0.175

Note: S-C method: Srinivasan and Chidambaram method; Conventional method: principal harmonic assumption

It is observed that Srinivasan and Chidambaram method gives an improved performance. Figure 4.14 shows a response in u for a servo problem. It is observed that the proposed method gives an improved performance. A similar performance is obtained for a regulatory problem $[(k_L\, G_L)_2 = 1]$ as shown in Fig. 4.15. The comparison of ISE and IAE values for the servo and regulatory performances is given in Table 4.5.

Table 4.5 *Performance comparison for parallel cascade control systems*

	Performance measure	Conventional method	S-C method $N = 5$
Servo	ISE	0.172	0.159
	IAE	0.3597	0.271
	ITAE	0.195	0.082
Regulatory	ISE	0.2069	0.116
	IAE	0.4882	0.2592
	ITAE	0.3789	0.1275

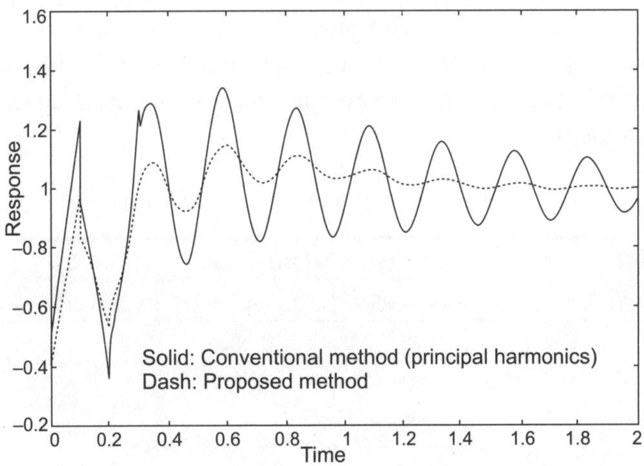

Fig. 4.14 *Response in u corresponding to Fig. 4.13. PID controller for the outer loop and PI for the inner loop*

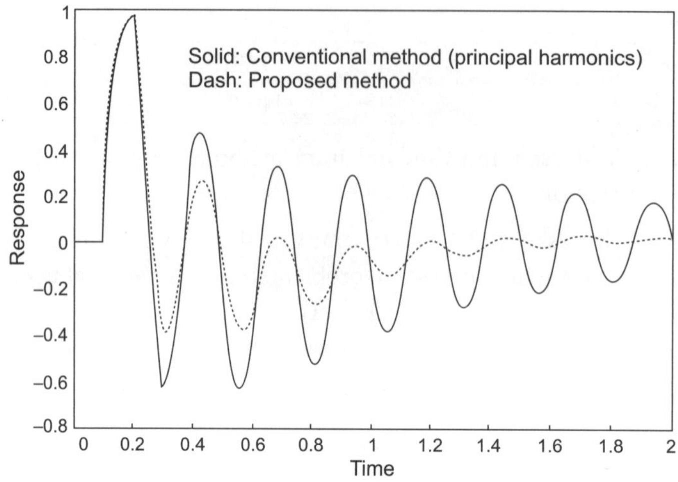

Fig. 4.15 *Regulatory response in y_1 for a disturbance in the inner loop using PID controller for the outer loop and PI for the inner loop*

The effect of measurement noise is studied by adding a random noise with Gaussian distribution with zero mean and a standard deviation of 0.2% in the inner loop and the corrupted signal is used for the feedback. The present method gives $k_{u1} = 1.18$ and $\omega_u = 27.3182$. A PID controller is designed using the Ziegler–Nichols method. The outer loop is kept under a PID controller and the relay is used in the inner loop (Fig. 4.16). Based on the oscillations obtained in y_2, value of k_u is obtained for the inner loop as 1.18. A PI controller is designed for the inner loop based on the value of k_u. The results for relay with noise are given in Table 4.6. The results are close to the results obtained without any noise. Figure 4.17 shows a response for a unit step change in the set point.

The effect of load is studied by adding a load of 0.02 in the inner loop. Transfer function for the load is assumed as that of the process transfer function. The load affected output signal is used for the feedback relay and as the system output for k_u and p_u identification. The effect of load is reported in Table 4.6. Figure 4.18 shows a response for a unit change in the set point with load present in the inner loop.

Table 4.6 *Controller design comparison for simultaneous relay for parallel cascade control system*

Parameter		Conventional method		Proposed method	
		Noise 0.2%	Load 0.02	Noise 0.2%	Load 0.02
Outer loop	k_{c1}	0.78	0.786	0.70	0.7
	τ_{I1}	0.115	0.12	0.115	0.12
	τ_{D1}	0.029	0.03	0.029	0.03
Inner loop	k_{c2}	0.595	0.58	0.53	0.52
	τ_{I2}	0.175	0.175	0.175	0.175

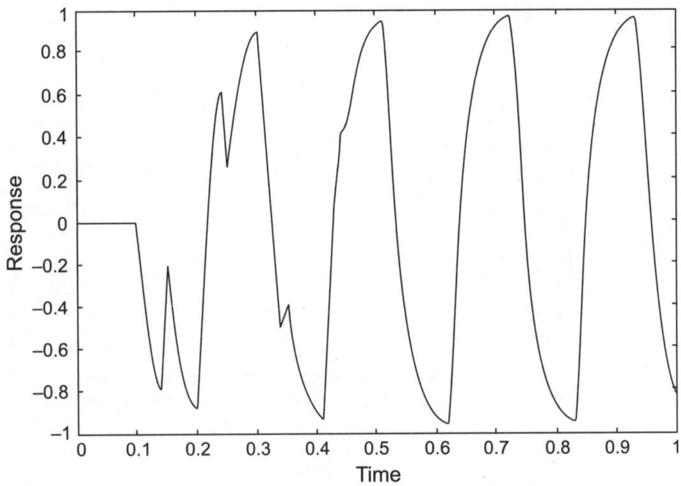

Fig. 4.16 *Response in y_2 using symmetric relay with noise in the inner loop and the outer loop on PID controller*

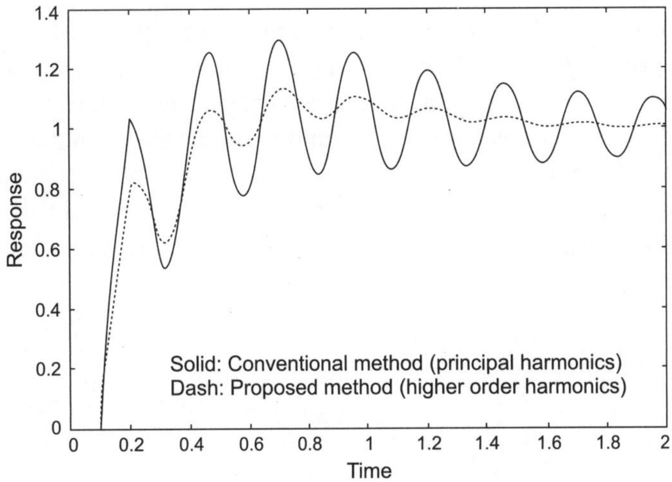

Fig. 4.17 *Servo response in y_1 with PI controller in the inner loop and the outer loop on PID settings (with 0.2% noise)*

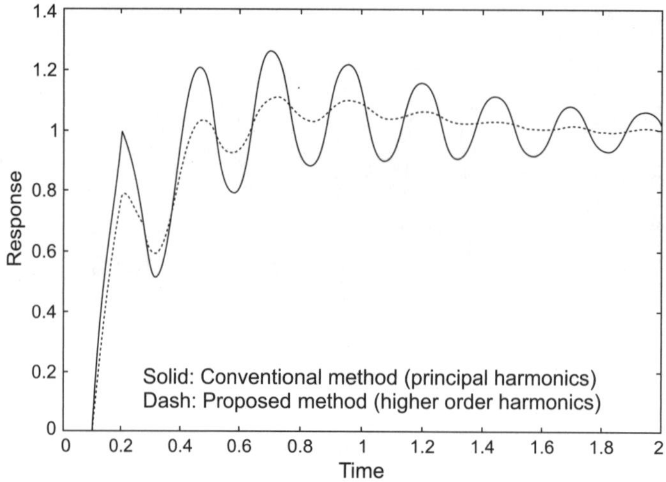

Fig. 4.18 *Servo response in y_1 with PI controller in the inner loop and the outer loop on PID settings (in presence of load)*

4.6 Conclusions

The modified analysis of relay autotuning proposed for a single feedback system discussed in Chapter 2 is extended to simultaneous tuning of (i) series cascade control system and (ii) a parallel cascade control system. The method effectively takes care of higher order harmonics. Based on the calculated value of ultimate controller gain and the noted period of oscillation, PI–PID controllers are designed by the Ziegler–Nichols tuning method. The performances of the PI–PID controllers are compared with the principal harmonics method of Saraf et al. (2003) The present methods give a better performance than that of the conventional analysis.

5 A Simple Method of Tuning Cascade Controllers

A simple method is proposed to design PID controllers (i) for a series cascade control system and (ii) for a parallel cascade control system. The method is based on equating coefficients of the corresponding powers of q and q^2 in the numerator to α_1 and α_2 times that in the denominator of the closed loop transfer function model for a servo problem. This method can be used only when the inner and outer loop transfer functions are known. If these transfer functions are not known, then an identification step needs to be carried out. The method is first applied to design a proportional (P) controller for the inner loop and then to design a proportional plus integral (PI) controller for the outer loop. Performances of the controllers are evaluated for the FOPTD models of the inner loop and outer loop.

5.1 Introduction

As stated earlier, cascade control is one of the most popular structures for process control. A cascade control system consists of a primary controller and a secondary controller (refer to Fig. 5.1). Cascade control scheme is used to improve the dynamics response of the closed loop system when the disturbance enters the inner loop or disturbances are present in the manipulated variable. The frequency response method (Edgar et al., 1982) is usually employed to design such controllers. The method involves trial and error graphical method. Krishnaswamy et al. (1990) proposed a tuning chart that predicts the primary controller settings by minimizing the ITAE criterion due to load disturbances on the secondary loop.

Lee et al. (1998b) proposed a method for synthesizing series cascade controllers. The method consists of first finding the ideal controllers that give desired closed loop responses and then finding the PID approximation of the ideal controllers using Maclaurin series. Their method is shown to be better than the ITAE method. However, the method has two tuning parameters (closed loop and outer loop time constants) whose values are to be selected by trial and error method. In the present work, a simple method is proposed for designing cascade controllers. The proposed method is an extension of the method presented by Chidambaram and Padmasree (2003) and Padmasree et al. (2004) for a single loop feedback system to a cascade control system. The proposed method is based on the servo problem. Most of the methods in the literature, such as IMC, pole placement and synthesis method, are based on the servo problem. The servo problem needs only $y(s)/u(s)$ information, while a controller based on the regulatory problem requires information on which load variable is considered (inner or outer loop) and its transfer function model.

As stated earlier, a parallel cascade control system consists of a primary controller and a secondary controller (Luyben and Luyben, 1997). Schematic diagram of parallel control scheme is shown in Fig. 4.10. Output of the primary controller resets the set point of the secondary loop. The secondary controller can be designed for disturbance rejection purpose, while the primary controller is for maintaining good servo responses. Output of the secondary controller is the manipulated variable that affects two output variables. Yu (1988) explored the disturbance rejection capability of parallel cascade control systems. Shen and Yu (1990) applied these results to the selection of the secondary measurement when a number of different disturbances are present. Brambilla and Semino (1992) introduced a non-linear filter between the controllers in order to partially decouple the two loops and improve the performances of the control system. Many processes have an inherently parallel cascade structure, i.e., the secondary output is not merely an intermediate process output as in series cascade structure (Pottmann et al., 1996). Cascade control scheme is used to improve the dynamics response of the closed loop system when the disturbance is entering the inner loop or disturbances are present in the manipulated variable.

Procedure for a parallel cascade control system is to tune the secondary controller first and then tune the primary controller with the secondary controller on automatic. In the present work, a simple method is proposed for designing parallel cascade controllers. It is an extension of the method presented by Chidambaram and Padmasree (2003) and Padmasree et al. (2004) for a single loop feedback system to a parallel cascade control system.

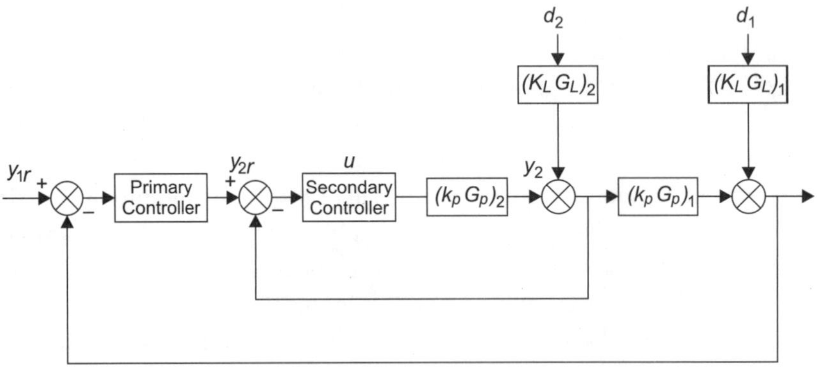

Fig. 5.1 *Series cascade control scheme*

5.2 Method for Series Cascade Control Systems

Let us consider the transfer function model of the process for the inner loop (Fig. 5.1) as an FOPTD (first order plus time delay) model:

$$y_2(s)/u(s) = \left[k_{p2}\exp(-D_2 s)/(\tau_2 s+1)\right] \tag{5.1}$$

The transfer function of the process in the outer loop is also an FOPTD model:

$$y_1(s)/y_2(s) = \left[k_{p1}\exp(-D_1 s)/(\tau_1 s+1)\right] \tag{5.2}$$

Usually we consider a simple proportional (P) controller for the inner loop and a proportional plus integral (PI) controller for the outer loop. This combination gives a good performance and contains only three tuning parameters which are to be calculated (Lee et al., 1998b). We first design the inner loop controller. The closed loop transfer function model for the inner loop is given by

$$y_2 / y_{2r} = K_2\exp(-\varepsilon q)/\left[q+1+K_2\exp(-\varepsilon q)\right] \tag{5.3}$$

where $K_2 = k_{c2}\,k_{p_2};\ \varepsilon = D_2/\tau_2;\ q = \tau_2 s.$ (5.4)

Time delay system responds only after the time delay (D), hence we cannot alter response between time $t = 0$ and D. In Eq.(5.3), we need not consider the term exp($-\varepsilon q$) further in the numerator as it will only shift the time axis. To make degree of polynomial in q in the numerator same as that of the denominator, the following steps are carried out:

In the denominator exp($-\varepsilon q$) is rewritten as exp($-0.5\varepsilon q$)/exp($0.5\varepsilon q$). Then the term exp($0.5\varepsilon q$) is taken to the numerator and also to the first term in the denominator. By substituting the Taylor's series approximation for the delay terms exp($0.5\varepsilon q$) and exp($-0.5\varepsilon q$), we get

$$\begin{aligned} y_2 / y_{2r} = K_2\left[1+0.5\varepsilon q+0.125\varepsilon^2 q^2+\cdots\right]/\left[(q+1)\left(1+0.5\varepsilon q+0.125\varepsilon^2 q^2+\cdots\right)\right. \\ \left. +K_2\left(1-0.5\varepsilon q+0.125\varepsilon^2 q^2-\cdots\right)\right]. \end{aligned} \tag{5.5}$$

Since the objective of the control system is to make y follow y_{2r}, the numerator polynomial can be made equal to denominator polynomial. For controller design, depending on the type of controller, we are equating the corresponding coefficient of q or q^2 or q^3 in the numerator to that of the denominator. This step means that for proportional controller we equate the coefficient of q only. For design of PI controller the coefficients of q and q^2 are related to that of the denominator. For the design of PID controllers, the coefficients of q, q^2 and q^3 are to be related. Hence, accordingly, the proper number terms in the Taylor's series approximation for exp($0.5\varepsilon q$) and exp($-0.5\varepsilon q$) are considered.

Since a proportional controller is used, there is bound to be an offset. Hence, the coefficient of q^0 in the numerator cannot be equated to the denominator. By equating the coefficient of q in the numerator to that in the denominator, controller gain is calculated as

$$k_{c2} = [(1/\varepsilon)+0.5]/k_{p2} \tag{5.6}$$

A similar procedure is adopted to design a controller for the outer loop. Now we have PI controller in the outer loop. The closed loop transfer function for the outer loop is obtained as

$$y_1 / y_{1r} = R_1\exp(-\eta s)/R_2 \tag{5.7}$$

$$R_1 = (P_1 q+P_2)\exp(0.5\psi q) \tag{5.8}$$

$$\begin{aligned} R_2 = \left[(q+1)\exp(0.5\varepsilon q)+K_2\exp(-0.5\varepsilon q)\right]\left[\phi q^2+q\right]\exp(0.5\eta q) \\ +(P_1 q+P_2)\exp(-0.5\eta q)\exp(0.5\varepsilon q)\right] \end{aligned} \tag{5.9}$$

where

$$\phi = \tau_1 / \tau_2;\ \tau_1^* = \tau_1 / \tau_2;\ \lambda = D_1 / \tau_2;\ \eta = (\varepsilon + \lambda);\ \psi = \eta + \varepsilon \tag{5.10}$$

$$K_1 = k_{c1} k_{p1};\ K_2 = k_{c2} k_{p2};\ P_1 = K_1 K_2;\ P_2 = P_1 / \tau_1^* \tag{5.11}$$

To make degree of the numerator equal to that of the denominator, we make use of the Taylor's series approximation for the exponential term. Since we equate the coefficients of q and q^2 in the numerator to that in the denominator, it is sufficient to retain the terms in the Taylor's series only up to the power of q^2.

Let us now consider the proposed method with two tuning parameters (α_1 and α_2). It is to be noted that there are no tuning parameters for the inner loop. Similar to the procedure carried out for designing the inner loop controller, on equating the coefficient of q, q^2 in the numerator to α_1 and α_2 times that of the corresponding coefficient in the denominator in Eq.(5.7), we get

$$P_1 A_1 + P_2 A_2 = \alpha_1 (1 + K_2) \tag{5.12}$$

$$P_1 A_3 + P_2 A_4 = \alpha_2 A_5 \tag{5.13}$$

where $A_1 = (1 - \alpha_1)$

$$A_2 = 0.5(\psi + \lambda \alpha_1)$$

$$A_3 = 0.5(\psi + \lambda \alpha_2)$$

$$A_4 = 0.125(\psi^2 - \lambda^2 \alpha_2)$$

$$A_5 = (\phi(1 + K_2) + 1 + 0.5\,\varepsilon - 0.5\varepsilon K_2 + 0.5\eta(1 + K_2)) \tag{5.14}$$

Once P_1 and P_2 are calculated, then k_{c1} and τ_I can be calculated from eqn. (5.11) as

$$k_{c1} = P_1 / (k_{c2} k_{p2} k_{p1}) \tag{5.15}$$

$$\tau_1 = (k_{c1} k_{p1} k_{c2} k_{p2}) \tau_2 / P_2 \tag{5.16}$$

5.3 Simulation Results

Let us consider the example system considered by Lee et al. (1998b) with the transfer function models as

$$k_{p1}\ G_{p1} = \exp(-10s)/(100s + 1) \tag{5.17}$$

$$k_{p2}\ G_{p2} = 2.0\ \exp(-2s)/(20s + 1) \tag{5.18}$$

Lee et al. (1998b) proposed the P setting for the inner loop and PI settings for the outer loop as $k_{c2} = 3.44$, $k_{c1} = 5.83$ and $\tau_{I1} = 105$, respectively. Krishnaswamy et al. (1990) proposed the corresponding settings as $k_{c2} = 2.978$, $k_{c1} = 7.3$ and $\tau_{I1} = 200$. There is no tuning parameter for the inner loop. For the outer loop, there are two tuning parameters. It is to be noted that for a single loop FOPTD system Padmasree et al. (2004) showed that equating corresponding coefficients (i.e., making $\alpha_1 = 1.0$ and $\alpha_2 = 1.0$) gives a good result. We call this method as method 1. Thus equating $\alpha_1 = 1.0$ and $\alpha_2 = 1.0$ for the above example gives the settings as $k_{c2} = 5.25$, $k_{c1} = 9.666$ and $\tau_{I1} = 105.91$ as shown in Table 5.1. The servo responses for a unit step change in the set point of y_{1r} are shown in Fig. 5.2. Similarly, the regulatory responses for a unit step change in the load (entering in the secondary loop at the outlet of the process) are compared in Fig. 5.3. The error comparisons of the different method under perfect parameters are shown in Table 5.2. However, it is found in the present work that such an approach gives a better regulatory performance than that of the servo response. In Tables 5.1 and 5.2, all the performance measures such as ISE, IAE and ITAE are given. Since the response takes considerably longer time, the ITAE measure will be appropriate here for comparison.

The first method in which the controller settings are calculated based on $\alpha_1 = 1.0$ and $\alpha_2 = 1.0$ gives an oscillatory servo response. This may be due to tight settings. Hence, a second method is proposed. Values of α_1 and α_2 are chosen so as to provide a good performance. The guidelines for α_1 and α_2 are studied especially for the case of FOPTD model. Extensive simulation studies are carried out to find the best ratio of α_2/α_1 in the sense of ITAE performance. The ratio α_2/α_1 is varied from 0.3 to 0.7 and that of α_1 from 0.5 to 1.2. It is observed that at $\alpha_2/\alpha_1 = 0.4$ and $\alpha_1 = 0.9$ for the servo response in y_1, the lowest ITAE is obtained. We call this modified method as method 2. For the outer loop, since the values of these parameters (α_1, α_2/α_1) vary from around 0.5 to 1, it may be easier to tune these parameters to calculate the controller settings.

It is found by simulation that for the present case study, use of $\alpha_1 = 0.9$ and $\alpha_2 = 0.4\ \alpha_1$, (giving $k_{c2} = 5.25$, $k_{c1} = 4.68$ and $\tau_{I1} = 104.3$) gives the best result. From Eq.(5.6), it is seen that k_{c2} does not depend on α_2. But, α_2 affects k_{c1}.

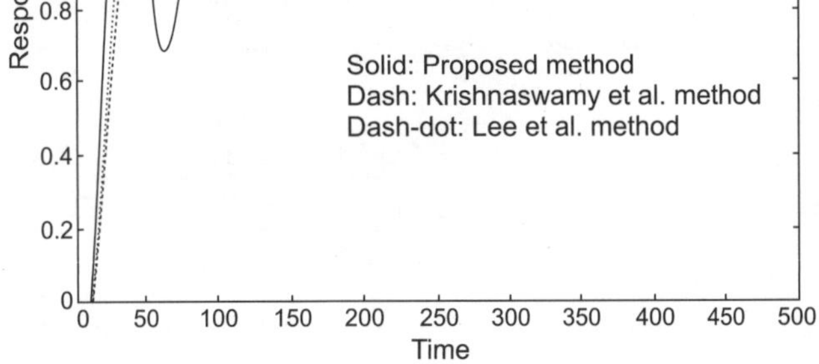

Fig. 5.2 *Comparison of different methods for servo response in y_1 using PI controller for the outer loop and P controller for the inner loop, under perfect parameters ($\alpha_1 = 1.0$ and $\alpha_2 = 1.0$ for series cascade control system)*

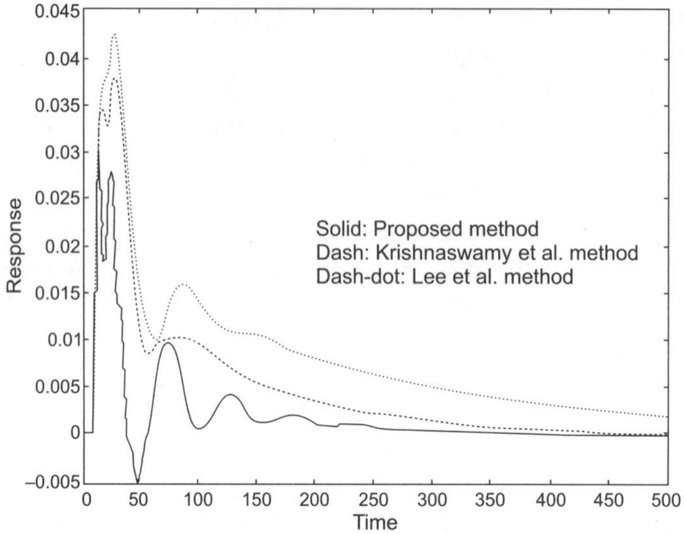

Fig. 5.3 *Comparison of different methods for regulatory response in y_1 for a disturbance in the inner loop (α_1 = 1.0 and α_2 = 1.0 for series cascade control system)*

Table 5.1 *P/PI Controller settings*

Controller parameters	Proposed method 1	Krishnaswamy et al.	Lee et al.
k_{c2}	5.25	2.978	3.44
k_{c1}	9.6667	7.3	5.83
τ_I	105.91	200	105

Note: For the proposed method: α_1 = 1.0 and α_2 = 1.0 (series cascade control system)

Table 5.2 *Performance comparison of different methods under perfect parameters*

Error	Proposed method 1		Krishnaswamy et al.		Lee et al.	
	Servo	Regulatory	Servo	Regulatory	Servo	Regulatory
ISE	22.194	0.0133	20.19	0.065	19.94	0.044
IAE	38.959	1.1258	38.48	4.1602	27.37	2.5954
ITAE	1590	99.3511	2920	651.1177	593.73	257.767

Note: For controller settings refer to Table 5.1(series cascade control system)

Therefore, the effect of reduction of the primary controller gain due to changes in α_1 and α_2 is observed on the outer loop only. In general, α_2 affects both k_{c1} and τ_I. However, for the particular values of α_2/α_1 = 0.4 and α_1 = 0.9, the value of τ_I is not affected. Basically we first design the inner loop controller and then the outer loop controller.

Controller settings by the different methods are given in Table 5.3. The servo responses for a unit step change in the set point of y_{1r} are shown in Fig. 5.4. Similarly regulatory responses for a unit step change in the load (entering in the secondary loop at the outlet of process) are compared in Fig. 5.5. In Fig. 5.5, the initial oscillations in the manipulated variable response are also observed for the Lee et al. (1998b) method. Table 5.4 gives performance comparison of the controlled system. The proposed method gives the best result.

Table 5.3 *Controller settings when $\alpha_1 = 0.9$ and $\alpha_2 = 0.4\ \alpha_1$ (series cascade control system)*

Controller parameters	Proposed method 2
k_{c2}	5.25
k_{c1}	4.68
τ_I	104.3

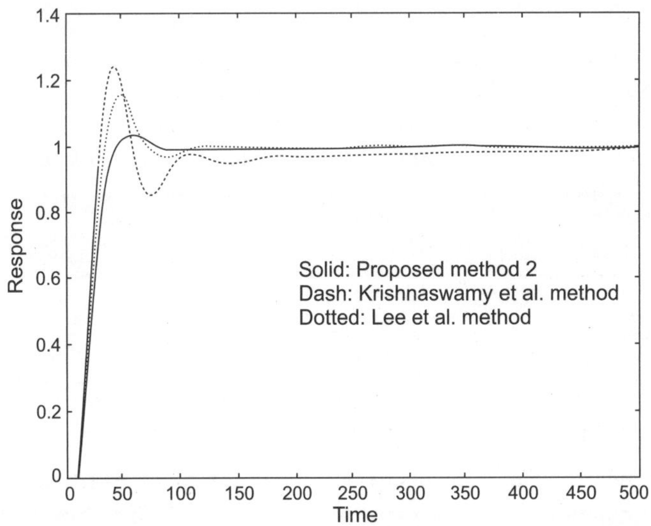

Fig. 5.4 *Comparison of different methods for servo response in y_1 using PI controller for the outer loop and P controller for the inner loop, under perfect parameters (series cascade control system)*

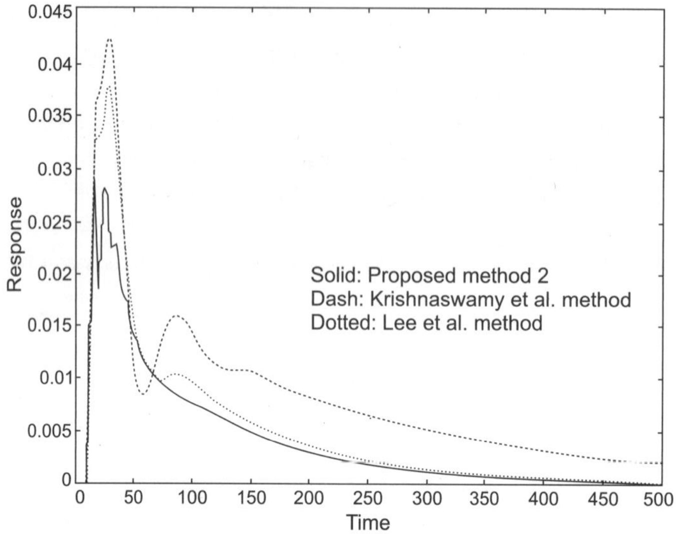

Fig. 5.5 *Comparison of different methods for regulatory response in y_1 for a disturbance in the inner loop under perfect parameters (series cascade control system)*

This method can be used only when the inner and outer loop transfer functions are known. If these transfer functions are not known, then an identification step needs to be carried out.

Table 5.4 *Performance comparison of different methods under perfect parameters*

Error	Proposed method 2		Krishnaswamy et al.		Lee et al.	
	Servo	Regulatory	Servo	Regulatory	Servo	Regulatory
ISE	20.05	0.025	20.19	0.065	19.94	0.044
IAE	25.99	2.104	38.48	4.1602	27.37	2.5954
ITAE	495.11	215.41	2920	651.1177	593.73	257.767

Note: Controller settings as given in Table 5.3 for the present method

The robustness of the proposed controller is evaluated for 20% perturbation of the inner loop process gain (in simulation the process gain used is 1.2 times of the value used in designing the controller). Fig. 5.6 shows the servo responses and Fig. 5.7 shows the regulatory responses. Table 5.5a shows performance comparison of the controlled system under uncertainty in k_{p2}. The slightly improved performance of the controller under perturbation in k_{p2} is obtained for the other controller design methods as well (refer to Tables 5.5a and 5.4).

The gain margin calculations (Table 5.7) show that all the three methods use a higher value of the gain margin. Hence, under this condition, the perturbed system will give a better performance. The present method 2 gives the best performance.

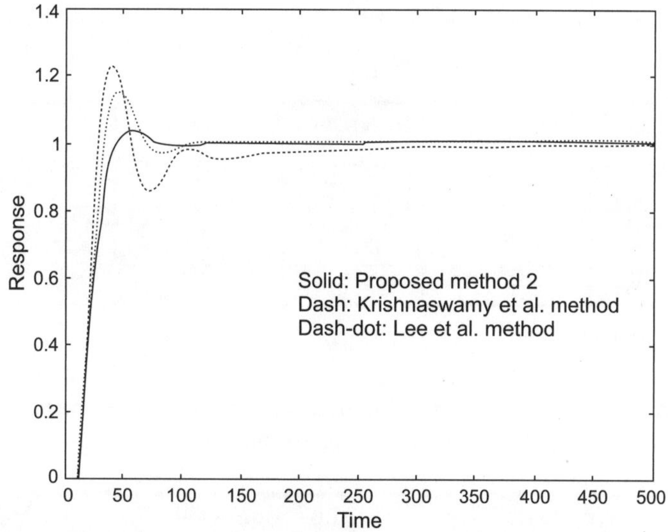

Fig. 5.6 *Servo response in y_1 using PI controller in the outer loop and P controller in the inner loop, with 20% uncertainty in k_{p2} in the process (series cascade control system)*

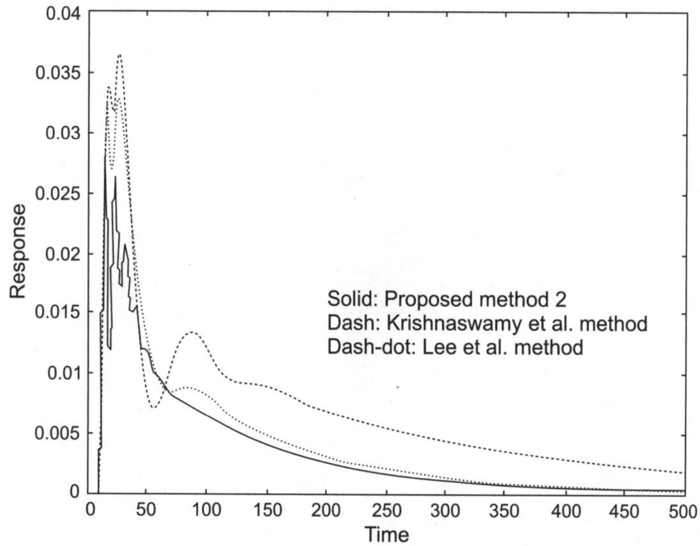

Fig. 5.7 *Regulatory response in y_1 for a disturbance in the inner loop with 20% uncertainty in k_{p2} in the process (series cascade control system)*

Table 5.5a *Performance comparison of the controlled system under uncertainty in k_{p2} 20% high (k_{p2} = 2.4 in the process and controller designed for k_{p2} = 2)*

Error	Proposed method 2		Krishnaswamy et al.		Lee et al.	
	Servo	Regulatory	Servo	Regulatory	Servo	Regulatory
ISE	19.65	0.0183	19.57	0.0456	19.37	0.0309
IAE	25.45	1.7535	37.397	3.4690	26.46	2.1629
ITAE	476.16	178.92	2830.8	541.3174	555.54	213.9037

(Series cascade control system)

Table 5.5b *Performance comparison of the controlled system under uncertainty in D_2 20% high (D_2 = 2.4 in the process and controller designed for D_2 = 2)*

Error	Proposed method 2		Krishnaswamy et al.		Lee et al.	
	Servo	Regulatory	Servo	Regulatory	Servo	Regulatory
ISE	20.14	0.0266	20.4728	0.0659	20.1053	0.0446
IAE	25.97	2.1043	38.8687	4.1611	27.5322	2.5955
ITAE	493.44	214.61	2935.8	649.8758	599.3067	256.77

(Series cascade control system)

Table 5.5c *Performance comparison of the controlled system under uncertainty in τ_2 20% high (τ_2 = 24 in the process and controller designed for τ_2 = 20)*

Error	Proposed method 2		Krishnaswamy et al.		Lee et al.	
	Servo	Regulatory	Servo	Regulatory	Servo	Regulatory
ISE	20.39	0.0283	20.9848	0.0722	20.5430	0.0496
IAE	26.51	2.1051	39.8310	4.1694	28.6465	2.5965
ITAE	516.37	207.45	2976.5	639.3069	657.3665	247.94

(Series cascade control system)

Simulation result shows that the present method is also robust (Tables 5.5b and 5.5c) for uncertainty in the time delay (Figs 5.8 and 5.9 for servo and regulatory responses respectively) and separately in the time constant (Figs 5.10 and 5.11 for servo and regulatory responses respectively). In general, a feedback loop tries to reduce the sensitivity of the model parameters on the closed loop performance. In cascade control systems, because of the additional feedback loop, the perturbation in the inner loop process gain and time delay will not have significant effect on the servo and regulatory problems.

The proposed controller is also evaluated for 20% perturbation of the outer loop process gain. Similar robust behavior is also observed for a perturbation in the outer loop process gain (Table 5.6a), separately in the process time delay (Table 5.6b) and in the time constant (Table 5.6c). For all the three design methods, the perturbation or variation in the outer loop parameters (k_{p1} or D_1 or τ_1) does not affect the regulatory response. The ITAE is considered here for performance evaluation. In the present method, the variation in time delay has much less effect on the performance than that of the variation in the time constant. The method gives an improved performance than that of the other methods.

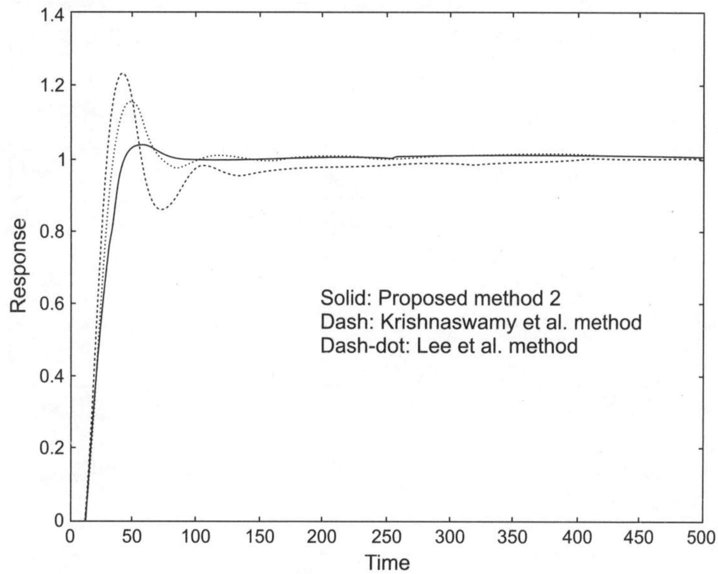

Fig. 5.8 *Servo response in y_1 with PI controller for the outer loop and P controller for the inner loop with 20% uncertainty in D_2 in the process*

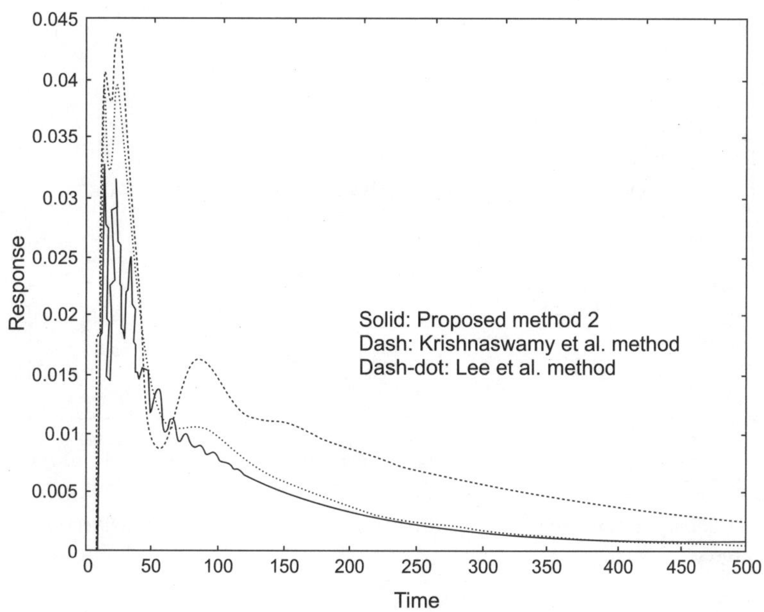

Fig. 5.9 *Regulatory response in y_1 for a disturbance in the inner loop with 20% uncertainty in D_2 in the process*

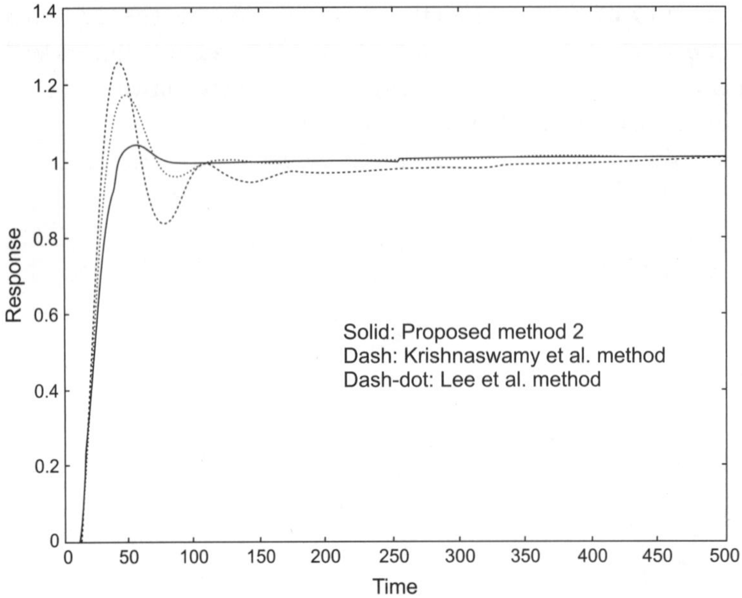

Fig. 5.10 *Servo response in y_1 using PI controller for the outer loop and P controller for the inner loop with 20% uncertainty in τ_2 in the process*

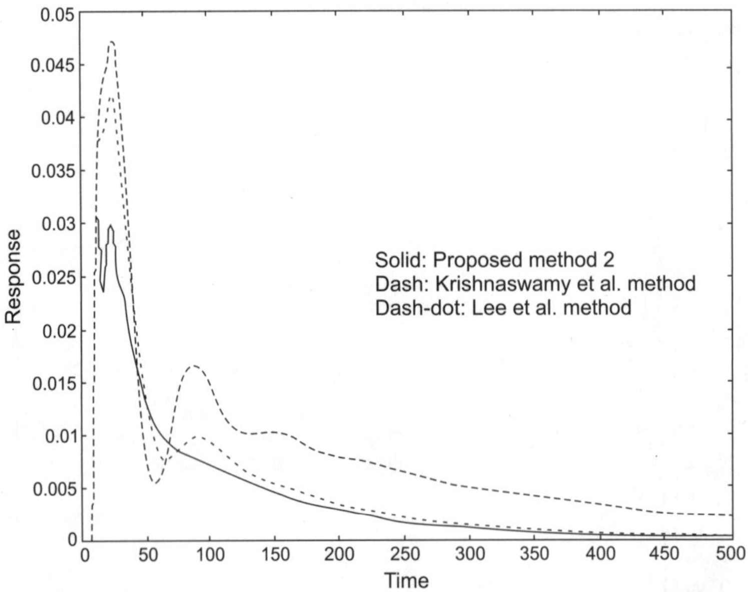

Fig. 5.11 *Regulatory response in y_1 for a disturbance in the inner loop with 20% uncertainty in τ_2 in the process*

Table 5.6a *Performance comparison of the controlled system under uncertainty in k_{p1} 20% higher in the process (k_{p1} = 1.2, α_1 = 0.9 and α_2 = 0.4 α_1)*

Error	Proposed method 2		Krishnaswamy et al.		Lee et al.	
	Servo	Regulatory	Servo	Regulatory	Servo	Regulatory
ISE	18.88	0.0292	21.38	0.0764	19.80	0.0515
IAE	25.33	2.105	40.82	4.1789	29.36	2.5963
ITAE	495.83	207.33	2764.5	641.125	741.79	249.32

(Series cascade control system)

Table 5.6b *Performance comparison of the controlled system under uncertainty in D_1 20% higher in the process (D_1 = 12, α_1 = 0.9 and α_2 = 0.4 α_1)*

Error	Proposed method 2		Krishnaswamy et al.		Lee et al.	
	Servo	Regulatory	Servo	Regulatory	Servo	Regulatory
ISE	22.15	0.0271	23.86	0.0698	22.76	0.0473
IAE	29.51	2.1043	44.6626	4.161	33.09	2.5955
ITAE	650.53	215.4813	3230.9	651.70	911.15	257.82

(Series cascade control system)

Table 5.6c *Performance comparison of the controlled system under uncertainty in τ_1 20% higher in the process (τ_1 = 120, α_1 = 0.9 and α_2 = 0.4 α_1)*

Error	Proposed method 2		Krishnaswamy et al.		Lee et al.	
	Servo	Regulatory	Servo	Regulatory	Servo	Regulatory
ISE	21.74	0.0242	20.36	0.0599	21.02	0.0404
IAE	30.81	2.1075	35.30	4.1685	29.44	2.5987
ITAE	928.44	217.5	2405.5	658.37	811.68	259.821

(Series cascade control system)

Table 5.7 *Gain margin (A_m) and phase margin (ϕ_m) with inner loop P controller and outer loop PI controller*

Method	Proposed method 2	Krishnaswamy et al.	Lee et al.
A_m	2.9587	2.3337	2.695
$\phi_m(°)$	60.89	55.98	57.78

(Series cascade control system)

5.4 Stability Analysis

Stability of the given controllers can be checked by calculating gain and phase margin. Ho et al. (1995; 1996) reported a method of calculating phase margin and gain margin of well-known PID tuning formulas. Settings that give a larger phase margin and gain margin are preferred. The outer loop process and controller transfer functions are denoted by $G_p(s)$ and $G_c(s)$ respectively. The loop transfer function is given by

$$G_p(s)G_c(s) = k_{c1}\left[1+\left(1/\tau_1 s\right)\right]\left(\left[k_{c2}k_{p2}\exp\left(-D_2 s\right)\right]/\left[\tau s+1+k_{c2}k_{p2}\exp\left(-D_2 s\right)\right]\right)$$
$$\left(k_{p1}\exp\left(-D_1 s\right)\right)/\left(\tau_1 s+1\right) \tag{5.19}$$

Frequency at which the Nyquist curve has a phase of '$-\pi$ (phase cross over frequency, ω_p)' is obtained by solving the following equation

$$\tan^{-1}\left(-1/\tau_I\omega_p\right) - D_1\omega_p - D_2\omega_p + \tan^{-1}\left(-\tau_1\omega_p\right) +$$
$$\tan^{-1}\left\{\left[\left(\tau_2\omega_p - k_2\sin\left(D_2\omega_p\right)\right]/\left[1+k_2\cos\left(D_2\omega_p\right)\right]\right\} = -\pi \tag{5.20}$$

The gain margin is obtained by the following equation

$$A_m = \left(\frac{\tau_I\omega_p}{k_1k_2}\right)\frac{\left(1+\tau_1^2\omega_p^2\right)^{0.5}\left(\left[1+k_2\cos\left(D_2\omega_P\right)\right]^2+\left[\tau_2\omega_P - k_2\sin\left(D_2\omega_P\right)\right]^2\right)}{\left(1+\tau_I^2\omega_p^2\right)^{0.5}\left(\left[1+k_2\cos\left(D_2\omega_P\right)\right]^2+\left[\tau_2\omega_P - k_2\sin\left(D_2\omega_P\right)\right]^2\right)^{0.5}} \tag{5.21}$$

The frequency at which the Nyquist curve has an amplitude of 1 is known as gain cross over frequency (ω_g) and is obtained by solving the following equation

$$k_1k_2 = \tau_I\omega_g\frac{\left(1+\tau_1^2\omega_g^2\right)^{0.5}\left([1+k_2\cos(D_2\omega_g)]^2+[\tau_2\omega_g - k_2\sin(D_2\omega_g)]^2\right)}{\left(1+\tau_I^2\omega_g^2\right)^{0.5}\left([1+k_2\cos(D_2\omega_g)]^2+[\tau_2\omega_g - k_2\sin(D_2\omega_g)]^2\right)^{0.5}} \tag{5.22}$$

The phase margin is given by the following equation:

$$\phi_m = \tan^{-1}\left(-1/\tau_1\omega_g\right) - D_1\omega_g - D_2\omega_g + \tan^{-1}\left(-\tau_1\omega_g\right) +$$
$$\tan^{-1}\left(\tau_2\omega_g - k_2\sin\left(D_2\omega_g\right)\right)/\left(1+k_2\cos\left(D_2\omega_g\right)\right) + \pi = 0 \tag{5.23}$$

In the present work, the phase margin and gain margin are calculated for the system $k_{p1}G_{p1} = \exp(-10s)/(100s+1)$ and $k_{p2}G_{p2} = 2.0\exp(-2s)/(20s+1)$ with the controller designed by different methods, and the results are reported in the Table 5.7. Controller designed by the present method gives the largest phase margin and hence is more stable and robust than the other methods. These are also shown by simulation study of the closed loop response as discussed earlier.

5.5 Method for Parallel Cascade Control Systems

Let us consider that the transfer function of the process in the inner loop is given by an FOPTD model: $k_{p2}\exp(-D_2s)/(\tau_2s+1)$.The transfer function of the process in the outer loop is also an FOPTD transfer function model $[k_{p1}\exp(-D_1s)/(\tau_1s+1)]$. Usually we consider a simple proportional controller for the inner loop and a PI controller for the outer loop (Fig. 5.12). The closed loop transfer function model for the inner loop is given by

$$y_2/y_{2r} = K_2e^{(-\varepsilon q)}/\left(q+1+K_2e^{(-\varepsilon q)}\right) \tag{5.24}$$

where,

$$K_2 = k_{c2}k_{p2};\ \varepsilon = D_2/\tau_2;\ q = \tau_2 s \tag{5.25}$$

In Eq.(5.24), we need not consider the term $\exp(-\varepsilon q)$ further in the numerator as it is going to shift the time axis only. By substituting the second order Pade's approximation for the delay term, we get

$$y_2/y_{2r} = K_2\left[1+0.5\varepsilon s+0.25\varepsilon^2s^2\right]/\left[(s+1)\left(1+0.5\varepsilon s+0.25\varepsilon^2s^2\right)\right.$$
$$\left.+K_2\left(1-0.5\varepsilon s+0.25\varepsilon^2s^2\right)\right] \tag{5.26}$$

Since the objective of the control system is to make y follow y_r, we can equate the coefficient of s in the numerator to α times that of the denominator.

$$D_2k_{c2}k_{p2} = 2[\alpha/(\alpha+1)](\tau_2+0.5D_2) \tag{5.27}$$

Here it is found by simulation that $\alpha = 0.67$ gives a better performance.

The closed loop transfer function for the outer loop is obtained as

$$y_1/y_{r1} = \left(R_1e^{0.5\psi q}e^{-\lambda q}\right)/R_2 \tag{5.28}$$

$$R_1 = \left(P_1q^2+\left(P_1+P_2\tau_2\right)q+P_2\tau_2\right)e^{0.5\psi q} \tag{5.29}$$

$$R_2 = \left[\left(\varphi q^3+(\varphi+1)q^2+q\right)e^{0.5\psi q}+\left(K_2\varphi q^2+K_2q\right)e^{-0.5(\varepsilon-\lambda)q}\right.$$
$$\left.+\left(P_1q^2+\left(P_1+P_2\tau_2\right)q+P_2\tau_2\right)e^{0.5(\varepsilon-\lambda)q}\right] \tag{5.30}$$

where

$$\varphi = \tau_1/\tau_2\ ;\ \tau_I^* = \tau_I/\tau_2\ ;\ \lambda = L_1/\tau_2\ ;\varepsilon = L_2/\tau_2;\ \psi = \varepsilon+\lambda \tag{5.31}$$

$$P_1 = k_{p1}k_{c1}k_{c2};\ P_2 = P_1/\tau_I^* \tag{5.32}$$

Similar to the procedure carried out for designing the inner loop controller, on equating the corresponding coefficient of q, q^2 of the numerator to α_1 and α_2 time of the denominator in Eq.(5.28), we get

$$P_1A_1+P_2\tau_2A_2 = \alpha_1A_3 \tag{5.33}$$

$$P_1 A_4 + P_2 \tau_2 A_5 = \alpha_2 A_6 \tag{5.34}$$

Where $A_1 = (1-\alpha_1)$ (5.35a)

$$A_2 = \left(1 + 0.5\psi - \alpha_1 + 0.5(\lambda - \varepsilon)\alpha_1\right) \tag{5.35b}$$

$$A_3 = \left(1 + K_2\right) \tag{5.35c}$$

$$A_4 = \left(1 + 0.5\psi - \alpha_2 + 0.6(\lambda - \varepsilon)\alpha_2\right) \tag{5.35d}$$

$$A_5 = \left(0.5\psi + 0.125\psi^2 + 0.5(\lambda - \varepsilon)\alpha_2 - 0.125(\lambda - \varepsilon)^2 \alpha_2\right) \tag{5.35e}$$

$$A_6 = \left(1 + \varphi + 0.5\psi + K_2\varphi - 0.5K_2(\varepsilon - \lambda)\right) \tag{5.35f}$$

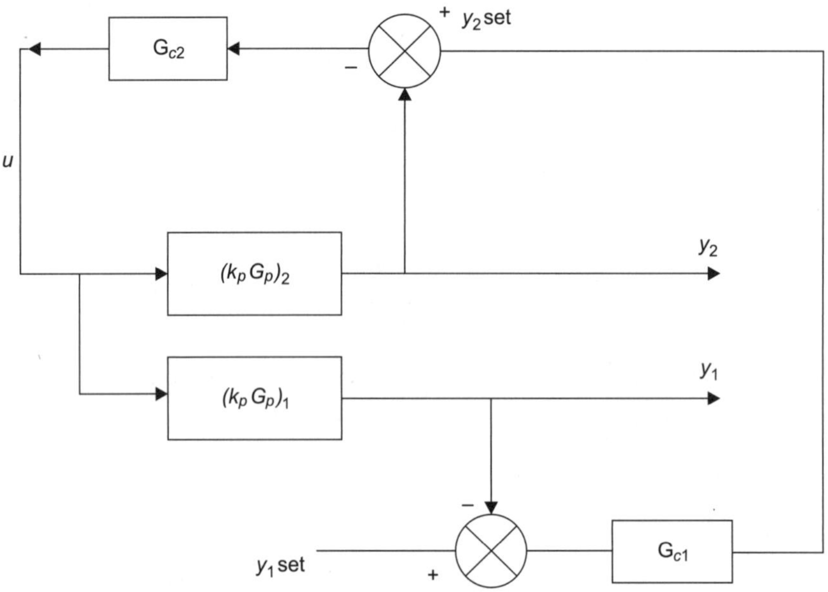

Fig. 5.12 *Parallel cascade control scheme*

Different values of α_1 are chosen and it is found by simulation that $\alpha_1 = 1.2$ and $\alpha_2 = 0.6\ \alpha_1$ give the best results. Once P_1 and P_2 are calculated, then k_{c1} and τ_{I1} can be calculated from Eq.(5.32) as

$$k_{c1} = P_1 / \left(k_{c2} k_{p2} k_{p1}\right) \tag{5.36}$$

$$\tau_{I1} = \left(k_{c1} k_{p1} k_{c2} k_{p2}\right) \tau_1 / P_2 \tag{5.37}$$

It should be noted that for a single loop FOPTD system, Chidambaram and Padmasree (2003) showed that equating corresponding coefficients (i.e., making $\alpha_1 = 1.0$ and $\alpha_2 = 1.0$) gives a good result. Similar results are reported for series cascade control in the earlier subsection. However, it is found in the present work that such an approach creates an instability problem.

So the method proposed by Chidambaram and Padmasree (2003), by equating the coefficient of q, q^2 in the numerator to α_1 and α_2 times that of the corresponding coefficient in the denominator, is found to give a good performance. For the secondary loop, if the value of α is assumed as 1 [similar to the method of Chidambaram and Padmasree (2003)] and tuning α_1 with $\alpha_2 = 0.6\alpha_1$ gives a slight oscillation in the output response. Hence, use of $\alpha = 0.8$ is suggested.

5.6 Simulation Results for Parallel Cascade Controllers

Let us consider the example system whose transfer functions models are given by:

Case study 1

$$\text{Inner loop } K_{p2}G_{P2} = \frac{2e^{-2s}}{20s+1}; \quad \text{Outer loop } K_{p1}G_{P1} = \frac{e^{-10s}}{100s+1} \tag{5.38}$$

Let us consider a proportional controller for the secondary loop. Since it has only one tuning parameter, it is easy to tune. We use a PI controller for the primary controller. Using the continuous cycling Ziegler–Nichols method, *P* setting for the inner loop and PI settings for the outer loop are obtained as: k_{c2} = 4.0875, k_{c1} = 4.3447 and τ_{I1} = 18.33, respectively. The present method gives the settings as: k_{c2} = 4.2, k_{c1} = 4.8038 and τ_{I1} = 8.9582 (Table 5.8). The servo responses for a unit step change in the set point of y_{r1} are shown in Figure 5.13. The performances of the present method are better than that of the Ziegler–Nichols method.

The regulatory responses for a unit step change in the load (entering in the secondary loop at the outlet of the process) are compared in Fig. 5.14. The present method gives a better performance. Table 5.9 gives ITAE values comparisons for servo and regulatory responses for a unit step change. The system output value is –0.25 without any controller.

To further verify the performance of this proposed method, a case study 2 with a different time delay is chosen.

Table 5.8 *Controller settings*

Case study	Controller parameters	Ziegler–Nichols	Proposed method [$\alpha_1 = 1.2, \alpha_2 = 0.6\,\alpha_1$]
1	k_{c2}	4.0875	4.2
	k_{c1}	4.3447	4.8038
	τ_I	18.33	8.9582
2	k_{c2}	2.1233	2.2
	k_{c1}	4.149	4.6894
	τ_I	13.33	21.2533

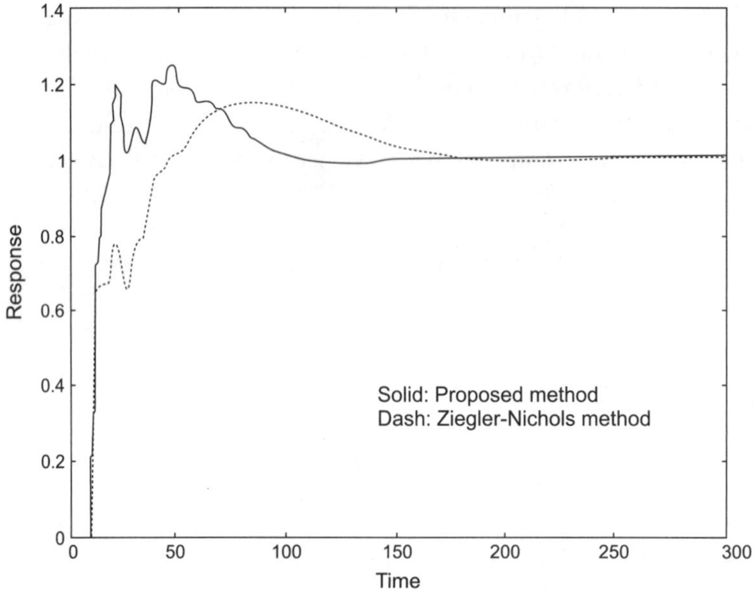

Fig. 5.13 *Comparison of servo response for case study 1*

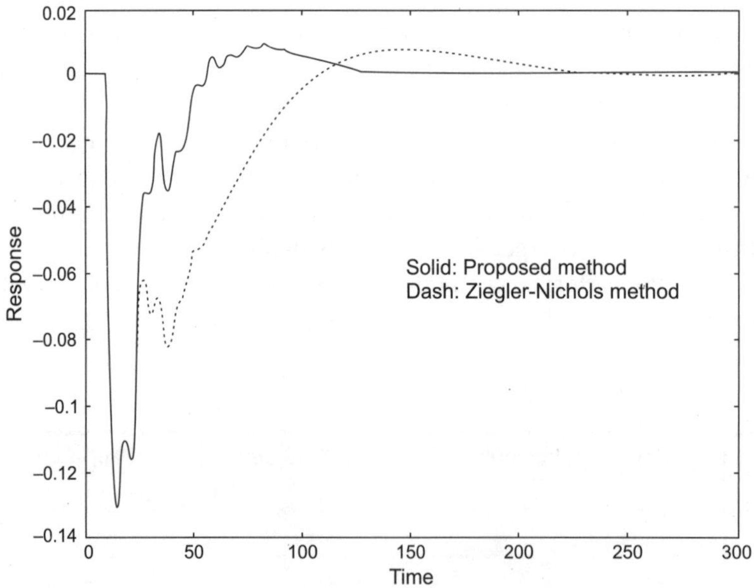

Fig. 5.14 *Comparison of regulatory response for case study (parallel cascade control system)*

Case study 2

$$\text{Inner loop}\quad K_{p2}G_{P2} = \frac{2e^{-4s}}{(20s+1)};\quad \text{Outer loop}\quad K_{p1}G_{P1} = \frac{e^{-20s}}{(100s+1)} \tag{5.39}$$

Using the Ziegler–Nichols method, P setting for the inner loop and PI settings for the outer loop are obtained as: k_{c2} = 2.1233, k_{c1}= 4.149 and τ_{I1} = 13.33, respectively. Since D_2/τ_2 and D_1/τ_1 are changing, α_1 and α_2 will also change. At α_1 = 1.2, the present method gives the settings as: k_{c2} = 2.2, k_{c1} = 4.6894 and τ_{I1} = 21.2533.

The servo responses for a unit step change in the set point of y_{r1} are shown in Fig. 5.15. Figure 5.16 shows the regulatory performance in y_1, for disturbance entering in the inner loop. The proposed method shows a better performance than that of the Ziegler–Nichols method

For this case study, robustness of the proposed controller is evaluated for a 20% perturbation of process gain in the process. The same controller settings as in Table 5.8 for case 1 are used. Figure 5.17 shows the servo response and Fig. 5.18 shows the regulatory response. The output reaches a value of –0.25, when there are no controllers. Simulation results show that the present method is robust. Similar robustness is observed for a perturbation in the time delay and separately in the process time constant. Table 5.10 shows the performance comparisons of the closed loop system under uncertainty in the model parameters.

Table 5.9 *Performance comparison of the controlled system under perfect parameters*

Case study	Error	Ziegler–Nichols		Proposed method	
		Servo	Regulatory	Servo	Regulatory
1	ISE	15.26	0.345	13.5589	0.1868
	IAE	30.8924	5.2704	23.1326	2.5808
	ITAE	1.4×10^3	288.663	693.5077	90.7747
2	ISE	31.67	0.4820	26.68	0.5034
	IAE	52.8151	5.9747	44.52	6.2024
	ITAE	2.674×10^3	397.5402	2.179×10^3	409.20

Table 5.10 *Performance comparison of the controlled system under uncertainty in k_{p2} (parallel cascade control system)*

Gain	Error	Ziegler–Nichols		Proposed method	
		Servo	Regulatory	Servo	Regulatory
k_{p2} = 2.4	ISE	16.7264	0.3120	14.0215	0.1596
	IAE	35.0334	5.3517	25.4475	2.6569
	ITAE	1.867×10^3	320.15	900.5495	106.1059

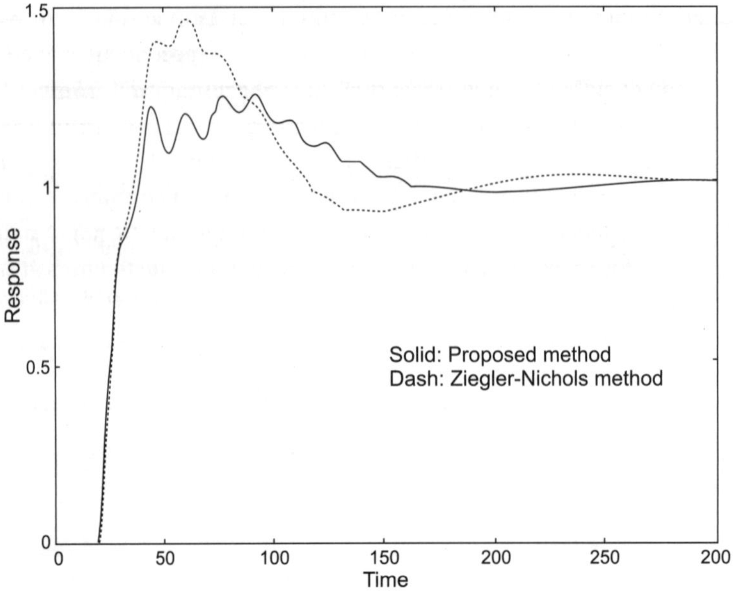

Fig. 5.15 *Comparison of servo response for case study 2 (parallel cascade control system)*

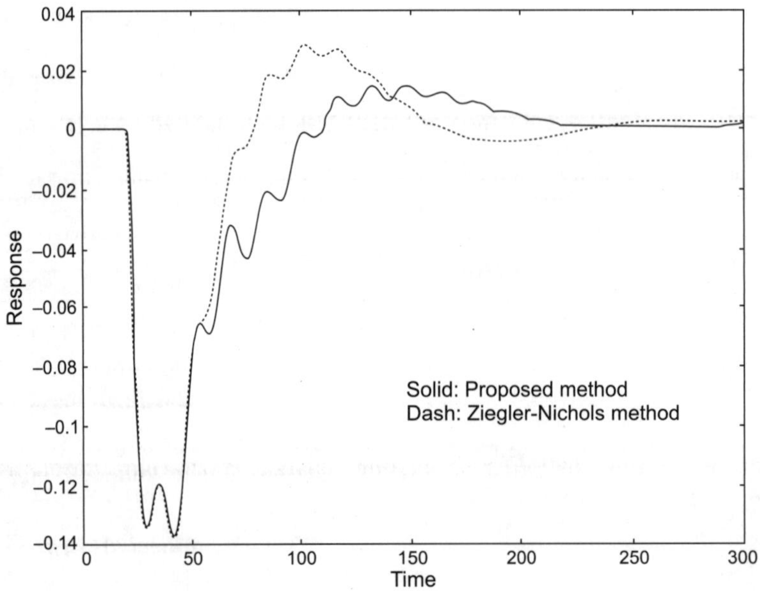

Fig. 5.16 *Comparison of regulatory response for case study 2 (parallel cascade control system)*

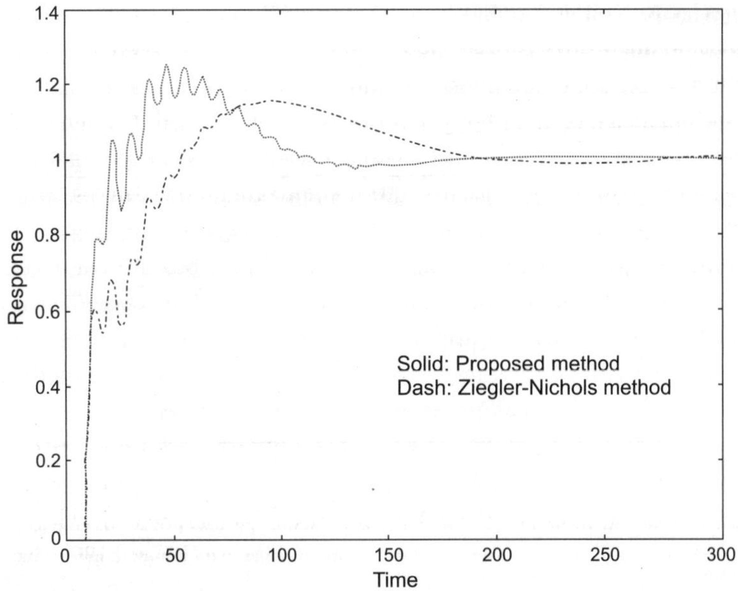

Fig. 5.17 *Servo response for $k_{p2} = 2.4$ for case study 1 (parallel cascade control system)*

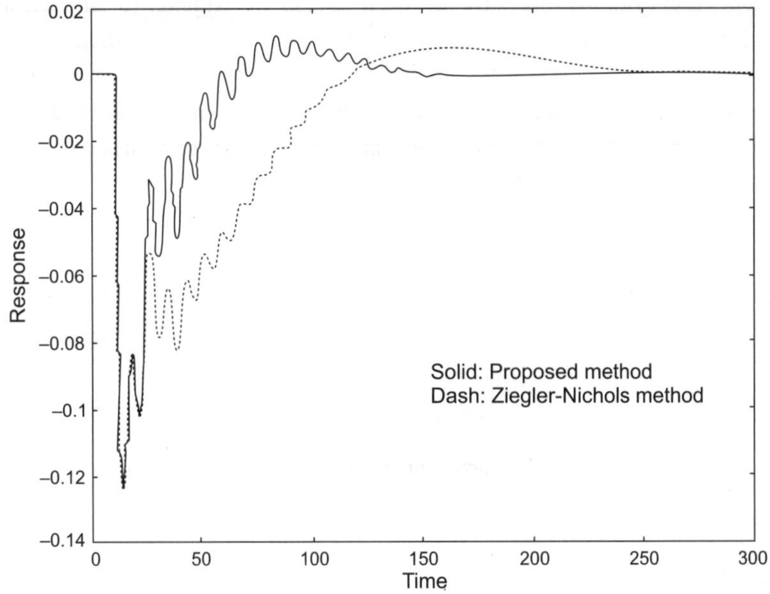

Fig. 5.18 *Regulatory response for $k_{p2} = 2.4$ for case study 1 (parallel cascade control system)*

5.7 Conclusions

A simple method, by relating the coefficients of q and q^2 of the numerator to that of the denominator of the closed loop transfer function model, is proposed to design controllers for (i) a series cascade control system and (ii) for a parallel cascade control system. For series cascade control systems, the servo and regulatory responses of the method are better when compared to the recent method of Lee et al. (1998b) and that of Krishnaswamy et al. (1990) The method has two tuning parameters within the range 0.2 to 1.3. It gives the best robust performance when there is an uncertainty in the model parameters. The stability of the controller is also analyzed theoretically. For parallel cascade control systems, the servo and regulatory responses of the method are better than that of the Ziegler–Nichols tuning method. The method also gives a robust performance when there is an uncertainty in the model parameters.

6 Improved Saturation Relay Test for Systems with Large Dead Time

Yu (1999) introduced the saturation relay feedback test for determining the ultimate gain and ultimate frequency of the system that are required for the design of PI/PID controllers. In the present chapter, two modifications are proposed to improve the efficiency of the saturation relay feedback test, especially for systems with a large dead time. In the first method, the safety factor is corrected based on the ideal relay feedback test. In the second method, higher order harmonics are included in the analysis of ideal relay. Simulation results are given to show the improvement of the proposed method over that of the Yu (1999) method for an FOPTD system and for a higher order system. The method is also compared with that proposed by Luyben (2001) for identifying the FOPTD model parameters.

6.1 Introduction

Li et al. (1991) pointed out that the error in estimation of ultimate gain ranges from −18% to +27% (with respect to the theoretical value) in ideal relay feedback test. To reduce the error in the estimation of ultimate gain, Yu (1999) suggested the saturation relay feedback test. In this method, the ultimate gain estimated from the ideal relay feedback test is multiplied with a safety factor to get slope of the saturation relay and the saturation relay feedback test is conducted to get better results for the ultimate gain and frequency.

The method proposed by Yu (1999) gave accurate results for small D/τ ratio (generally $D/\tau < 2$). However, he did not report any result for large D/τ ratio (generally $D/\tau > 5$). In the present study, error involved in the estimation of ultimate gain for large D/τ ratio is analyzed. To get an improved result, two modifications are proposed, mainly focusing on the selection of proper slope for the saturation relay. In first method, the safety factor is corrected based on the trend of the system output waveform during the ideal relay feedback test. In the second method, higher order harmonics are included in the analysis of the ideal relay, based on the improved conventional relay autotune method proposed by Srinivasan and Chidambaram (2004).

Luyben (2001) pointed out that additional information can be obtained from relay feedback test namely, the shape of the response. Thyagarajan and Yu (2003) proposed a method that uses a conventional ideal relay feedback test and gave a procedure based on the shape information to find parameters for the corresponding model. In the present work, the methods proposed by Srinivasan and Chidambaram (2004) and Luyben (2001) for identifying the FOPTD model parameters are compared.

6.2 Importance of Saturation Relay Slope

Since the square wave output of the ideal relay results in an abrupt change in the slope of the relay at zero point, $[e(t) = 0]$, Yu (1999) suggested the saturation relay, which provides a smooth transition of the slope of the relay around the zero point. The saturation relay is characterized by its height and slope (Fig. 6.1). For a feedback loop with a saturation relay, the output of the relay is similar to a sine wave with extreme cuts, based on the height of the relay (Yu, 1999). Hence, the saturation relay feedback test provides more accurate results than that of an ideal relay.

To conduct a saturation relay feedback test, proper slope has to be chosen for the saturation relay. If the slope is too high, the relay output is similar to a square wave and, as a result, the accuracy of the feedback test is lost. As a general thumb rule, lesser the slope value, greater the accuracy of the feedback test.

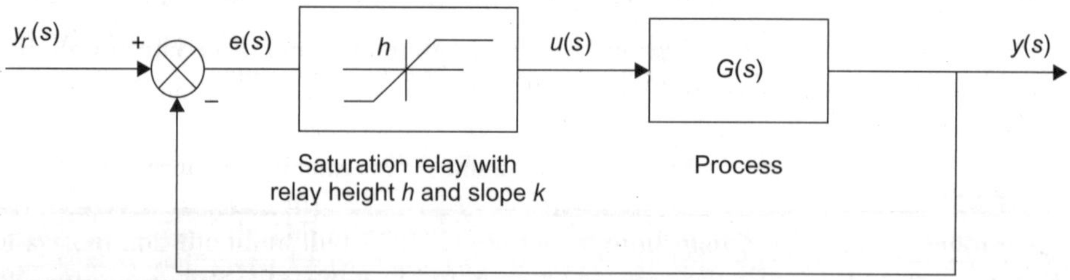

Fig. 6.1 *Block diagram for saturation relay feedback system*

However, if the slope of the saturation relay is less than the ultimate controller gain of the actual system, the feedback test fails to produce sustained oscillations (Yu, 1999). Thus, success of the feedback test depends mainly on the selection of a proper slope for the saturation relay.

6.3 Method 1

In this method, the slope of the saturation relay is found from the ideal relay feedback test. Since the ultimate controller gain from the ideal relay test is only an approximate value, a safety factor is necessary to ensure that the slope selected for the saturation relay is greater than the ultimate controller gain of the actual system. Yu (1999) suggested a safety factor of 1.4 to be multiplied with the approximate ultimate controller gain, from the ideal relay feedback test to get the slope of the saturation relay. Yu (1999) presented the analysis for only smaller time-delay to time-constant ratio ($D/\tau < 1$). In the case of large D/τ ratio (generally $D/\tau > 5$), the ultimate gain obtained from the ideal relay feedback test is always over-estimated (Li et al., 1991; Srinivasan and Chidambaram, 2004). Hence, the safety factor cannot be more than unity.

The safety factor, 1.4, actually increases the slope of the saturation relay unnecessarily and causes an additional error in the saturation relay feedback test. Instead, the ultimate gain found from the ideal relay feedback test can be directly assigned as the slope of the saturation relay, without a need for any safety factor. The system output waveform during the ideal relay feedback test provides the clue to find whether D/τ ratio is smaller or larger. If the system output waveform is triangular, then the D/τ ratio is small (Friman and Waller, 1997; Srinivasan and Chidambaram, 2004). If the system output waveform is similar to a square waveform, then the D/τ ratio is large (Friman and Waller, 1997; Srinivasan and Chidambaram, 2004).
Hence, procedure for the proposed method 1 is as follows.

1. Perform the ideal relay feedback test.
2. Find the approximate ultimate controller gain (k_u), based on principal harmonics only.

   $$k_u = 4h/(\pi a) \tag{6.1}$$

   Where 'a' is the observed amplitude of the system's output waveform
3. Observe the trend of system output waveform.
4. If the waveform is triangular, take 1.4 as the safety factor (S_f) as suggested by Yu (1999). If the waveform is closer to a square waveform, take $S_f = 1.0$
5. Select the slope of the saturation relay (k), from the approximate k_u found by the ideal relay feedback test.

   $$\text{Slope, } k = S_f\ k_u \tag{6.2}$$

6. Perform the saturation relay feedback test.
7. Find the ultimate controller gain and ultimate frequency, based on principal harmonics only.

   $$k_u = (2h/\pi)\left[\sin^{-1}(a_1/a)/a_1 + (a^2 - a_1^2)^{0.5}/a^2\right] \tag{6.3}$$

where $a_1 = h/k$ from saturation relay.

$$\omega_u = 2\ \pi/P_u \tag{6.4}$$

where P_u is the period of oscillation of the system's output waveform.

Though the method involves two tests, improvement in the accuracy of the estimation of ultimate controller gain and ultimate frequency in fact pays for the additional amount of work. Hence, the error caused by the improper selection of slope value is avoided in the proposed method 1. The safety factor, which is chosen by observing the system output waveform in the ideal relay feedback test, prevents selecting an unnecessary high slope value for the saturation relay.

Simulations are carried out for an FOPTD model, with $k_p = 1$, $\tau = 1$ and D ranging from 1, 5, 10, to 30. Theoretical values of ultimate gain and ultimate frequency are calculated and tabulated in Table 6.1. Estimations of the ultimate gain based on the ideal relay feedback test by Yu (1999) method and the proposed method 1 are also given in Table 6.1.

6.4 Method 2

Though the error in the estimation of the ultimate controller gain is reduced significantly by selecting a proper safety factor, still there is a possibility of reducing it further. The actual cause for the error in the ultimate gain estimation is the misinterpretation of a waveform as a sine waveform, during the analysis of the ideal relay. Hence, the error can be considerably reduced, only if higher order harmonics are included during the analysis of the ideal relay test.

Srinivasan and Chidambaram (2004) proposed a method to include higher order harmonics to improve the ideal relay feedback test. However, in that method, the number of higher order harmonics that must be included varies with the D/τ ratio and it has to be carefully chosen in order to get the accurate results. However, in the proposed method, only three harmonic components are used to characterize the output of the ideal relay. This will increase the accuracy of the ideal relay feedback test. Further, selection of the slope for the saturation relay, based on ideal relay feedback test improves. As a net result, this will definitely help in reducing the error in the estimation of ultimate controller gain. Moreover, there is no confusion in selecting the appropriate number of higher order harmonics that has to be included, as the ideal relay feedback test is the initial test and the results of this test are further improved by the saturation relay feedback test.

Hence, procedure for method 2 is as follows.

1. Perform the ideal relay feedback test.
2. Observe the trend of the system output waveform.
3. If the waveform is triangular,
   - find approximate ultimate controller gain (k_u), based on principal harmonics only.
   - take 1.4 as the safety factor (S_f) as suggested by Yu (1999).
4. If the waveform is similar to a square waveform,
   - find improved ultimate controller gain (k_u), based on higher order harmonics (n = 1, 3, 5) only (Luyben, 2001).

     $$k_u = 4h / (\pi a_C) \quad (6.5)$$

     where $$a_c = y(t^*)/(1 - 1/3 + 1/5);\ t^* = \pi/(2\omega_u) \quad (6.6)$$

   - take 1 as the safety factor (S_f) for conducting the saturation relay test.
5. Using the above selected safety factor and k_u, calculate the slope of the saturation relay (k):

   $$\text{Slope, } k = S_f k_u \quad (6.7)$$

6. As in method 1.
7. As in method 1.

Simulations are carried out for an FOPTD model, with $k_p = 1$, $\tau = 1$ and D ranging from 1, 5, 10, to 30. The simulation results show that method 2 has the best accuracy in the estimation of ultimate controller gain (Table 6.1).

Table 6.1 *Theoretical value and estimations of the ultimate gain by relay tests for the FOPTD systems*

L/τ	k_u	ω_u	Ideal relay feedback test	Yu method	Proposed method 1	Proposed method 2
1	2.262	2.028	2.019 [10.73]	2.231[1.379]	2.231 [1.379]	2.231[1.38]
5	1.132	0.530	1.282 [13.24]	1.260 [11.33]	1.149 [1.50]	1.128 [0.36]
10	1.040	0.286	1.273 [22.41]	1.203 [15.65]	1.126 [8.29]	1.069 [2.74]
15	1.019	0.196	1.273 [24.93]	1.203 [18.03]	1.126 [10.51]	1.066 [4.62]
20	1.011	0.149	1.273 [25.92]	1.202 [18.96]	1.126 [11.39]	1.066 [5.42]
25	1.007	0.12	1.273 [26.40]	1.202 [19.42]	1.126 [11.81]	1.066 [5.83]
30	1.005	0.101	1.273 [26.28]	1.202 [19.67]	1.126 [12.05]	1.066 [6.05]

Note: Safety factor (S_f) = 1.4 for Yu method; $S_f = 1$ for the method 1 and method 2 (% error in the estimation of ultimate gain)

6.5 Simulation Study

To verify the effectiveness of the proposed methods, four examples are studied.

6.5.1 Case study 1

Consider an FOPTD system with $k_p = 1$, $\tau = 1$ and $D = 10$. The controller is designed based on ultimate gain and frequency of the identified model. The designed PI controller is simulated on the actual system. Closed loop responses of the controllers by Yu (1999) method and the present two methods are evaluated on the actual system. The comparison shows that method 2 gives a closed loop response closer to that of the actual system. Hence, method 2 is recommended. The servo response is found to be closer to that of the actual system (refer to Fig. 6.2). The controller is designed by the Ziegler–Nichols method based on ultimate gain and frequency of the identified model. The designed PI controller is simulated on the actual system. Performance of the controller designed on the identified model should be closer to that designed on the actual system.

Luyben (2001) suggested for large D/τ values ($D/\tau = 10$), the IMC tuning method gives best results than the Ziegler–Nichols tuning method. For the present example system, IMC tuning method is also used. The controller parameters using IMC method are given in Table 6.2.

As brought out by Luyben (2001), it is found in the present study also, the IMC method gives an improved performance than that of the Ziegler–Nichols for systems with large delay. After designing controllers method by the Yu (1999) method, proposed method 1, method 2 and the Luyben (2001) method, the controllers are evaluated for robustness. A perturbation of 25% in process gain is given. Simulation result shows that all the methods are robust (Table 6.2). Method 2 gives result close to that of the actual system.

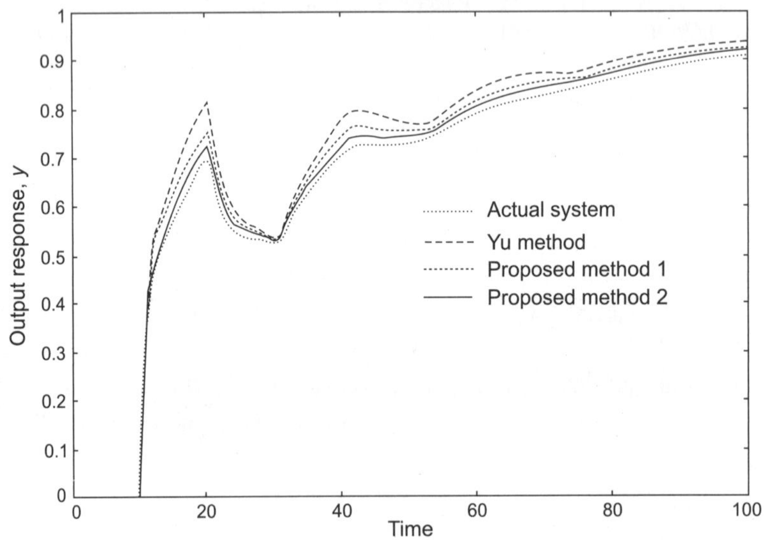

Fig. 6.2 *Comparison of closed loop responses for the FOPTD system (with $D/\tau = 10$) using the controllers from various methods*

Table 6.2 *PI controller settings for FOPTD system with $k_p = 1$, $\tau = 1$ and $D = 10$*

Parameter	Theoretical value	Yu method	Proposed method 1	Proposed method 2	Luyben method
k_u	1.0402	1.203	1.1264	1.0687	1.2732
ω_u	0.2863	0.2909	0.2904	0.2934	0.2939
k_c	0.47[0.35]	0.54 [0.45]	0.51 [0.41]	0.48[0.36]	0.57[0.43]
τ_I	18.29[6.0]	17.99 [6.68]	18.03 [6.41]	17.85[6.02]	17.83[5.85]
ISE	18.49[12.61]	16.26[11.85]	17.12[12.08]	17.75[12.52]	15.6[11.80]
ISE***	15.21[11.78]	14.42[11.93]	14.77[11.74]	15.09[11.76]	14.3[12.27]

Note: ISE*** for FOPTD system when 25% perturbation in k_p
$k_c = 0.45\,k_u$; $\tau_I = 2\pi/(1.2\omega_u)$; $S_f = 1.4$ for Yu method; $S_f = 1$ for the proposed methods. [The PI controller parameters and the ISE values by using IMC method are reported in the bracket]

6.5.2 Effect of noise

Effect of noise is also considered. It is observed that in presence of noise, the ultimate gain and frequency (and hence the controller settings) are not affected for all the three methods. An example with $k_p = \tau = 1.0$ and $D = 10$ (i.e., $D/\tau = 10$) is considered. Noise with zero mean Gaussian distribution and standard deviation of 0.5% is added to the output of the system. The corrupted signal is used in the relay identification test. The amplitude and period of oscillations are identified by taking average at the peak locations. k_u and ω_u are calculated using the Yu (1999) method, modified method 1 and modified method 2. These calculated values of k_u, a and p_u are given in Table 6.3. The ultimate frequency is 0.2938 for all the three methods. Figure 6.3 shows system output response with and without noise. After introducing the noise, it is observed that the initial dynamics and response are shifted on time axis. While calculating k_u and p_u the initial dynamics is not used. Hence, in presence of noise, the ultimate gain and frequency (and hence the controller settings) are not affected for all the three methods.

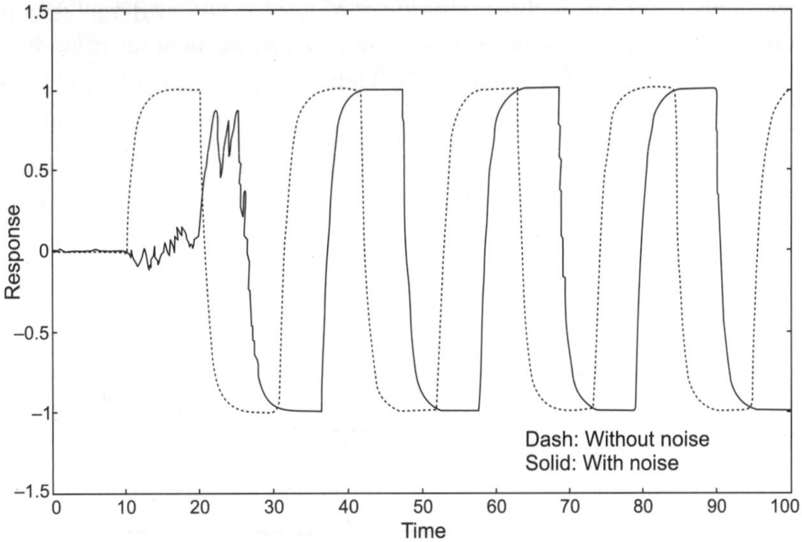

Fig. 6.3 *Process output with measurement noise*

Table 6.3 *Identified k_u and p_u while using noise*

Parameters	Yu method	Method 1	Method 2
A	0.999	1.0	0.999
p_u	21.4443	21.4443	21.443
ω_u	0.293	0.293	0.293
k_u	1.2034	1.1263	1.0478

6.5.3 Case study 2

Consider an SOPTD system with unit process gain, $\tau_1 = 2$, $\tau_2 = 4$ and $D = 40$. The above-described procedure for the FOPTD system is followed for this higher order system (Table 6.4). The ideal relay test shows a waveform close to that of a square form. Hence, a scaling factor of 1 is considered for the proposed methods. Again, the comparison shows that the method 2 follows the actual behavior of the system very closely (refer to Fig. 6.4 and Table 6.4). From these simulation results, it is clear that the modified methods are more accurate than the conventional saturation relay feedback test.

Table 6.4 *PI controller settings for SOPTD system with $k_p = 1$, $\tau_1 = 2$, $\tau_2 = 4$ and $D = 40$*

Parameter	Theoretical value	Yu method	Proposed method 1	Proposed method 2
k_u	1.0465	1.203	1.1265	1.0687
ω_u	0.0685	0.0694	0.0692	0.0698
k_c	0.4709	0.5414	0.5069	0.4809
τ_I	76.4378	75.4465	75.6646	75.0142
ISE	75.6721	67.3882	70.8979	73.5425

Note: $k_c = 0.45\ k_u$; $\tau_I = 2\pi/(1.2\omega_u)$; $S_f = 1.4$ for Yu method; $S_f = 1$ for the proposed method 2

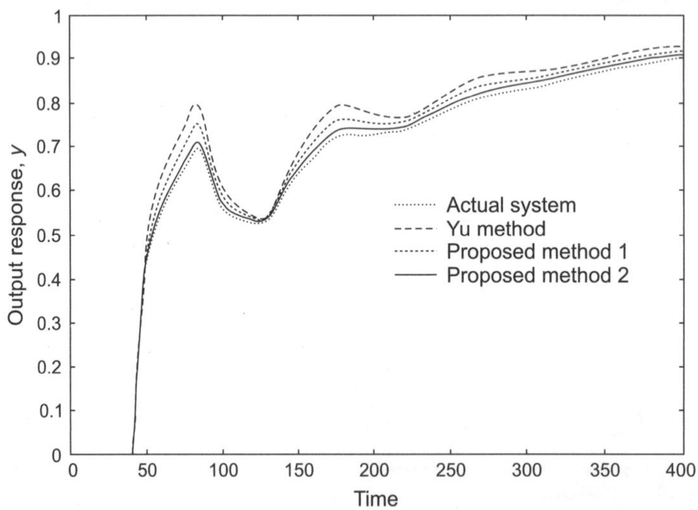

Fig. 6.4 *Comparison of closed loop response for SOPTD system (with $k_p = 1$, $\tau_1 = 2$, $\tau_2 = 4$ and $D = 40$) using the controllers from various methods*

6.5.4 Case studies 3 and 4

Consider a third order transfer function model with large time delay considered by Luyben (2001)

$$G(s)=0.125\exp(-10s)/(s+1)^3 \tag{6.8}$$

Using proposed method 2, the ultimate gain and ultimate frequency are calculated. The method suggested by Srinivasan and Chidambaram (2004) for identifying the FOPTD model parameters is applied and the results are compared with that proposed by Luyben (2001). An FOPTD model is also analytically derived using the amplitude criterion and phase angle criterion for the actual system and keeping the process gain (k_p) as same. The identified FOPTD model parameters, for Srinivasan and Chidambaram (2004) and Luyben (2001) method, match well with that of the actual system. After the identification step, the IMC tuning method is used for the PI controller design. The details are given in Table 6.5. The closed loop response on the actual system is evaluated. The servo and regulatory responses are shown in Figs 6.5 and 6.6 and the ISE values are given in Table 6.5. The controller based on method 2 gives a response close to that based on the actual system.

Another example reported by Lee et al. (1996) for the coupled tanks system with $D/\tau = 5.48$ is also considered.

$$G(s)=1.09\exp(-8.5s)/(1.55s+1) \tag{6.9}$$

The actual process parameters and identified parameters using the Luyben (2001) method and the proposed method 2 are given in Table 6.5. The identified parameters using method 2 follows that of the actual system closely. PI controllers are designed by the IMC method and the performance on the actual system is evaluated.

The servo and regulatory responses are shown in Figs 6.7 and 6.8 and the ISE values are given in Table 6.5. The controller based on method 2 gives a response close to that based on the actual system. For case studies 3 and 4, the response shows that the performance of method 2 is close to that of the actual system.

Table 6.5 *PI controller settings using IMC method for case studies 3 and 4*

Case study	Parameters	Actual*	Luyben method	Proposed method 2
3	k_p	0.125	0.121	0.125
	τ	1.78	2.98	1.85
	L	11.26	9.95	11.18
	λ	19.14	16.915	19.00
	k_c	3.097	3.8867	3.132
	τ_I	7.41	7.955	7.44
	ISE	14.36	13.58	14.32

Contd.

Table 6.5 *Contd.*

4	k_p	1.09	0.9993	1.0893
	τ	1.55	1.8639	1.5438
	L	8.50	7.9023	8.5024
	λ	14.45	13.433	14.455
	k_c	0.3682	0.433	0.368
	τ_I	5.8	5.815	5.795
	ISE	10.93	10.45	10.93

Note: Actual*: Actual FOPTD identified using phase criteria and gain criteria (i.e., making k_p, k_u and p_u of the original model and reduced model same)

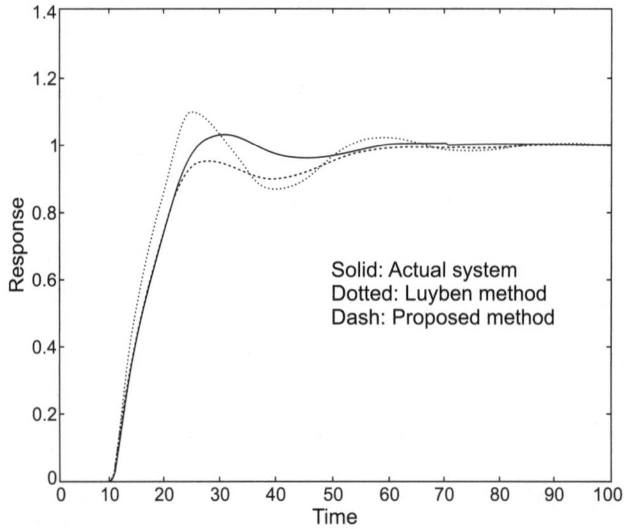

Fig. 6.5 *Comparison of closed loop servo response using PI controller for case study 3*

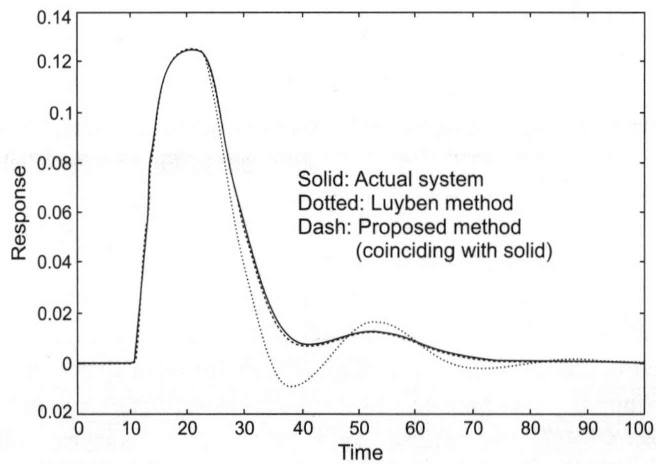

Fig. 6.6 *Comparison of closed loop regulatory response using PI controller for case study 3*

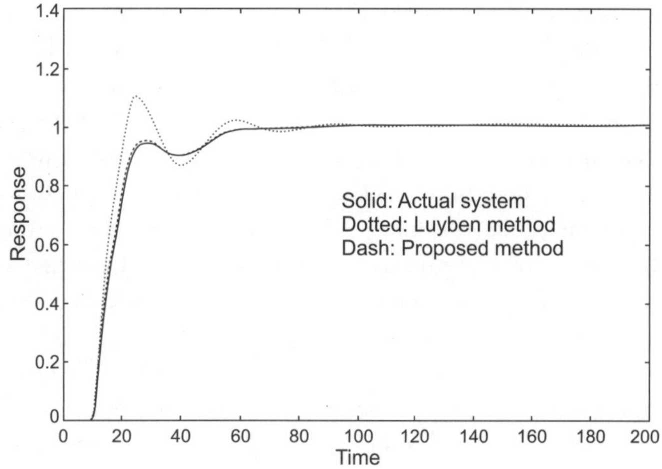

Fig. 6.7 *Comparison of closed loop servo response using PI controller for case study* 4

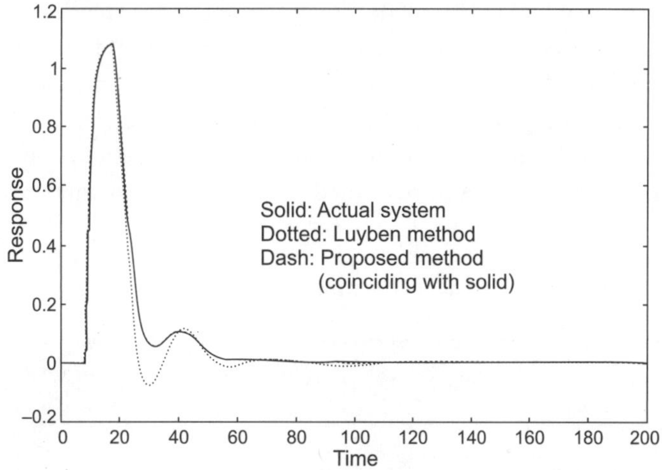

Fig. 6.8 *Comparison of closed loop regulatory response using PI controller for case study* 4

6.6 Non-Relay Identification Methods

Now we will consider some of the non-relay methods used for closed loop identification. Apart from the Ziegler–Nichols continuous cycling method and the Cohen–Coon open loop reaction curve method (Cohen and Coon, 1953), there are at least four methods available in the literature for closed loop identification of stable FOPTD transfer models. Yuwana and Seborg (1982) proposed a simple method that uses the closed loop step response using *P* controller and Pade approximation of the dead time element to evaluate the parameters of a first order process model. This method does not force the system to marginal stability, and because of the closed loop identification, the model parameters are less sensitive to disturbance and measurement noise occurring during the identification stage. Jutan and Rodriguez (1984) modified the Yuwana and Seborg (1982) algorithm by using a higher order approximation for the process model delay. They used least squares method to optimize the parameters in their approximation, which also provided an analytical solution to the closed loop dynamics.

Lee (1989) matched the dominant poles of the closed loop system with a second order process response in order to determine the process parameters. This modification enables the method to be used for the process with large dead times. Chen (1989) modified the Yuwana and Seborg (1982) method by determining the process ultimate frequency of oscillation and the controller gain directly from the closed loop system. Then these data were used to identify a delay plus first order model.

Let us consider a case study 4 example (FOPTD system) with $k_p = 1.09$, $D = 8.5$ and $\tau = 1.55$. For the present problem using $k_c = 0.5$, a closed loop response for a unit step change in the set point is shown in the Fig. 6.9. From the response, the values of y_{p1}, y_{p2}, y_{m1} and Δt are 0.5427, 0.4012, 0.2548 and 10.3820, respectively. Using these values, the identified model parameters k_p, L and τ are calculated from the relevant equations, and the results are given in Table 6.6. Based on the identified model parameters, a PI controller is designed using the IMC tuning formulae proposed by Luyben (2001) (refer to Table 6.6). The designed controllers are simulated on the actual system. As stated earlier, the performance of the controller designed on the identified model should be closer to that designed on the actual model. The servo response comparison shows that the proposed method 2 gives a response closer to that of the actual system (refer to Fig. 6.10 for response comparison). Hence, in method 2, using the saturation relay is recommended. Out of the non-relay methods, Lee (1989) method gives better results.

Table 6.6 *Comparison of identified model parameters, the designed PI controller parameters using IMC method and IAE values of the closed loop system for case study 4*

Parameters	Actual	Lee	Chen	Yuwana-Seborg	Jutan-Rodriguez	Proposed method 2
k_p	1.09	1.085	1.085	1.085	1.085	1.0893
τ	1.55	2.68	4.15	4.1861	1.47	1.5438
L	8.50	7.18	7.14	7.694	11.78	8.5024
λ	14.45	12.21	12.13	13.079	20.03	14.455
k_c	0.3682	0.473	0.586	0.565	0.338	0.368
τ_I	5.8	6.27	7.717	8.032	7.363	5.795
IAE*	14.45	13.89	15.72	15.09	18.86	14.44

Note: *IAE value calculated up to 100 seconds with a sampling time 0.01 seconds

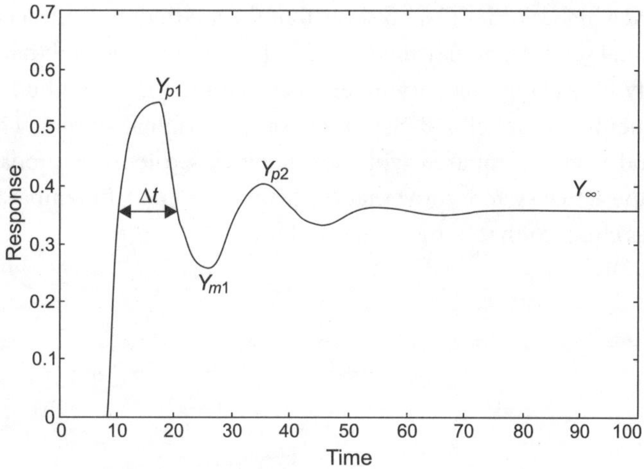

Fig. 6.9 *Typical closed loop under damped response for a step change in the set point using* $k_c = 0.5$

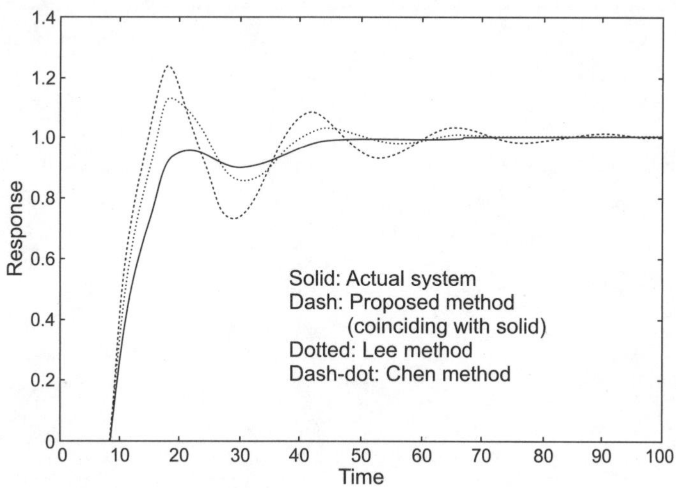

Fig. 6.10 *Comparison of closed loop servo response using PI controllers from various methods for case study 4*

6.7 Conclusions

Conventional saturation relay method proposed by Yu (1999) yields about 15–19% error in the estimation of ultimate gain for systems with large D/τ ratio ($D/\tau > 5$). To reduce the error in the estimation of ultimate gain, two modified methods are proposed, focusing on the selection of a proper slope for the saturation relay. When these methods are applied to systems with large D/τ ratio, the corresponding error is about 8–12% and 2–6% respectively. Case studies for an

FOPTD system and a higher order system show that the methods provide a better performance than the conventional saturation relay method. The present method and that by Luyben (2001) are used to identify FOPTD model parameters. Based on the IMC method, PI controllers are designed for the identified model and also based on the original system. The performance of the present method is also compared with non-relay closed loop methods. The closed loop performances on the actual system show that the Srinivasan and Chidambaram (2004) method gives performances closer to that of the actual system.

7 Identification of FOPTD Model using Single Symmetrical Relay Test

Using a single symmetric relay feedback test, a method is proposed to identify all the three parameters of a first order plus time delay (FOPTD) model. On identifying a higher order dynamics system by an FOPTD model, the conventional method identifies a negative time constant (Li et al., 1991) due to the error in neglecting higher order dynamics in the system output. In the present work, all the parameters of an FOPTD model are estimated with adequate accuracy. Four simulation examples are given. The estimated model parameters of an FOPTD model are compared with those obtained by Li et al. (1991) and also those with the exact model parameters of the system. The performance of the controller designed on the identified model is compared with that identified by Li et al. (1991) and with that of the actual process. The method gives results close to that of the actual system. Simulation results for stable and unstable systems are given.

7.1 Introduction

Identification of transfer function models from experimental data is essential for model based controller design. Often derivation of a rigorous mathematical model is difficult due to the complex nature of chemical processes. Hence, system identification is a valuable tool to identify low order models, based on the input-output data. The relay feedback is a single-shot experiment and the magnitude of oscillations can be varied. From the principal harmonics approximation, the ultimate gain (K_u) and ultimate frequency (ω_u) are found. The ultimate gain (K_u) and ultimate frequency (ω_u) are found from the principal harmonics approximation as

$$k_u = 4h/(\pi a) \tag{7.1a}$$

$$\omega_u = 2\pi/P_u \tag{7.1b}$$

Where 'h' is the magnitude of the relay, 'a' is the output amplitude and 'P_u' is the period of oscillations in the system output.

Luyben (1987) suggested the use of relay testing for identifying a transfer function model. Using the values of K_u and ω_u in the phase angle and amplitude criteria for the first order plus time delay (FOPTD) model, we get the following two equations relating the three parameters:

$$-D\omega_u - \tan^{-1}(\tau\omega_u) = -\pi \tag{7.2a}$$

$$k_u k_p \big/ \left(1+\tau^2\omega_u^2\right)^{0.5} = 1 \tag{7.2b}$$

Since only k_u and ω_u are available, the additional information such as steady-state gain or time delay should be known *a priori* in order to fit a typical transfer function model such as an FOPTD model. Many methods are proposed for obtaining the additional information. Luyben (2001) and Thyagarajan and Yu (2003) proposed a method that includes the effect of shape factor in identifying the process model parameters. However, they used equation (7.1a), which neglects higher order harmonics of the system. Hence, there is a need to propose a method by which models with more than two parameters can be identified accurately using a single symmetrical relay feedback test without *a priori* knowledge of process gain and/or time delay.

Srinivasan and Chidambaram (2003) proposed a modified asymmetrical relay feedback method to get improved estimates of the parameters of the FOPTD model. Using a single asymmetrical relay test, a method is suggested for formulating additional equations to evaluate all the three parameters of the FOPTD model. The asymmetric relay method requires an extra parameter (γ = displacement in the relay height) whose value is to be selected appropriately so that the calculation of the process gain and the estimate of ω_u are carried out accurately (Shen et al. 1996a). In the symmetric relay test such problems are not present. In this chapter, a method is proposed to estimate the three parameters of an FOPTD model using a single symmetrical relay feedback method. Three simulation studies are given to check the proposed identification of the FOPTD model parameters.

7.2 Problem Description

Consider an FOPTD model

$$G(s) = y(s)/u(s) = k_p \exp(-Ds)/(\tau s+1) \tag{7.3}$$

Where, 'k_p' (process gain), 'D' (time delay) and 'τ' (time constants) are the parameters to be estimated.

7.3 Relay Test

Consider a relay feedback system (Fig. 7.1) where $G(s)$ is the process transfer function, y is the output, y_r is the set point, e is the error and u is the manipulated variable. A feedback loop with a symmetrical relay is used. The relay switches between $+h$ and $-h$ as shown in Fig. 7.1 for a positive gain process.

7.4 Estimation of k_p, D and τ

From the definition of Laplace transform, we get

$$y(s) = \int_0^\infty y(t)\exp(-st)\,dt \tag{7.4}$$

The integral $y(s)$ can be evaluated for $s = 8/t_s$ where 't_s' is the time at which few (say 3) repeated cycles of oscillation appear in the output after the initial transients die out. The reason for taking $s = 8/t_s$ is that, for $t > t_s$, because of a very small value of the term exp($-st$), the contribution by the subsequent terms is negligible while evaluating the integral. Let the resulting integral value be denoted as $y(s_1)$. Similarly

$$u(s) = \int_0^{\infty} u(t)\exp(-st)\,dt \tag{7.5}$$

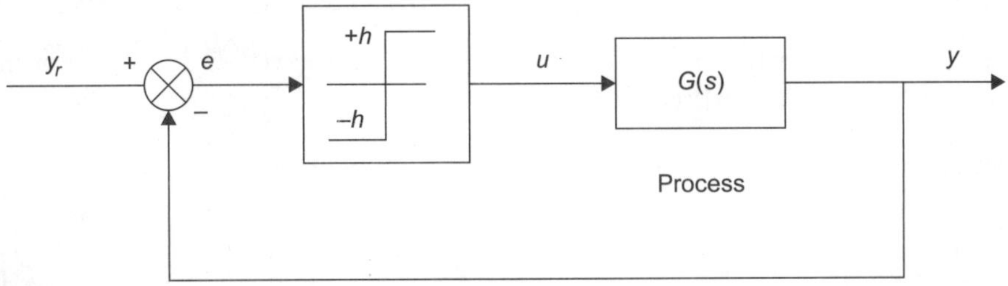

Fig. 7.1 *Block diagram for symmetric relay feedback system*

Equation (7.5) is also evaluated using $s = 8/t_s$. With the numerical values of $y(s_1)$ and $u(s_1)$, using the assumed model, the following equation is formulated:

$$y(s_1)/u(s_1) = k_p \exp(-Ds_1)/(\tau s_1 + 1) \tag{7.6}$$

Padmasree and Chidambaram (2001) first used this equation while analyzing the closed loop response of a PID controlled system. Here it is used for relay autotuning.

Since $y(t)$ and $u(t)$ are piece-wise continuous and periodic, the Laplace transform of $y(t)$ and $u(t)$ can be written as (Kreyszig, 1996)

$$y(s) = \frac{1}{[1 - \exp(-Ps)]} \int_0^P y(t)\exp(-st)\,dt \tag{7.7}$$

$$u(s) = \frac{1}{[1 - \exp(-Ps)]} \int_0^P u(t)\exp(-st)\,dt \tag{7.8}$$

From Eq.(7.7) and Eq.(7.8), we get

$$G(s) = y(s)/u(s) = \left[\int_0^P y(t)\exp(-st)\,dt\right] \Big/ \left[\int_0^P u(t)\exp(-st)\,dt\right] \tag{7.9}$$

where $P = 2\pi/\omega$ and ω is the frequency of oscillations observed in the output. By substituting $s = j\omega$ and using the Euler's expansion for exp($-j\omega t$) in Eq.(7.9), we get

$$G(j\omega) = y(j\omega)/u(j\omega) = (c_1 - jd_1)/(c_2 - jd_2) \tag{7.10}$$

where

$$c_1 = \int_0^P y(t)\cos(\omega t)\,dt \tag{7.11a}$$

$$d_1 = \int_0^P y(t)\sin(\omega t)dt \tag{7.11b}$$

$$c_2 = \int_0^P u(t)\cos(\omega t)dt \tag{7.11c}$$

$$d_2 = \int_0^P u(t)\sin(\omega t)dt \tag{7.11d}$$

Equations (7.11a) to (7.11d) can be evaluated numerically using $y(t)$ and $u(t)$ data of the relay test. Hence Eq.(7.10) can be written as

$$G(j\omega) = p + jq \tag{7.12}$$

where,

$$p = (c_1c_2 + d_1d_2)/(c_2^2 + d_2^2) \tag{7.13a}$$

$$q = (d_2c_1 - d_1c_2)/(c_2^2 + d_2^2) \tag{7.13b}$$

By substituting $s = j\omega$ in Eq.(7.3), we get

$$G(j\omega) = k_p \exp(-Dj\omega)/(\tau j\omega \pm 1) \tag{7.14a}$$

In above equation + sign is used for a stable system and – sign is used for an unstable system. On substituting $\exp(-Dj\omega) = \cos(D\omega) - j\sin(D\omega)$ in Eq.(7.14a), we get

$$p + jq = \left[k_p\left(\cos(D\omega) - j\sin(D\omega)\right)\right]/(\tau j\omega \pm 1) \tag{7.14b}$$

On cross-multiplying and separating the coefficients of the real and imaginary terms we get

$$\pm p - q\omega\tau - k_p\cos(D\omega) = 0 \tag{7.15}$$

$$p\tau\omega \pm q + k_p\sin(D\omega) = 0 \tag{7.16}$$

From the relay test, the three linear algebraic equations (7.6), (7.15) and (7.16) can be solved, to estimate the three unknown parameters: k_p, D and τ. In Eq.(7.15) and Eq.(7.16), the frequency of oscillation of the system output is used. It is to be noted that for the analysis of an asymmetric relay test Srinivasan and Chidambaram (2003) used Eqs (7.15) and (7.16) and calculated the process gain from $u(t)$ and $y(t)$ data as

$$k_p = \left[\int y(t)dt\right]/\left[\int u(t)dt\right] \tag{7.17}$$

The integration is to be carried out for one period of oscillation. In the present method because of symmetric relay, Eq.(7.17) cannot be used. Hence, in the symmetric relay autotuning, Eq.(7.6) is used. We also do not use the equation $k_u = 4h/(\pi a)$ and $\omega_u = 2\pi/P_u$. Thus using single symmetrical relay test, all the three parameters of an FOPTD system are calculated.

7.5 Simulation Study

7.5.1 Case study 1

Consider the transfer function given by Li et al. (1991)

$$G(s) = \exp(-2s)/[(10s+1)(s+1)] \tag{7.18}$$

Application of the ideal relay method by Li et al. (1991) gave the model parameters of FOPTD as $k_p = -0.501$, $D = 2$ and $\tau = -5.03$ (negative values of time constant and gain). According to Li et al. (1991), in such cases, the FOPTD model is not valid and it is recommended to fit a second order plus time delay (SOPTD) model. Whereas the proposed method is able to identify an FOPTD model ($k_p = 1.03$, $D = 2.84$ and $\tau = 11.98$) with the symmetrical relay test. Srinivasan and Chidambaram (2004) obtained the FOPTD model with $k_p = 1.017$, $D = 2.872$ and $\tau = 11.71$. Table 7.1 gives all the intermediate values required for the present calculation of the FOPTD model parameters. Based on the identified parameters, a suitable PID controller is designed. The PID controller is tuned using the Ziegler–Nichols continuous cycling method (Ziegler and Nichols, 1942). The controller parameters and the closed loop performance comparison are given in Table 7.2. The integral square error (ISE) of the proposed model is 4.4798 whereas that of the actual system is 4.4667. The proposed method gives better parameter estimates. The process output and the step response are as shown in Figs 7.2a and 7.2b. The step response of the proposed model is closer to that of the actual system.

Table 7.1 *Intermediate values for the examples considered*

Case study	a	ω_u	P_u	t_s	$y(s_1)$	$u(s_1)$	p	q
1	0.1943	0.5978	10.51	24.785	0.0445	0.4850	−0.1415	−2.29e−4
2	0.1808	0.8572	7.33	40.66	0.1061	0.4808	−0.1156	0.0029
3	0.6468	2.033	3.09	25.74	0.2306	−0.1686	−0.4375	−0.0448

Table 7.2 *Controller parameters based on identified model and actual process parameters*

Case study	Method	$(k_c k_p)_{max}$	ω_u	k_c	τ_I	τ_D	ISE
1	Actual	7.0719	0.5985	4.2431	5.2490	1.3123	4.4667
	Proposed	7.2767	0.6010	4.3660	5.2217	1.3123	4.4798
	Li et al. *	-	-	Not	applicable	-	-
2	Actual	8.5024	0.8443	5.1015	3.7208	0.9302	3.4935
	Proposed	8.0943	0.8453	5.1333	3.7167	0.9292	3.4959
	Li et al.	6.9502	0.8576	4.2208	3.6633	0.9158	3.5946
3	Actual	2.5366	2.3311	2.2837	2.6096	0.2563	2.3902
	Proposed	2.7119	2.3272	2.2470	2.8507	0.2630	2.3971
	Li et al.	1.8240	1.9810	1.9544	3.3259	0.2557	3.1817

Note: *Li et al. method gives a negative time constant. For the case studies 1 and 2, the tuning rule applied is the Ziegler–Nichols closed loop tuning method [$k_c = 0.6\,k_u$; $\tau_I = P_u/2$ and $\tau_D = P_u/8$]. For case study 3 tuning rules are given in Table 7.3

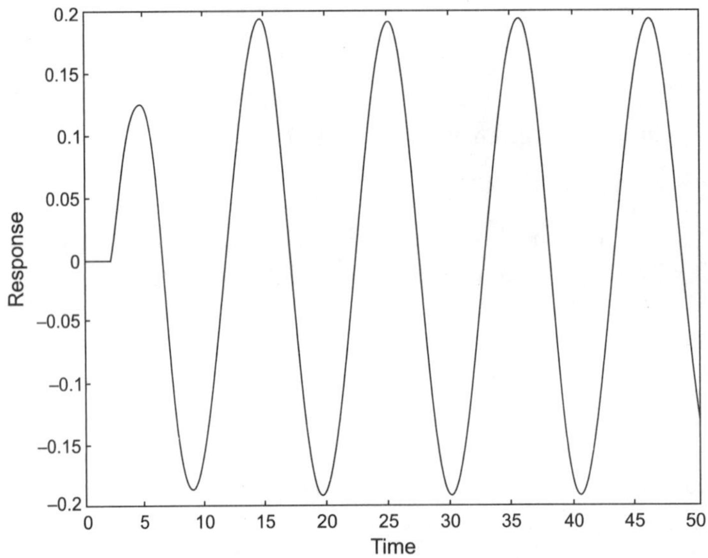

Fig. 7.2a *Process output symmetrical relay response (case study 1)*

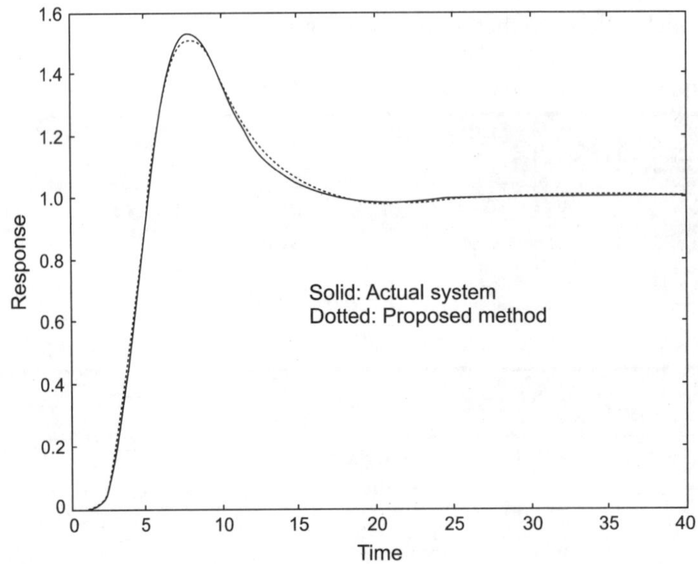

Fig. 7.2b *Closed loop response comparisons for case study 1*

7.5.2 Case study 2

Consider the transfer function

$$G(s) = \exp(-2s)/(10s+1) \tag{7.19}$$

The present method gives the model parameters as $k_p = 0.9467$, $D = 2.00$ and $\tau = 9.5028$. The method proposed by Srinivasan and Chidambaram (2003) gives the model parameters as $k_p = 1.03$, $D = 2.3$ and $\tau = 10.3$, whereas the ideal relay method of Li et al. (1991) gives $k_p = 0.988$, $D = 2.0$ and $\tau = 8.02$, respectively. In the Li et al. method, from the initial response, the value of delay (D) is noted. A significant improvement is obtained by the present method. Based on the identified parameters, a suitable PID controller is designed. As stated earlier, the PID controller is tuned using the Ziegler–Nichols continuous cycling method (Ziegler and Nichols, 1942). The controller parameters and the closed loop performance comparison are given in Table 7.2. The integral square error (ISE) of the proposed model is 3.4959 whereas that of the Li et al. model is 3.5946. The process output and the step response are as shown in Figs 7.3a and 7.3b. The step response of the proposed model is closer to that of the actual process than the Li et al. model.

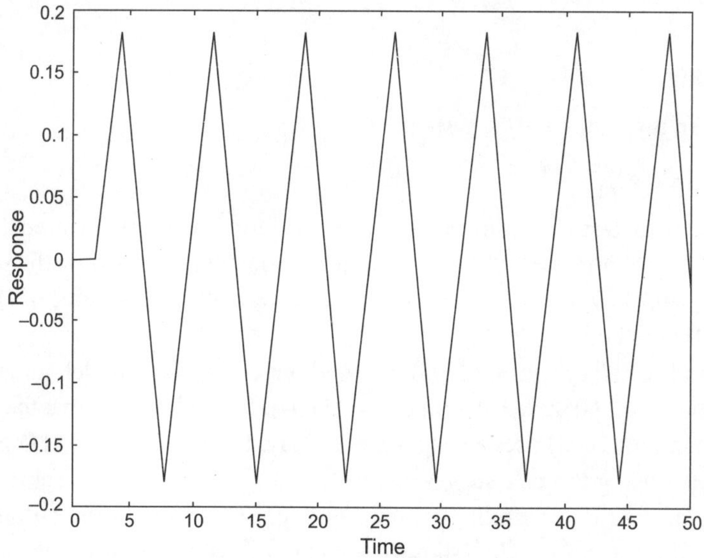

Fig. 7.3a *Process output symmetrical relay response (case study 2)*

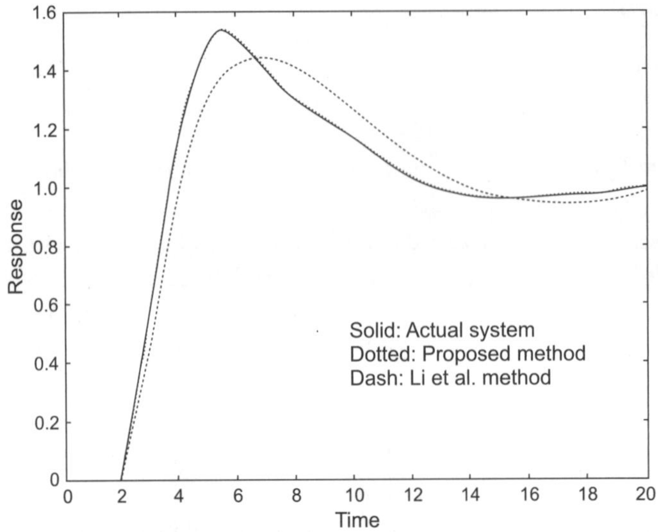

Fig. 7.3b *Closed loop response comparisons for case study 2 (solid and dot lines coincide)*

7.5.3 Case study 3

Let us consider an unstable FOPTD system:

$$G(s) = \exp(-0.5s)/(s-1) \tag{7.20}$$

The symmetrical relay test is conducted. The proposed method gives the model parameters as $k_p = 1.0638$, $D = 0.5127$ and $\tau = 1.0832$. The model parameters obtained by the method of Li et al. (1991) are: $k_p = 0.94$, $D = 0.5$ and $\tau = 0.77$. It is to be noted that the value of D is assumed in the Li et al. (1991) method.

Srinivasan and Chidambaram (2003) method gives similar model parameters as the proposed method ($k_p = 1.0052$, $D = 0.61$ and $\tau = 1.0094$). It is to be noted that the Srinivasan and Chidambaram method (2003) uses an asymmetric relay. Based on the FOPTD model, the PID controller is tuned using the rule suggested by Padmasree et al. (2004) (Table 7.3). Table 7.2 gives the values of the PID controller parameters. Figure 7.4a shows the output response for the symmetric relay test and Fig. 7.4b shows closed loop servo response of the unstable FOPTD system considered. Figure 7.4a shows the identified model by the proposed method is closer to that of the actual system.

Table 7.3 *Tuning formulae used for case study 3 (Padmasree et al., 2004)*

Parameter	Equation $[e = (q/t)]$	Range
$k_c k_p$	$1.2824\, \varepsilon^{-0.8325}$	$0.01 \le \varepsilon \le 1.2$
τ_I/τ	$5.573\varepsilon - 0.0063$	$0.01 \le \varepsilon \le 0.5$
	$0.483 \exp(3.3739\varepsilon)$	$0.5 \le \varepsilon \le 1.2$
τ_D/τ	$0.507\varepsilon + 0.0028$	$0.01 \le \varepsilon \le 1.2$

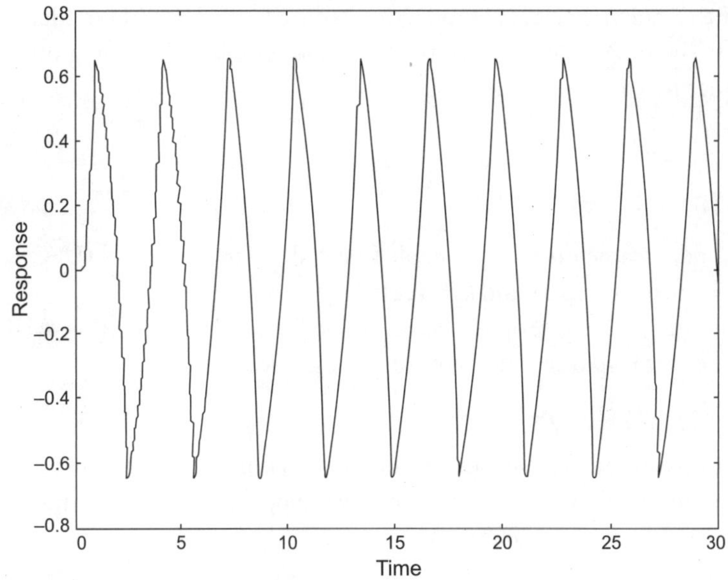

Fig. 7.4a Process output symmetrical relay response (case study 3)

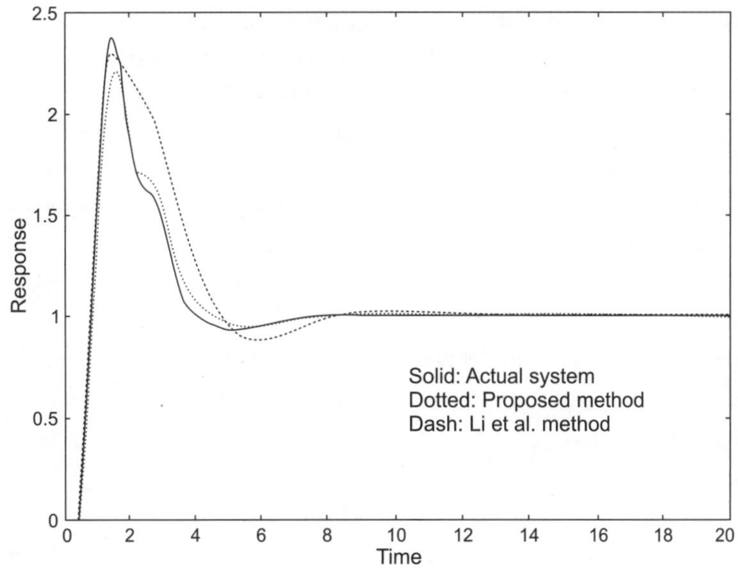

Fig. 7.4b Closed loop response comparisons for case study 3

7.5.4 Case study 4

Let us consider an open loop process transfer function of a system as

$$y(s)/u(s) = G(s) = \exp(-0.5s)/((2s-1)(0.5s+1)) \tag{7.21}$$

Let us apply the symmetric relay method proposed by Li et al. (1991) to identify an unstable FOPTD model. The time delay is measured from the response, and using the phase and gain margins the other two parameters are obtained as

$$D = \tan^{-1}(\tau\omega_u)/\omega_u \tag{7.22}$$

$$k_p = \left[1 + (\tau\omega_u)^2\right]^{0.5} / k_u \tag{7.23}$$

Using this method, parameters of an unstable first order plus time delay (UFOPTD) model are obtained as: $\tau = 0.5762$, $D = 0.51$ and $k_p = 0.4193$

Thyagarajan and Yu (2003) proposed a method to identify the approximate UFOPTD model using the shape of the response. The time constant is calculated as

$$2\exp(t_{a/2}/\tau) - \exp(t_a/\tau) = 1 \tag{7.24}$$

Here t_a is the time required to reach peak amplitude, $t_{a/2}$ is the time required to reach one half of the peak amplitude. Once τ is known, the other two model parameters (D and k_p) are obtained using the ultimate properties (k_u and ω_u) as

$$D = \tan^{-1}(\tau\omega_u)/\omega_u \tag{7.25}$$

$$k_p = \left(1 + (\tau\omega_u)^2\right)^{0.5} / k_u \tag{7.26}$$

For the example system considered, the intermediate values are $t_{a/2} = 0.4085$, $t_a = 1.1448$, peak amplitude = 0.4464 and period of oscillation is 1.1383. The identified model parameters using this method are $\tau = -0.9436$, $D = -0.7214$ and $k_p = 0.5145$.

Majhi and Atherton (1999) proposed a method of identification using a symmetric relay feed back test. The process gain k_p of UFOPTD model can be obtained by shifting process output signal to the left, and then measuring the area of output and input signals for half period of cycle as

$$k_p = \frac{a_y}{a_u} \tag{7.27}$$

Once k_p is known, the other two model parameters are obtained as

$$\tau = \frac{-p_u/2}{\ln\left[\dfrac{k_p h + a}{k_p h - a}\right]} \tag{7.28}$$

$$D = (p_u/2)\ln\left[\frac{k_p h + a}{k_p h}\right] \Big/ \ln\left[\frac{k_p h + a}{k_p h - a}\right] \tag{7.29}$$

For the example system considered, for $h = 1$, the intermediate values are, amplitude = 0.4464, $p_{u/2} = 2.76$, $a_y = 0.6162$ and $a_u = 1.761$. The identified model parameters using this method are: $k_p = 0.3499$, since $k_p < a$, τ and D are found to have imaginary values. Luyben (2001) proposed an identification method that gives an approximate FOPTD model using the shape of the relay response. The method is not effective for open loop unstable process. The steps involved in this method are as follows:

(i) Calculate the time (t_1) required to reach peak amplitude and time (t_2) to half peak; (ii) Calculate the shape factor $F = 4(t_1 - t_2)/P_u$; (iii) Using the correlation given for D/τ versus F, calculate the ratio D/τ; and (iv) using gain and phase margin, the model parameters k_p, τ and D are calculated.

Using this method, we get $t_1 = 33.23$, $t_2 = 32.494$, $F = 0.5336$ and hence $D/\tau = 0.4045$. The model parameters identified are $\tau = -10.4947$, $D = -1.3066$ and $k_p = 4.203$. This method is not giving meaningful results.

Using the proposed method, the intermediate values are calculated as $t = 38.75$, $s_1 = 0.2063$, $y(s_1) = -0.0613$, $u(s_1) = 0.0558$, the real part $p = -0.3485$ and imaginary part $q = -0.0244$ at amplitude 0.4464 and frequency of oscillation 1.1383. The model parameters identified are: $k_p = 0.8192$, $\tau = 1.8628$ and $D = 0.9314$. After identification of the UFOPTD model, the controllers are designed using tuning rules proposed by Padmasree and Chidambaram (2005) for the UFOPTD model. The controller parameters for the proposed method are $k_c = 2.7877$, $\tau_I = 4.8612$ and $\tau_D = 0.4774$, similarly for the Li et al. (1991) method are $k_c = 3.3855$, $\tau_I = 5.5139$ and $\tau_D = 0.2602$ (Table 7.4). Figure 7.5 shows close loop performance for unit step change in set point on the actual system. The closed loop performance of the proposed method is better than that of the Li et al. (1991) method.

Table 7.4 *Identified parameters of FOPTD model for case study 4*

Method	k_p	τ	θ	k_c	τ_I	τ_D	IAE	ISE	ITAE
Li et al.	0.4193	0.5762	0.51	3.39	5.51	0.26	5.46	4.67	19.94
Proposed method	0.8192	1.8628	0.9314	2.79	4.86	0.48	4.04	3.22	10.88

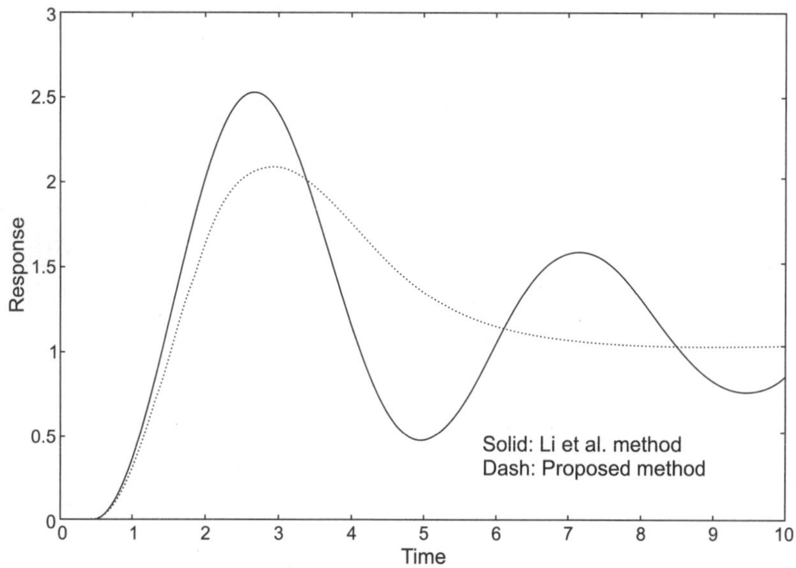

Fig. 7.5 *Comparison of the closed loop servo response of the controllers on the actual system*

7.6 Conclusions

A method is proposed for estimating the parameters of an FOPTD transfer function model using a single symmetric relay test. The advantages of this method are: single relay feedback test and the three parameters that can be identified. Simulation results show that, for the three case studies, the method gives improved parameter estimates. The closed loop performance of the identified system is closer to that of the actual system.

8 Autotuning of PID Controllers for Unstable FOPTD Systems

Using a single symmetric relay feedback test, a method is proposed to identify all the three parameters of a first order plus time delay (FOPTD) unstable model. It is found by simulation that the relay autotune method gives –23% error in the calculation of $k_{u,}$ when $D/\tau = 0.6$. In the present work, a method is proposed by incorporating the higher order harmonics to explain the error in the calculation of k_u and to estimate all the parameters of an unstable FOPTD system. Two simulation results are given on unstable first and second order plus time delay transfer function models. The estimated values of the parameters of the unstable FOPTD model are compared with the methods of Majhi and Atherton (2000) and Thyagarajan and Yu (2003). PID controllers are designed for the identified model and for the actual system. The proposed method gives results close to that of the actual system. For a second order plus time delay system, the latter two methods fail to identify the FOPTD model. Simulation results are also given for a non-linear bioreactor system. The PID controller designed on the model identified by the present method gives a performance closer to that of the controller designed on the locally linearized model.

8.1 Introduction

As stated earlier, the relay feedback method has become very popular because it is time efficient as compared to the conventional method. The amplitude (a) and period of oscillation (p_u) are noted from the sustained oscillation of the system output. The ultimate gain (k_u) and ultimate frequency (ω_u) are calculated from the principal harmonics approximation as

$$k_u = 4h/(\pi a) \tag{8.1}$$

$$\omega_u = 2\pi/\ p_u \tag{8.2}$$

where, 'a' is the amplitude, 'h' is the relay height and 'p_u' is the period of oscillations in the system output.

Luyben (1987) suggested the use of relay testing for identifying a transfer function model. Using k_u and ω_u in the phase angle and amplitude criteria for an unstable FOPTD model, the following two equations relating three model parameters are obtained:

$$k_u k_p \Big/ \left(\tau^2 \omega_u^{\ 2} + 1\right)^{0.5} = 1 \tag{8.3}$$

$$-D\omega_u + \tan^{-1}(\tau\omega_u) = 0 \tag{8.4}$$

Since only k_u and ω_u are available, additional information such as the steady-state gain or the time delay should be known *a priori* in order to fit a typical transfer function model such as an unstable FOPTD. The above Eqs (8.3) and (8.4) assume that the higher order harmonics are neglected.

As stated earlier in Chapter 7, Majhi and Atherton (1999) proposed an improved method for calculating the FOPTD model parameters by symmetric relay tuning. In this method, the output response is aligned with the input response by shifting to the left. The area under the curve for the process output response over a half period of the cycle is calculated (a_y). The corresponding net area for the process input response over the half period of the output response is also calculated (a_u). The steady-state gain k_p is estimated as

$$k_p = a_y / a_u \tag{8.5}$$

After obtaining k_p, the time delay (D) and time constant (τ) are calculated as

$$\tau = \frac{p_u / 2}{\ln\left(\dfrac{k_p h + a}{k_p h - a}\right)} \tag{8.6}$$

$$D = (p_u / 2) \ln\left(\frac{k_p h + a}{k_p h}\right) / \ln\left(\frac{k_p h + a}{k_p h - a}\right) \tag{8.7}$$

Thyagarajan and Yu (2003) proposed a method of identifying an FOPTD unstable model based on the shape of the response of the process using a symmetric relay. In method, the output response is aligned with the input response by shifting to the left. Then, the time to peak amplitude, the peak amplitude and the period of oscillation are noted. The time delay is considered as the time to the peak value. From the derived analytical expression of the process output response of an unstable FOPTD system for a symmetric relay input, the time constant and gain are calculated as

$$\tau = (p_u/2) / \ln\left[1/\left(2e^{-D/\tau} - 1\right)\right] \tag{8.8}$$

$$k_p = a / \left[h\left(e^{D/\tau} - 1\right)\right] \tag{8.9}$$

It is to be noted that, for higher order systems, the recorded time to peak value from the response (D) will not match with that of the actual time delay of the process.

Li et al. (1991) reported that the model identified by the symmetry relay autotune method gives an error as high as 27% to –18% in the value of k_u for stable FOPTD systems. Srinivasan and Chidambaram (2004) proposed a method of considering higher order harmonics, to explain the reported error of 27% to –18% in k_u calculations for stable systems. Srinivasan and Chidambaram (2004) also suggested a method to estimate the model parameters of stable FOPTD system. In the present work, Srinivasan and Chidambaram method is extended to an unstable FOPTD system using a single symmetric relay test.

8.2 Problem Description

Let us consider an unstable first order plus time delay (FOPTD) system:

$$y(s)/u(s) = k_p \exp(-Ds)/(\tau s - 1) \tag{8.10}$$

where 'k_p', the process gain, 'D', the time delay and 'τ', the time constant are the parameters to be estimated. Consider a relay feedback system (Fig. 8.1) where $G(s)$ is the process transfer function, 'y' is the output, 'y_r' is the set point, 'e' is the error and 'u' is the manipulated variable. Here, the feedback loop uses a symmetric relay.

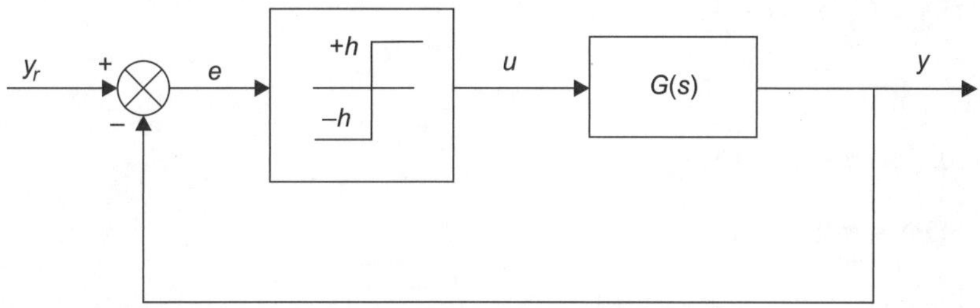

Fig. 8.1 *Block diagram for symmetric relay feedback system*

8.3 Consideration of Higher Order Harmonics

From the Fourier series analysis, it can be easily shown (Aström and Hägglund, 1984) that a relay consists of many sinusoidal waves of odd multiples of fundamental frequency 'ω' and with the amplitude $4h/(n\pi)$ (n = 1, 3, 5, ...). The input to the process thus consists of many sine waves. For an FOPTD system, the output wave is also a sinusoidal wave with different amplitude and frequency. Here $y(t)$ is the combined response due to many sine waves:

$$y(t) = \left[a_1 \sin(\omega_u t + \phi_1) + (1/3)\, a_3 \sin(3\omega_u t + \phi_3) + \cdots\right] \tag{8.11}$$

where

$$a_1 = 1\Big/\left[1 + (\tau\omega_u)^2\right]^{0.5}; \quad a_3 = 1\Big/\left[1 + (3\tau\omega_u)^2\right]^{0.5}; \quad a_5 = 1\Big/\left[1 + (5\tau\omega_u)^2\right]^{0.5}; \ldots \tag{8.12}$$

$$\phi_1 = -D\omega_u - \pi + \tan^{-1}(\tau\omega_u) = -\pi \tag{8.13a}$$

$$\phi_3 = -3D\omega_u - \pi + \tan^{-1}(3\tau\omega_u); \quad \phi_j = -j\,D\omega_u - \pi + \tan^{-1}(j\tau\omega_u) \tag{8.13b}$$

Equation (8.11) can be written as

$$y(t) = a_1\left[\sin(\omega_u t + \phi_1) + (1/3) b_3 \sin(3\omega_u t + \phi_3) + \cdots\right] \tag{8.14}$$

where

$$b_3 = \left\{\left[1 + (\tau\omega_u)^2\right]\Big/\left[1 + (3\tau\omega_u)^2\right]\right\}^{0.5}; \quad b_5 = \left\{\left[1 + (\tau\omega_u)^2\right]\Big/\left[1 + (5\tau\omega_u)^2\right]\right\}^{0.5};, \text{ etc.} \tag{8.15}$$

If $\tau\omega_u$ is assumed to be very large (∞), then $y(t)$ will consists of only the fundamental harmonics. In Eq.(8.11), the terms containing $a_3 \sin(3\omega_u t + \phi_3)$, $a_5 \sin(5\omega_u t + \phi_5)$, etc. will be neglected. As stated earlier, this assumption gives a large error in the calculation of k_u and hence in the estimated model parameters of FOPTD model. In what follows, we will consider the higher order dynamics for the calculation of k_u and in the identification of the model parameters. Since for the unstable FOPTD systems, the value of D/τ should be less than 1 (Venkatashankar and Chidambaram, 1994) for stabilization using a proportional controller and hence using a relay, let us derive an approximate evaluation of $y(t)$ for the limiting cases of larger $\tau\omega_u$.

For the limiting case of larger $\tau\omega_u$, Eq.(8.13a) becomes

$$\phi_1 = -D\omega_u + (\pi/2) = 0 \tag{8.16}$$

Hence, $D\omega_u = 0.5\pi$ (8.17)

$$\phi_3 = -3D\omega_u - \pi + \tan^{-1}(3\tau\omega_u) \tag{8.18a}$$

$$= -3D\omega_u - \pi/2 = -2\pi \tag{8.18b}$$

Similarly, $\phi_5 = -6\pi/2; \ldots; \phi_N = -(N+1)\pi/2$ (8.18c)

Hence, Eq.(8.14) can be written as

$$y(t) = a_1\left[-\sin(\omega_u t) - (1/3)\; b_3 \sin(3\omega_u t) - \cdots\right] \tag{8.19}$$

where $b_3 = \left\{\left[1+(\tau\omega_u)^2\right]/\left[1+(3\tau\omega_u)^2\right]\right\}^{0.5}$; $b_5 = \left\{\left[1+(\tau\omega_u)^2\right]/\left[1+(5\tau\omega_u)^2\right]\right\}^{0.5}$ (8.20a)

For larger values of $\tau\omega_u$, the value of 1 can be neglected when compared to the value of $\tau\omega_u$ and hence the values of b_1, b_3 etc. can be approximated as

$$b_3 = 1/3;\; b_5 = 1/5;\; b_j = 1/j,\text{ etc.} \tag{8.20b}$$

Hence, Eq.(8.19) becomes

$$y(t) = a_1\left[-\sin(\omega_u t) - (1/9)\sin(3\omega_u t) - (1/25)\sin(5\omega_u t) - \cdots\right] \tag{8.21}$$

Here the value of 'a_1' is to be calculated. The value of 'a_1' is not the amplitude what we observe from the output oscillation. Let us estimate the error involved in k_u by using only the principal harmonics in the analysis of relay testing. From the output oscillations, it is possible to calculate $y(t)$ at any time 't'. 'ω_u' is the frequency of observed oscillations. From Eq.(8.21), we get

$$a_1 = -y(t)/\Sigma\left[\sin(i\omega_u t)/i^2\right] \tag{8.22}$$

Let us consider the time (t^*) at which

$$\omega_u t^* = 0.5\pi. \tag{8.23}$$

Then Eq.(8.21) becomes

$$y(t^*) = a_1\left[1 + (1/9) + (1/25) + (1/49) + (1/81) + \cdots\right] \tag{8.24}$$

In the above equation, let 'N' be the number of terms considered. Using the limiting value for the summation term ($0.125\pi^2$), we get from Eq.(8.22):

$$a_1 = 0.810\, y(t^*) \tag{8.25}$$

The time at which the value of $y(t^*)$ is to be noted from the response is given by $t^* = 0.5\pi/\omega_u$. In the conventional method, we have to use only the first term in Eq.(8.22) [i.e., $a = y(t^*)$]. The consideration of all higher order harmonics gives $a_1 = 0.81y(t^*)$. This limiting value shows that a maximum error –19% in k_u is obtained by using the conventional analysis (principal harmonics method). The observed amplitude (a) is always greater than the corrected amplitude (a_1). Later it will be shown that the actual maximum error in k_u will be of –23% [percentage error in ultimate gain is given by $(k_{u(\text{principal harmonics})} - k_{u(\text{exact})})/k_{u(\text{exact})}$]. The calculated value of k_u by the conventional method is always lesser than the actual value.

In deriving the equation for the ultimate gain of the controller, it is assumed that all the higher order harmonics of the relay output are filtered by the system and allows only the fundamental frequency of oscillation (Aström and Hägglund, 1984). The simulation result for several D/τ ratios indicate that the system is not filtering out all the higher order harmonics. For $D/\tau = 0.6$, the relay autotune method gives –23% error in the calculation of k_u by the principal harmonics assumption (Table 8.1). In the proposed work, we suggest a method to incorporate adequate higher order harmonics to get improved estimates of the ultimate gain. It is better to incorporate $N = 3$ rather than $N = 1$ (which is principal harmonics assumption). It is also observed that, depending upon the D/τ ratio, the system will filter out some of the higher order harmonics (not all). Hence, incorporating all N's will also lead to an error in the estimation of the ultimate gain.

Table 8.1 *Effect of including higher order harmonics on k_u*

D/τ	$N = 1$ k_u	$N = 3$ k_u	$N = 5$ k_u	$N = 7$ k_u	$N = 9$ k_u	$N = \infty$ k_u	k_u exact	% error k_u
0.60	1.550	2.133	2.194	2.220	2.228	2.287	2.019	–23.23
0.50	1.969	2.726	2.804	2.837	2.848	2.924	2.536	–22.36
0.40	2.591	3.515	3.615	3.658	3.672	3.770	3.315	–21.84
0.20	5.862	7.266	7.473	7.562	7.590	7.792	7.229	–18.91
0.1	12.519	14.715	15.134	15.314	15.372	15.777	15.07	–16.93
0.06	21.877	24.627	25.312	25.618	25.826	26.415	25.54	–14.34
0.04	33.683	35.967	36.905	37.338	37.669	38.583	38.63	–12.81
0.02	70.735	72.756	74.458	75.339	76.241	77.636	77.90	–9.19

Note: % error in k_u is calculated as $(k_{u(N=1)} - k_{u(\text{exact})})/k_{u(\text{exact})}$
Relay gives oscillations only up to $D/\tau = 0.69$

The values of a_1 are calculated for different number of terms (N) considered in Eq.(8.24). After calculating a_1 values, the values of k_u are calculated (corresponding to N). Table 8.1 shows the effect of including higher order harmonics for various D/τ ratios, ranging from 0.6 to 0.02. Table 8.2 gives the details of the intermediate calculations. For the lower values of D/τ, using only the fundamental frequency in the analysis of relay testing, the proposed method predicts a maximum error of +23% in the amplitude (a). The actual number of higher order harmonics to be considered depends on dynamics of the process. When there is no initial dynamics, the system may be of the FOPTD system and a value of $N = 3$ is suggested. However, if the system response shows any initial dynamics, then the system may be of second order or higher order, and a value of $N = 7$ or $N = 9$ is recommended. Table 8.2 shows that for smaller values of D/τ for FOPTD systems, a larger value of 'N' (number of higher order harmonics) is to be used to get an accurate value of k_u. It is found that when $0.1 < (D/\tau) \leq 0.69$, the value of $N = 3$ is recommended and when $D/\tau = 0.1$ or below, $N = 5$ or $N = 7$ is recommended. Once we decide on the value of N to be used, we can get a more accurate value of k_u. The corrected k_u is calculated as $k_u = 4h/(\pi\ a_1)$, where a_1 is calculated from Eq.(8.24). Knowing the corrected k_u and using the measured value of ω_u, appropriate equations are used to calculate model parameters (k_p, D and τ) of an unstable FOPTD model. Let us now discuss how to derive the equations to identify the model parameters of an unstable FOPTD system.

Table 8.2 *Details of calculations for Table* 8.1

D/τ	a_0 observed amplitude	ω_u frequency	t^* ($=0.5^*\pi/\omega_u$)	$y(t^*)$	a_1 calculated amplitude
0.6	0.8214	1.3483	1.1650	0.6870	0.5968 (N=3)
0.5	0.6464	2.0268	0.7750	0.5375	0.4669 (N=3)
0.4	0.4913	2.9224	0.5375	0.4169	0.3622 (N=3)
0.2	0.2172	6.9813	0.2250	0.2017	0.1752 (N=3)
0.1	0.1017	14.96	0.105	0.0996	0.0841 (N=5)
0.06	0.0582	25.1327	0.0625	0.0595	0.0503 (N=5)
0.04	0.0378	39.2699	0.04	0.0408	0.033 (N=9)
0.02	0.018	78.5398	0.02	0.0202	0.0164 (N=9)

8.4 Estimation of k_p, D and τ

From the definition of Laplace transform (Kreyszig, 1996), we get

$$y(s) = \int_0^\infty y(t)\exp(-st)dt \tag{8.26}$$

The above integral can be evaluated for a particular value of s (say s_1). It is suggested to use the value $s_1 = 8/t_s$, where 't_s' is the time at which three repeated cycles of oscillation appear in the output. The reason for taking $s_1 = 8/t_s$ is that, for $t > t_s$, because of a very small value of the term $\exp(-s_1 t)$, the contributions by subsequent terms is negligible while evaluating the integral. Let the above resulting integral value be denoted as $y(s_1)$. Similarly,

$$u(s) = \int_0^\infty u(t)\exp(-st)dt \tag{8.27}$$

Equation (8.27) is also evaluated using $s_1 = 8/t_s$. With the numerical values of $y(s_1)$ and $u(s_1)$, using the assumed model [Eq.(8.1)], an equation is formulated as given below:

$$y(s_1)/u(s_1) = k_p \exp(-Ds_1)/(\tau s_1 - 1) \tag{8.28}$$

On cross-multiplying, we get

$$[y(s_1)/u(s_1)][\tau s_1 - 1] - k_p \exp[-Ds_1] = 0 \tag{8.29}$$

From the amplitude criterion, we get

$$k_p k_u \Big/ \left[(\tau\omega_u)^2 + 1\right]^{0.5} = 1 \tag{8.30}$$

k_u used in Eq.(8.30) is the corrected k_u giving

$$v = \tau\omega_u = \left[(k_p k_u)^2 - 1\right]^{0.5} \tag{8.31}$$

From the phase angle criterion, we get

$$D = \tan^{-1}(v)/\omega_u \tag{8.32}$$

Equation (8.29) becomes

$$[y(s_1)/u(s_1)][(v/\omega_u)s_1 - 1] - k_p \exp\left[-(s_1/\omega_u)\tan^{-1}(v)\right] = 0 \tag{8.33}$$

Solving numerically, Eq.(8.33), we get the value for v. Since

$$v = \left[(k_p k_u)^2 - 1\right]^{0.5} \tag{8.34}$$

using $v = \tau\omega_u$ and Eq.(8.32) we get τ and D.

8.5 Simulation Study

8.5.1 Case study 1

Let us consider an unstable FOPTD system as

$$G(s) = \exp(-0.2s)/(s-1) \tag{8.35}$$

With the symmetric relay test of height ±1 and at $N = 3$, the intermediate values are calculated as $k_u = 7.2664$, $s_1 = 1.6327$, $y(s_1) = 0.0173$, $u(s_1) = 0.0155$ and $\omega_u = 6.9813$. The proposed method at $N = 3$ gives model parameters as $k_p = 0.9614$, $D = 0.2044$ and $\tau = 0.9903$. At $N = 1$, the intermediate values are calculated as $k_u = 5.8621$, $s_1 = 1.6327$, $y(s_1) = 0.0173$, $u(s_1) = 0.0155$ and the frequency as 6.9813. Model parameters obtained at $N = 1$ (conventional method) are $k_p = 1.3853$, $D = 0.2073$ and $\tau = 1.1544$. Based on the FOPTD model identified, the PID controller is tuned using the tuning rules suggested by Padmasree et al. (2004) (refer to Table 7.3 for the tuning rules). Table 8.3 shows the model parameters identified for the unstable FOPTD for D/τ ranging from 0.6 to 0.2. Table 8.4 gives the identified model and the controller settings.

Figure 8.2a shows the system output response for the symmetric relay test ($D/\tau = 0.2$) and Fig. 8.2b shows closed loop servo response of the unstable FOPTD system considered. It shows that the identified model by the proposed method gives a response closer to that of the actual system. Based on the method proposed by Thyagarajan and Yu (2003), the time to peak amplitude (delay) is obtained as 0.2. Using Eq.(8.8) and Eq.(8.9), the model parameters are identified as $k_p = 0.9855$ and τ=1.0041.

Table 8.3 *Model parameters identified for unstable FOPTD systems*

	$D/\tau = 0.6$			$D/\tau = 0.5$			$D/\tau = 0.4$			$D/\tau = 0.2$		
	k_p	τ	D	k_p	τ	D	k_p	τ	D	k_p	τ	D
Proposed	0.9782	1.35	0.79	1.0089	1.25	0.58	0.9841	1.1332	0.4372	0.9614	0.99	0.2044
Actual	1.0	1.0	0.6	1.0	1.0	0.5	1.0	1.0	0.4	1.0	1.0	0.2

Note: Proposed: Proposed method; Actual: Actual parameters

Table 8.4 *Comparison of identified model parameters and controller parameters for case study* 1

	k_p	τ	D	k_c	τ_I	τ_D	IAE
Actual	1	1	0.2	4.8968	1.1084	0.1042	0.815
Conventional (N = 1)	1.3853	1.1544	0.207	3.8665	1.1481	0.1083	0.9773
Proposed (N = 3)	0.9614	0.9903	0.2044	4.9616	1.1330	0.1064	0.815
Thyagarajan and Yu	0.9855	1.004	0.2	4.9854	1.1084	0.1092	0.817
Majhi and Atherton	0.9244	0.9397	0.1983	5.0659	1.0993	0.1032	0.799

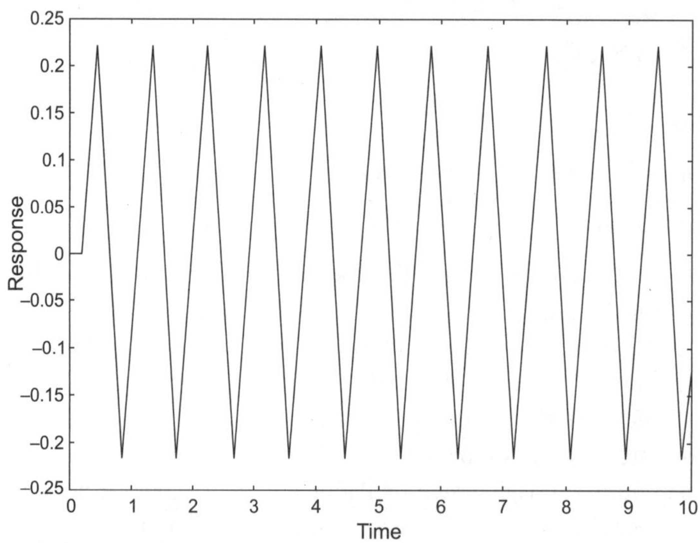

Fig. 8.2a *Relay oscillations for (D/τ) = 0.2 for case study 1*

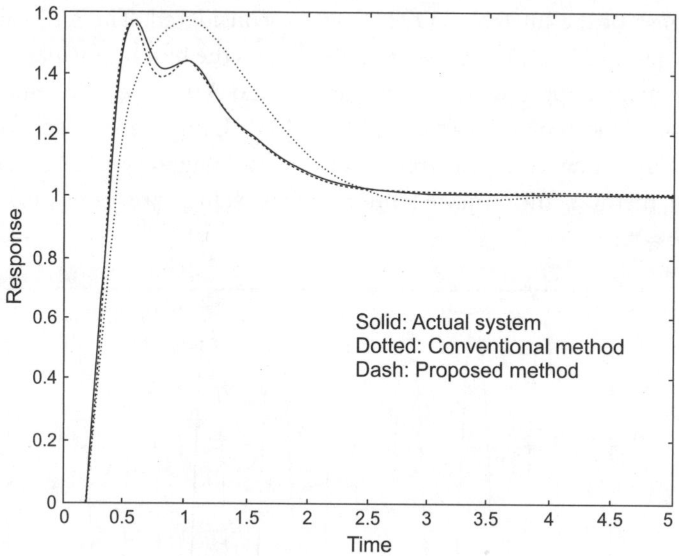

Fig. 8.2b *Comparison of set point responses for case study 1*

For the example considered, based on the method proposed by Majhi and Atherton (2000), the required intermediate values are noted as observed amplitude $a = 0.2172$, $a_y = 0.0501$ and $a_u = 0.0542$. The identified parameters are $k_p = 0.9244$, $D = 0.1983$ and $\tau = 0.9397$. All the three methods discussed above give the model parameters close to that of the actual system (refer Table 8.4).

8.5.2 Case study 2

Let us consider a process transfer function possessing a stable pole and an unstable pole, considered by Majhi and Atherton (1999; 2000):

$$G(s) = \exp(-0.5s)/\left[(0.5s+1)(2s-1)\right] \tag{8.36}$$

The value for the ultimate frequency and ultimate controller gain are found by the stability analysis as 1.2614 and 3.2085, respectively.

The symmetric relay with relay height ±1 is conducted. Figure 8.3 shows the output response. Figure 8.3 shows initial dynamics and the sustained oscillations. The number of terms in Eq.(8.24) is selected as $N = 7$. Using $N = 1$, the intermediate values are obtained as $k_u = 2.8522$, $s_1 = 0.2065$, $y(s_1) = -0.0613$, $u(s_1) = 0.0558$ and frequency = 1.1383. The proposed method (for $N = 1$) gives the model parameter as $k_p = 0.8266$, $D = 0.9951$ and $\tau = 1.8757$. At $N = 7$, the required intermediate values are calculated as $k_u = 3.5292$, $s_1 = 0.2065$, $y(s_1) = -0.0613$, $u(s_1) = 0.0558$ and $\omega_u = 1.1383$. The value of k_u by the proposed method using $N = 7$ is found closer to that of the actual system. The proposed method gives the model parameters as $k_p = 0.7534$, $D = 1.0412$ and $\tau = 2.1642$. Based on the FOPTD model, the PID controller is tuned using the tuning rule suggested by Padmasree et al. (2004) (refer to Table 7.3 for details). Figure 8.4 shows the closed loop servo response of the unstable FOPTD system considered. The IAE value is 4.4493 for using $N = 1$, and 4.0 for $N = 7$. The present method gives the best performance.

Based on the method proposed by Thyagarajan and Yu (2003), we noted the values of $p_u = 5.52$, $a = 0.4464$ and time to peak value $D = 1.1381$. Using Eq.(8.8) and Eq.(8.9), the value of process gain and time constant are calculated as imaginary values. Using Majhi and Atherton (2000) method, the required intermediate values are calculated as $a_y = 0.7855$, $a_u = 1.7544$ and $h = \pm 1$.

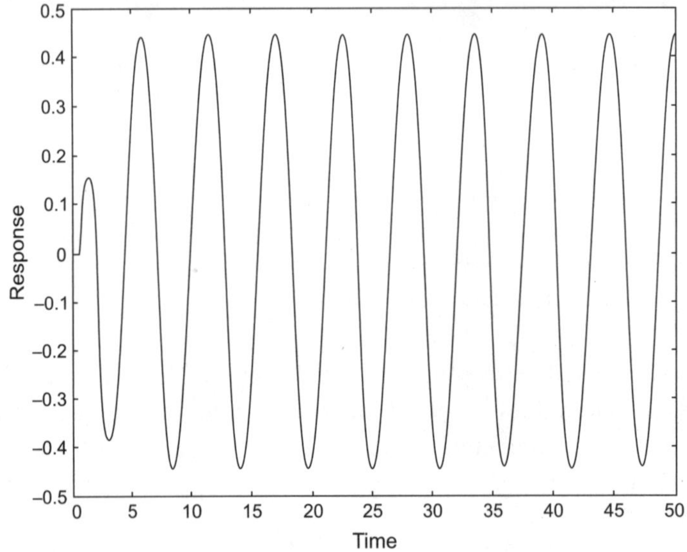

Fig. 8.3 *Relay oscillations for case study 2*

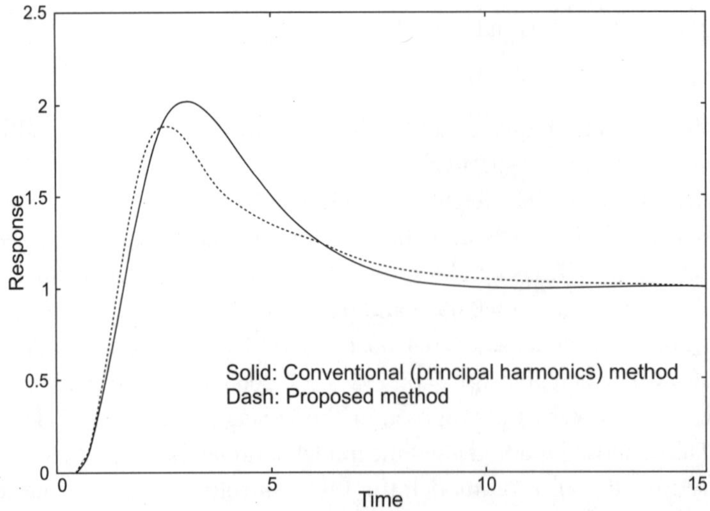

Fig. 8.4 *Comparison of set point responses for case study 2*

The process gain k_p is calculated as 0.4459. Since the identified value of k_p is less than that of the measured amplitude (a = 0.4464), D and τ are obtained as imaginary values [refer to Eq.(8.6) and Eq.(8.7)].

8.5.3 Effect of measurement noise on identification

The effect of measurement noise on the accuracy of the estimation of the parameters is considered for case study 2. The measurement noise with a zero mean Gaussian distribution and a standard deviation of 0.5% and separately of 1% is added to the output of the system. In the identification test, the corrupted signal is used in the feedback control and for the system output. Once the initial dynamics are died out, the amplitude and period of oscillation are calculated by taking the average values at various peak locations. Table 8.5 gives the identified parameters using the proposed methods: Thyagarajan and Yu method and Majhi and Atherton method. It is to be noted that using Majhi and Atherton method gives the identified value of k_p less than that of the observed amplitude (a). The denominators of the right-hand side of Eq.(8.6) and Eq.(8.7) become negative. Hence, D and τ values are obtained as imaginary values [refer to Eq.(8.6) and Eq.(8.7)]. Using Thyagarajan and Yu method, the calculated time delay is larger. Hence, k_p and τ values are obtained as imaginary values [refer to Eq.(8.8) and Eq.(8.9)].

Using the proposed method for parameter identification, the effect of 0.5% standard deviation noise is not significant (Table 8.5). As the noise level increases, the parameter estimation deteriorates. It is desirable to use a filter to remove the noise. For higher order systems, Thyagarajan and Yu method and Majhi and Atherton method give imaginary values for k_p and τ, and D and τ, respectively.

Table 8.5 *Effect of measurement noise for case study 2. $G(s) = exp(-0.5s)/((0.5s+1)(2s-1))$*

Method	σ	k_p	D	τ	k_c	τ_I	τ_D	IAE
Conventional method	Without noise	0.8266	0.9951	1.8757	2.6297	5.4258	0.5098	4.22
Proposed method	Without noise	0.7534	1.0412	2.1642	3.1299	5.7894	0.5339	4.05
	σ = 0.5%	0.7595	1.0549	2.2041	3.1183	5.8655	0.541	4.09
	σ = 1%	0.5332	0.8807	1.3524	3.4372	5.8783	0.4503	3.86
Thyagarajan and Yu method	Without noise	*	1.1381	*	*	*	*	*
	σ = 0.5%	*	1.225	*	*	*	*	*
	σ = 1%	*	1.192	*	*	*	*	*

Contd.

Table 8.5 *Contd.*

Majhi and Atherton method	Without noise	0.4459	**	**	**	**	**	**
	$\sigma = 0.5\%$	0.4346	**	**	**	**	**	**
	$\sigma = 1\%$	0.452	**	**	**	**	**	**

Note: * For Thyagarajan and Yu method, the obtained D value is larger. Hence, k_p and τ are obtained as imaginary values [refer to Eqs (8.8) and (8.9)]
**In Majhi and Atherton method, since the identified value of k_p is less than that of the measured amplitude, the values for D and τ are obtained as imaginary values [refer to Eqs (8.6) and (8.7)]

8.5.4 Effect of load on model parameter identification

The effect of load on parameter identification is considered for case study 1 and case study 2. A load of 0.02 and separately of 0.04 is introduced. The transfer function for the load is assumed to be that of the process transfer function. The affected output signal is used for the feedback relay and as the system output for model identification purposes. The effect of load on model identification for case studies 1 and 2, is reported in Table 8.6.

Using the Majhi and Atherton method for case study 2, the identified value of k_p is found to be less than that of the measured amplitude. The denominators of the right-hand side of Eqs (8.6) and (8.7) become negative. Hence, D and τ are imaginary values [refer to Eqs (8.6) and (8.7)]. Similarly, using the Thyagarajan and Yu method for case study 2, the calculated time delay is larger. Hence, k_p and τ values are obtained as imaginary values [refer to Eqs (8.8) and (8.9)]. For case study 1, the proposed method and the Thyagarajan and Yu method, there is no significant effect of the load on the parameter estimates.

Table 8.6 *Effect of load on identification for case studies* 1 *and* 2

Case study	Load	Proposed method			Thyagarajan and Yu			Majhi and Atherton		
		k_p	τ	D	k_p	τ	D	k_p	τ	D
1	0	0.9614	0.9903	0.2044	0.9855	1.004	0.2	0.9244	0.9397	0.1983
	0.02	0.8846	0.8869	0.2021	1.0122	1.004	0.2	0.9164	0.9056	0.1973
	0.04	0.8367	0.8169	0.2001	0.989	0.9536	0.1987	0.6687	0.6302	0.1857
2	0	0.7534	2.1642	1.0412	*	*	1.1381	0.4459	**	**
	0.02	0.5697	1.372	0.8819	*	*	1.1704	0.4468	**	**
	0.04	0.5611	1.2833	0.8558	*	*	1.2095	0.4517	**	**

Note: * case study 2: using Thyagarajan and Yu method for estimation, the obtained D value is larger. Hence, k_p and τ are obtained as imaginary values [refer to Eqs (8.8) and (8.9)]
** case study 2: using Majhi and Atherton method for estimation, the obtained k_p value is less than that of the observed amplitude of the system output. Hence, the values of D and τ are obtained as imaginary values [refer to Eqs (8.6) and (8.7)]

8.6 Simulation Study of an Unstable Non-linear Bioreactor

The proposed method of identification of an unstable FOPTD system is applied to a non-linear continuous bioreactor that exhibits output multiplicity. The dimensionless model equations are given by Agarwal and Lim (1984):

$$dX/dt = (\mu - D)X \tag{8.37}$$

$$dS/dt = (S_t - S)D - (\mu X/\gamma) \tag{8.38}$$

where

$$\mu = (\mu_m S)/(km + S + K_f S^2) \tag{8.39}$$

Here 'X', 'S', 'S_f' are the dimensionless concentration of cell, substrate and feed substrate, respectively. 'D' is the dilution rate and 'μ' is the specific generation rate. The model parameters are given by Agarwal and Lim (1984) as

$$\begin{array}{lll} \gamma = 0.4\%\ g/g; & S_f = 0.4\%\ g/g; & \mu_m = 0.53h^{-1} \\ km = 0.12\%\ g/g; & k_f = 0.4545\%\ g/g & \end{array} \tag{8.40}$$

The steady-state solution of Eq.(8.37) to Eq.(8.39) gives the following three multiple steady states

$$[X, S]_1 = [0, 4] \text{ (Wash out condition)} \tag{8.41}$$

$$[X, S]_2 = [0.9951, 1.5122] \text{ (Unstable steady-state)} \tag{8.42}$$

$$[X, S]_3 = [1.5301, 0.1746] \text{ (Stable steady-state)} \tag{8.43}$$

Initially, the system is assumed to be at the unstable steady-state condition. At time $t = 0$, $X = 0.9951$ and $S = 1.5122$. At this condition, the dilution rate (D) is 0.3 l/hr. The dilution rate is considered as a manipulated variable in order to control the cell mass concentration (X) at the unstable steady-state at $X = 0.9951$. A delay of one hour is considered in the measurement of X. For the given condition of unstable operating point, the local linearized model is obtained as an unstable FOPTD with the following parameters: $k_p = -5.898$, $\tau = 5.888$ and $D = 1.0$. A symmetrical relay with relay height (h) = 0.03 is conducted. Relay output response (deviation value from the steady-state point of X = 0.9951) is shown in Fig. 8.5. The identified unstable FOPTD model parameter values by the present method are $k_p = -5.5903$, $\tau = 5.6125$ and $D = 1.0152$. Using Thyagarajan and Yu method, the identified unstable model parameters are $k_p = -5.9501$, $\tau = 5.895$ and $D = 1.0$. Even though the delay recorded from peak amplitude location from the relay response matches with that of the linearized model, the shape of the system output differs significantly from that of the identified unstable FOPTD transfer function model (refer to Fig. 8.5 for the system response for the relay input) used by Thyagarajan and Yu (2003).

The response of the actual system gives a straight-line portion whereas the response used by Thyagarajan and Yu (2003) had an initial curved response. The difference in the behavior is due to the non-linearity of the bioreactor. A closed loop response of the non-linear model with designed PID controller is evaluated for a step change in X from 0.9951 to 1.294. The identified model parameters and controller parameters are given in Table 8.7. A good response is obtained as shown in Fig. 8.6. For the present problem, using the Majhi and Atherton (1999) method, the process gain is calculated as $k_p = -2.5$. After calculating k_p, the values of D and τ are obtained using Eq.(8.6) and Eq.(8.7) as $D = 0.8505$ and $\tau = 2.3265$.

The PID settings calculated on each of the identified models are given in Table 8.7. The locally linearized model (Jacob and Chidambaram, 1996) is also used for designing PID controllers (Table 8.7). The controller settings are simulated on the non-linear model equations of the bioreactor. The closed responses are shown in Fig. 8.6. The response of the controller based on the proposed identified model is very close to that of the actual system. The response based on the identified model of Majhi and Atherton (2000) gives an oscillatory response.

Table 8.7 *PID settings for the identified unstable FOPTD model parameters of bioreactor*

Method	k_p	τ	D	k_c	τ_I	τ_D
Linearized	–5.898	5.888	1.0	–0.9513	5.5363	0.5235
Proposed	–5.5903	5.6125	1.0152	–0.9524	5.6228	0.5304
Thyagarajan and Yu	–5.9501	5.895	1.0	–0.9439	5.5363	0.5235
Majhi and Atherton	–2.5	2.3265	0.8505	–1.1855	4.7255	0.4377

Note: PID settings: Padmasree et al. method for unstable FOPTD system
Equation used for PID settings (refer to Table 7.3 for the tuning rules, Padmasree et al., 2004)

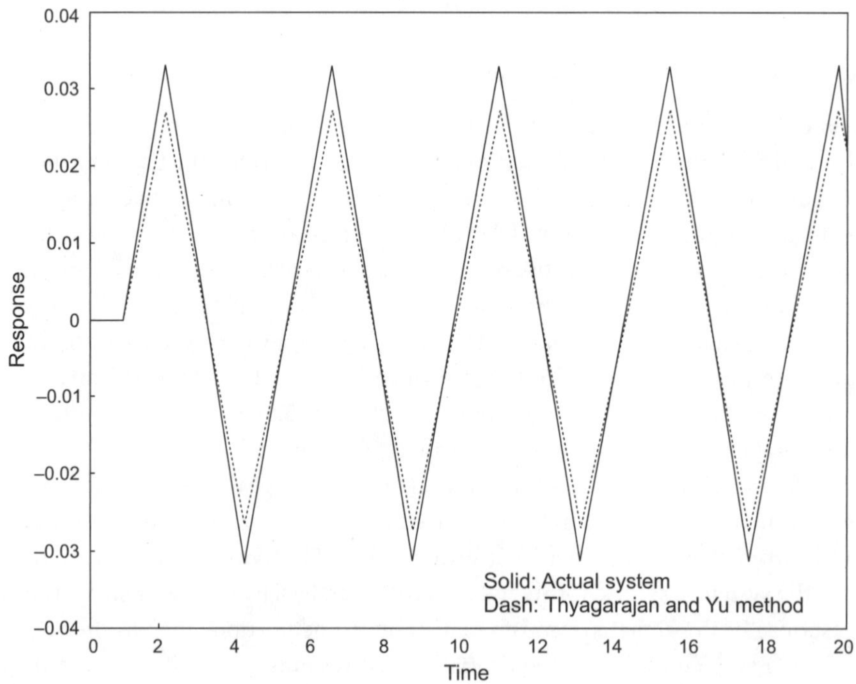

Fig. 8.5 *Comparison of actual response (deviation variable) with that obtained by the Thyagarajan and Yu method (for non-linear bioreactor problem)*

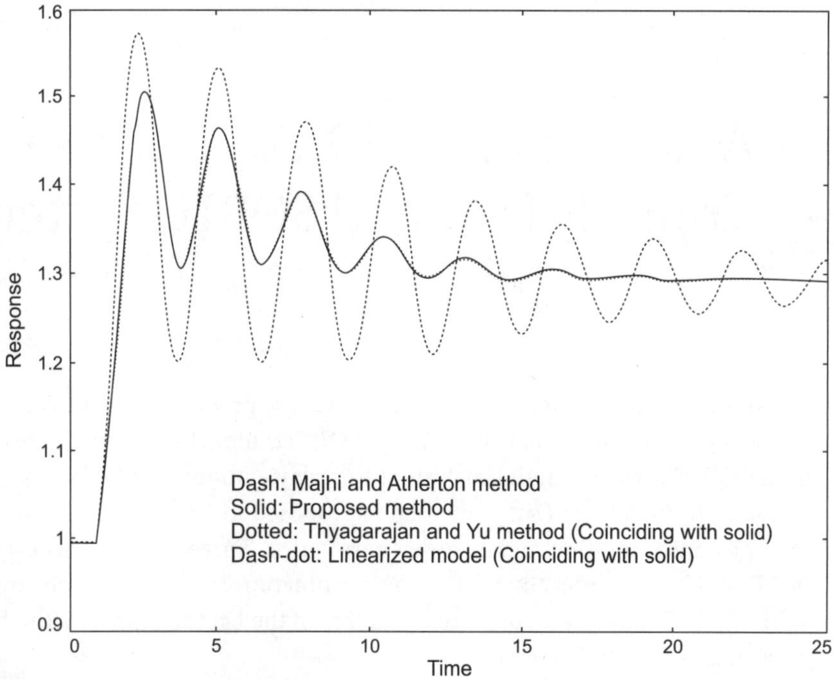

Fig. 8.6 *Comparison of the responses of the non-linear bioreactor using the controller settings from different identification methods*

8.7 Conclusions

A method is proposed to consider higher order harmonics for estimating the parameters of an unstable FOPTD transfer function model using a single symmetric relay test. The advantages of this method are single relay feedback test and that all the three parameters can be identified. Simulation results show that, for two transfer function case studies and a non-linear bioreactor system, the method gives improved parameter estimates and hence an improved closed loop response. For FOPTD systems, the performance of the method is compared with that of the Majhi and Atherton (2000) and the Thyagarajan and Yu (2003) methods. The method gives the best performance. The conventional method of analyzing the relay output for the calculation of ultimate controller gain gives a maximum error of –23% in k_u. For the unstable FOPTD systems, the value of k_u calculated is always less than that of the actual system. For higher order systems, the Thyagarajan and Yu and the Majhi and Atherton methods give imaginary values for the FOPTD model parameters. The FOPTD model parameters calculated by the proposed method gives a good closed loop performance.

9 Autotuning of PID Controllers for Critically Damped SOPTD Systems

Using a single symmetric relay feedback test, two methods are proposed to identify all the three parameters of a stable second order plus time delay (SOPTD) model with equal time constants [$k_p \exp(-Ds)/(\tau s+1)^2$]. The conventional analysis of the relay autotune method gives 27% error in the calculation of k_u for a large D/τ. A method is proposed to improve the accuracy in the calculation of k_u by incorporating higher order harmonics. Three simulation examples are given on SOPTD and higher order time delay transfer function models. The estimated model parameters of the SOPTD model are compared with that of the Li et al. (1991) method and that of the Thyagarajan and Yu (2003) method.

9.1 Introduction

Using relay autotune method, certain higher order model systems when approximated to an FOPTD model gives a negative time constant (Li et al., 1991). In such cases, Li et al. (1991) suggested to identify an SOPTD model instead of an FOPTD model. The SOPTD model can incorporate higher order process dynamics better than the FOPTD model. A controller designed based on the SOPTD model gives a better closed loop response than the one designed on an FOPTD model. It is better to have an SOPTD model with equal time constants, since only three parameters are to be identified.

In the present work, the SOPTD model with equal time constants is considered

$$y(s)/u(s) = k_p \exp(-Ds)/(\tau s+1)^2 \tag{9.1}$$

where 'k_p' the process gain, 'D' the time delay and 'τ' the time constant, are the parameters to be estimated. Li et al. (1991) used the relay feedback method to identify the SOPTD model parameters. Based on simulation studies, Li et al. (1991) reported that when the ratio of the time delay to time constant is larger, the model identified by the relay autotune method gives 27% error in k_u calculations. Similarly, for smaller D/τ ratios, the relay autotune method gives an error of −11% in the k_u calculations. This error is due to the assumptions of principal harmonics of the system output.

Thyagarajan and Yu (2003) proposed a method of identifying an SOPTD model using a symmetric relay test. As reported by Thyagarajan and Yu (2003), if the response consists of sinusoidal oscillations with exponentially increasing magnitude that reach to steady-state after some cycles, the process can be considered as a second order process with equal time constants and with a small D/τ value. In this method, the value of peak amplitude (a_0) and period of oscillation (p_u) are noted. The time constant and gain are calculated as

$$-\varepsilon\tau\omega_u - 2\tan^{-1}\left(\tau\omega_u\right) = -\pi \tag{9.2}$$

$$k_p = \left[1 + \left(\tau\omega_u\right)^2\right] \Big/ k_u \tag{9.3}$$

where $\varepsilon = D/\tau$ and k_p is process steady-state gain. As reported by Thyagarajan and Yu (2003), for most of the SOPTD with small delay and equal time constant examples, default setting of ε is 0.001. However, in certain cases, the model parameters obtained using the default value of $\varepsilon = 0.001$ is not satisfactory. This is because ε depends on the rate at which the oscillations develop. For $\varepsilon = 0.001$, 0.005 and 0.01, a linear equation is given relating ε and normalized time constant (τ_G/P_u), where τ_G is the global time constant, characterized by the global relay feedback responses (defined by peaks of cycling). The linear model gives a refined value of ε, which is used in Eq.(9.2) and Eq.(9.3) to identify the time constant and gain. For situations with $\varepsilon > 0.01$, i.e., when the relay output is not exponentially developed, their method is not applicable. All the simulation examples considered by Thyagarajan and Yu (2003) had ε values lesser than or equal to 0.01.

Panda and Yu (2005) proposed a method of identifying the model parameters of the critically damped SOPTD system by using shape factor. They reported to have solved two simultaneous non-linear algebraic equations to get D and τ. Lee et al. (2007) used the integral of relay responses to estimate the model parameters of FOPTD and SOPTD models with or without zero. In this method, the gain of the system is to be calculated using a separate experiment and the method considers the effect of up to third order harmonics. Majhi (2007) presented a method of estimating the time constant, time delay and non-minimum phase zero of second order transfer function model with a zero. This method needs numerical solutions of three simultaneous non-linear algebraic equations and also requires the knowledge of process gain. Liu et al. (2008) presented a method based on relay tuning method to calculate the three parameters of the system given by Eq.(9.3). This method requires solution of a nonlinear algebraic equation. Liu and Gao (2009) proposed a method of estimating the model parameters of Eq.(9.1) using a biased relay test.

Srinivasan and Chidambaram (2004) proposed a method to consider higher order harmonics of the symmetric relay oscillations for an FOPTD model. They also proposed a method to calculate the model parameters accurately for the FOPTD system. The method and the result are simple to use. Vivek and Chidambaram (2005a) extended the method to identify an unstable FOPTD system. Vivek and Chidambaram (2005b) proposed a method to identify all the parameters of an FOPTD system using a symmetric relay autotuning method. In the present work, these two methods are extended to SOPTD systems with equal time constants.

9.2 Problem Description

Let us consider an open loop stable SOPTD system with equal time constants [refer Eq.(9.1)]. Here k_p, D and τ are the parameters to be estimated. Consider a relay feedback system (refer Fig.

9.1), where $G(s)$ is the process transfer function, 'y' is the output, 'y_r' is the set point, 'e' is the error and 'u' is the manipulated variable. Here a symmetric relay is considered.

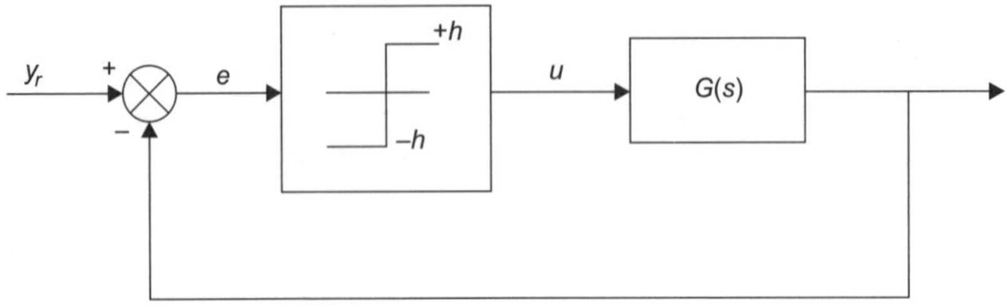

Fig. 9.1 *Block diagram for symmetric relay feedback system*

9.3 Consideration of Higher Order Harmonics

From the Fourier series analysis, it can be easily shown (Åström and Hägglund, 1984) that a relay consists of many sinusoidal waves of odd multiples of fundamental frequency 'ω' and with the amplitude $4h/(n\pi)$ (n = 1, 3, 5, …). The input to the process thus consists of many sine waves. For an SOPTD system, the output wave is also a sinusoidal wave with different amplitude and same frequency. .Here, $y(t)$ is the combined response due to many of the sine waves:

$$y(t)=\left[a_1\sin(\omega_u t+\phi_1)+(1/3)\,a_3\sin(3\omega_u t+\phi_3)+(1/5)\,a_5\sin(5\omega_u t+\phi_5)+\cdots\right] \tag{9.4}$$

where

$$a_1=1\Big/\left[1+(\tau\omega_u)^2\right];\ a_3=1\Big/\left[1+(3\tau\omega_u)^2\right];\ a_5=1\Big/\left[1+(5\tau\omega_u)^2\right];\ldots,\text{etc.} \tag{9.5}$$

$$\phi_1=-D\omega_u-2\tan^{-1}(\tau\omega_u)=-\pi \tag{9.6}$$

$$\phi_3=-3D\omega_u-2\tan^{-1}(3\tau\omega_u);\ \phi_j=-jD\omega_u-2\tan^{-1}(j\tau\omega_u) \tag{9.7}$$

Equation (9.4) can be written as

$$y(t)=a_1\left[\sin(\omega_u t+\phi_1)+(1/3)b_3\sin(3\omega_u t+\phi_3)+(1/5)b_5\sin(5\omega_u t+\phi_5)\ldots\right] \tag{9.8}$$

where

$$b_3=\left[1+(\tau\omega_u)^2\right]\Big/\left[1+(3\tau\omega_u)^2\right];\quad b_5=\left[1+(\tau\omega_u)^2\right]\Big/\left[1+(5\tau\omega_u)^2\right];\ \ldots,\text{etc.} \tag{9.9}$$

If $\tau\omega_u$ is assumed very large, then $y(t)$ will consist of only the fundamental harmonics. In Eq.(9.10), the terms containing $a_3\sin(3\tau\omega_u+\phi_3)$, $a_5\sin(5\tau\omega_u+\phi_5)$, etc. will be neglected. As stated earlier, this assumption gives a large error in the calculation of k_u and hence an error in the estimated model parameters of the SOPTD model. In what follows, we will consider the higher order dynamics for the calculation of k_u and in the identification of the model parameters. Let us derive approximate evaluation of $y(t)$ for the limiting cases of smaller $\tau\omega_u$ and separately for larger $\tau\omega_u$. Wherever the relay oscillations are close to a rectangular waveform, the results for smaller $\tau\omega_u$ is to be used. If the relay oscillations are close to a sinusoidal waveform, then the standard equation considering fundamental frequency of oscillation is used.

For SOPTD system with equal time constants, the phase angle criterion is given by

$$\phi_1 = -D\omega_u - 2\tan^{-1}(\tau\omega_u) = -\pi \tag{9.10}$$

For the limiting case of smaller $\tau\omega_u$, we have $\tan^{-1}(m\tau\omega_u) = m \tan^{-1}(\tau\omega_u)$

$$\phi_3 = -3D\omega_u - 2\tan^{-1}(3\tau\omega_u) = -3\pi \tag{9.11a}$$

Similarly, $\phi_5 = -5\pi; \ldots; \phi_N = -N\pi$ (9.11b)

Hence, Eq.(9.8) becomes

$$y(t) = a_1\left[-\sin(\omega_u t) - (1/3)b_3 \sin(3\omega_u t) - (1/5)b_5 \sin(5\omega_u t) - \cdots\right] \tag{9.12}$$

where $b_3 = \left[1+(\tau\omega_u)^2\right] / \left[1+(3\tau\omega_u)^2\right]; \; b_5 = \left[1+(\tau\omega_u)^2\right] / \left[1+(5\tau\omega_u)^2\right]$ (9.13)

For smaller values of $\tau\omega_u$, the value of $\tau\omega_u$ can be neglected when compared to 1 and hence the values of b_1, b_3, etc. can each be approximated to 1.

Hence, Eq.(9.12) becomes

$$y(t) = a_1\left[-\sin(\omega_u t) - (1/3)\sin(3\omega_u t) - (1/5)\sin(5\omega_u t) - \cdots\right] \tag{9.14}$$

Here the value of 'a_1' is to be calculated. The value of 'a_1' is not the amplitude what we observe from the output oscillation. Let us estimate the error involved in k_u by using only the principal harmonics in the analysis of the relay testing. From the output oscillations, it is possible to calculate $y(t^*)$ at any time 't^*'. ω_u is the frequency of observed oscillations. From Eq.(9.14), we get

$$a_1 = -y(t^*) / \Sigma\left[\sin(i\omega_u t)/i\right] \tag{9.15}$$

Let us consider the time (t^*) at which

$$\omega_u t^* = 0.5\pi. \tag{9.16}$$

Then Eq.(9.14) becomes

$$y(t^*) = a\left[1 - (1/3) + (1/5) - (1/7) + (1/9) + \cdots\right] \tag{9.17}$$

Here, 'a' is the modulus of a_1. In Eq.(9.14), let 'N' be the number of terms considered in Eq.(9.14). Using the limiting value for the summation term (0.25π), we get from Eq.(9.17)

$$a = 1.273\, y(t^*) \tag{9.18}$$

The time at which the value of $y(t^*)$ is to be noted from the response is given from Eq.(9.16) as

$$t^* = 0.5\, \pi/\omega_u. \tag{9.19}$$

From the relay oscillation test, the value of ω_u is noted and t^* is calculated using Eq.(9.19). The value of the process output $y(t^*)$ is noted at t^*. Then by selecting proper number of terms in Eq.(9.17), the calculated amplitude is obtained.

In the conventional method, we use only the first term in Eq.(9.17) [i.e., $a = y(t^*)$]. Whereas consideration of all the higher order harmonics gives $a = 1.273\ y(t^*)$. This limiting value shows that a maximum error of 27% in k_u is obtained by using the conventional analysis (principal harmonics method). The observed amplitude (a_0) is always less than the corrected amplitude (a). Later it will be shown that the actual maximum error in k_u will be of 27% [percentage error in ultimate gain is given by $(k_{u(\text{principal harmonics})} - k_{u(\text{exact})})/\ k_{u(\text{exact})}$]. The calculated value of k_u by the conventional method is always greater than the exact values.

The method is verified for an SOPTD system with various D/τ ratios (keeping $D = 4$ and changing τ from 1 to 0.1, refer Table 9.1, for model parameters used for simulation). As stated earlier, in the proposed work, we suggest a method to incorporate adequate number of higher order harmonics to get an improved estimate of the ultimate gain. It is better to incorporate $N = 3$ or $N = 5$ than $N = 1$ (which is principal harmonics assumption). However, the present study shows that the value of $N = 5$ gives better results on k_u. It is also observed that, depending on the D/τ ratio, the system will filter out some of the higher order harmonics (not all). Hence, by incorporation of terms for all N may lead to an error in the estimation of the ultimate gain (Table 9.1). Table 9.1 shows the effect of including higher order harmonics, for various D/τ ratios, ranging from 4 to 40. Table 9.2 gives the details of the calculation. The actual number of the higher order harmonics to be considered also depends on the dynamics of the process. When there is no initial dynamics, a value of $N = 5$ is suggested. If the system response shows any initial dynamics, then $N = 7$ or $N = 9$ is recommended.

Table 9.1 *Effect of including higher order harmonics on k_u*

D/τ	$N = 1$ k_u	$N = 3$ k_u	$N = 5$ k_u	$N = 7$ k_u	$N = 9$ k_u	$N = \infty$ k_u	k_u exact	% error in k_u
4	1.33	1.258	1.211	1.191	1.179	1.139	1.28	3.91
6	1.283	1.164	1.121	1.103	1.096	1.055	1.158	10.79
8	1.274	1.128	1.087	1.068	1.058	1.022	1.09	16.88
10	1.273	1.113	1.073	1.055	1.045	1.0	1.07	18.97
20	1.273	1.1036	1.063	1.045	1.035	1.0	1.02	24.80
40	1.273	1.1035	1.063	1.045	1.035	1.0	1.0	27.3

Note: % error in $k_u = (k_{u\ \text{principal harmonics}} - k_{u\ \text{exact}})/k_{u\ \text{exact}}$

Table 9.2 *Details of calculations for Table* 9.1

D/τ	a_0 observed amplitude	ω_u frequency	$t^* = 0.5\ \pi/\omega_u$	$y(t^*)$	a calculated amplitude
4	0.9543	0.55	2.8298	0.8777	1.012
6	0.9919	0.614	2.5575	0.9475	1.093
8	0.9987	0.649	2.42	0.9778	1.1711
10	0.9998	0.672	2.3351	0.9907	1.1866

Contd.

Table 9.2 *Contd.*

20	1.0	0.724	2.1675	0.999	1.1965
40	1.0	0.753	2.0849	1.0	1.1977

Knowing the corrected value of k_u and using the measured value of ω_u, the appropriate equations are used to calculate the model parameters (k_p, D and τ) of the SOPTD model. Let us consider now the other limiting case, i.e., where $\tau\omega_u$ is larger.

In what follows, we will consider the higher order dynamics for the calculation of k_u and in the identification of the model parameters. Let us derive approximate evaluation of $y(t)$ for the limiting cases of the larger $\tau\omega_u$. The phase angle criterion is given by

$$\phi_1 = -D\omega_u - 2\tan^{-1}(\tau\omega_u) = -\pi \tag{9.20}$$

For larger $\tau\omega_u$ value, $\tan^{-1}(\tau\omega_u) \cong \pi/2$. Substituting in Eq.(9.20) gives
$D\omega_u = 0$ (since D and ω_u will have some value, the product cannot be equal to zero).

Let us consider $D\omega_u = q$ (9.21)

Hence, $\phi_3 = -3D\omega_u - 2\tan^{-1}(3\tau\omega_u) = -3q - \pi$ (9.22)

Similarly, $\phi_5 = -5q - \pi; \ldots; \phi_m = -mq - \pi$ (9.23)

Hence, Eq.(9.8) can be written as

$$y(t) = a_1\left[-\sin(\omega_u t) - (1/3)b_3 \sin(3(\omega_u t - q)) - \cdots\right] \tag{9.24}$$

where $b_3 = \left[1 + (\tau\omega_u)^2\right] / \left[1 + (3\tau\omega_u)^2\right];\ b_5 = \left[1 + (\tau\omega_u)^2\right] / \left[1 + (5\tau\omega_u)^2\right]$ (9.25)

For larger values of $\tau\omega_u$, the value of 1 can be neglected when compared to the value of $\tau\omega_u$ in Eq.(9.25) and hence the values of b_1, b_3, etc. can be approximated as

$b_3 = 1/9;\ b_5 = 1/25;\ b_j = 1/j^2$, etc. (9.26)

Hence, Eq.(9.26) becomes

$$y(t) = a_1\left[-\sin(\omega_u t) - (1/27)\sin(3(\omega_u t - q)) - (1/25)\sin(5(\omega_u t - q)) - \cdots\right] \tag{9.27}$$

Here the value of 'a_1' is to be calculated. The value of 'a_1' is not the amplitude what we observe from the output oscillation. Let us estimate the error involved in k_u by using only the principal harmonics in the analysis of relay testing. From the output oscillations, it is possible to calculate $y(t)$ at any time 't'. 'ω_u' is the frequency of observed oscillations. From Eq.(9.27), we get

$$a_1 = -y(t) / \Sigma\left[\sin(i(\omega_u t - q)) / i^3\right] \tag{9.28}$$

Let us consider the time (t^*) at which

$$\omega_u t^* = 0.5\pi. \tag{9.29}$$

Figure 9.2 shows the value of $D\omega_u$ (denoted as q) versus $\tau\omega_u$ using Eq.(9.20). Figure 9.2 shows an asymptotic behavior at larger value of $\tau\omega_u$. The value of q at $\tau\omega_u = 50$ can be considered as a limiting value for q for the present work. This value of q is obtained as 0.035.

Using this q value in Eq.(9.27), we get

$$y(t^*) = a_1\,(1.0319) \tag{9.30}$$

The time at which the value of $y(t^*)$ to be noted from the response is given by $t^* = 0.5\pi/\omega_u$. In the conventional method, we have to use only the first term in Eq.(9.27) [i.e., $a = y(t^*)$]. The consideration of all higher order harmonics gives $a_1 = 0.9691\ y(t^*)$. This limiting value shows a maximum error of –3.19% in k_u. It is obtained by using the conventional analysis (principal harmonics method). The observed amplitude (a) is always greater than the corrected amplitude (a_1). The calculated value of k_u by the conventional method is always lesser than the actual values. In the above analysis for $q = 0$, 3.1% error in k_u is obtained.

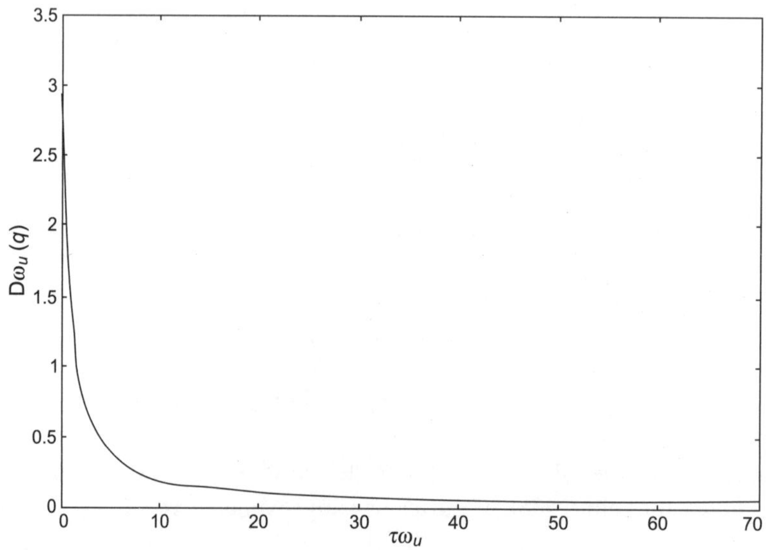

Fig. 9.2 q versus $\tau\omega_u$ corresponding to Eq.(9.22)

As reported earlier, for smaller D/τ ratios, the relay autotune method gives –10% error in the k_u calculations (Li et al., 1991). The above analysis gives –3.19% error in k_u calculations. Now let us consider the rearrangement of b_3 and b_5 terms in Eq.(9.25) as

$$b_3 = \left[1+(\tau\omega_u)^2\right]\Big/\left[1+(3\tau\omega_u)^2\right] = \left[1/9\right]+\left[1/\left(9\tau^2\omega_u^2\right)\right];$$

$$\text{and } b_5 = \left[1/25\right]+\left[1/\left(25\tau^2\omega_u^2\right)\right] \tag{9.31}$$

Substituting $b_3, b_5, \ldots b_j$ in Eq.(9.24) we get

$$y(t) = a_1\left[-\sin(\omega_u t) - (1/3)\left\{(1/9)+\left[1/\left(9\tau^2\omega_u^2\right)\right]\right\}\sin\left(3(\omega_u t - q)\right) - \cdots\right] \tag{9.32}$$

from Eq.(9.5) $(\tau\omega_u)^2 = (1-a_1)/a_1$ (9.33)

Using $(\tau\omega_u)^2$ in the Eq.(9.32), we get

$$y(t)=a_1\left[-\sin(\omega_u t)-(1/3)\{(1/9)+[a_1/9(1-a_1)]\}\sin(3(\omega_u t-q))-\cdots\right] \tag{9.34}$$

On further simplification, we get

$$\begin{aligned}y(t)=a_1&\left[-\sin(\omega_u t)-(1/27)\ \sin(3(\omega_u t-q))-(1/125)\ \sin(5(\omega_u t-q))-\cdots\right]\\&+\left[a_1^2/(1-a_1)\right]\left[-(1/27)\ \sin(3(\omega_u t-q))-(1/125)\ \sin(5(\omega_u t-q))-\cdots\right]\end{aligned} \tag{9.35}$$

Using limiting value for q as 0.0351, Eq.(9.35) is written as

$$y(t^*)=a_1(1.0319)+\left[a_1^2/(1-a_1)\right](0.0319) \tag{9.36}$$

Using a_1 ranging from 0 to 1.0, a linear model used to relate $y(t)$ and a_1 gives

$$y(t^*)=-1.1a_1+0.0052 \tag{9.37}$$

Rearranging Eq.(9.37), we get

$$a=0.909\left[y(t^*)-0.0052\right] \tag{9.38}$$

The consideration of all higher order harmonics gives $a\cong 0.909\, y(t^*)$. This limiting value shows that a maximum error –10% in k_u is obtained by using the conventional analysis (principal harmonics method). Let us now discuss how to formulate the equations to identify the model parameters of SOPTD systems with equal time constants.

9.4 Estimation of k_p, D and τ using k_u

From the definition of the Laplace transform (Kreyszig, 1996), we get

$$y(s)=\int_0^\infty y(t)\exp(-st)dt \tag{9.39}$$

The above integral can be evaluated for a particular value of s (say s_1). It is suggested to use the value $s_1 = 8/t_s$, where 't_s' is the time at which three repeated cycles of oscillation appear in the output. The reason for taking $s_1 = 8/t_s$ is that, for $t > t_s$, because of a very small value of the term $\exp(-s_1 t)$, contributions by subsequent terms is negligible while evaluating the integral. Let the above resulting integral value be denoted as $y(s_1)$. Similarly,

$$u(s)=\int_0^\infty u(t)\exp(-st)dt \tag{9.40}$$

The above equation is also evaluated using $s_1 = 8/t_s$. With the numerical values of $y(s_1)$ and $u(s_1)$, using the assumed model Eq.(9.1), an equation is formulated as given below:

$$y(s_1)/u(s_1)=k_p\ \exp(-Ds_1)/(\tau s_1+1)^2 \tag{9.41}$$

On cross-multiplying, we get

$$\left[y(s_1)/u(s_1)\right](\tau s_1+1)^2-k_p\exp[-Ds_1]=0 \tag{9.42}$$

From the amplitude criterion, we get

$$k_p k_u\Big/\left[(\tau\omega_u)^2+1\right]=1 \tag{9.43}$$

k_u used in the above equation is the corrected k_u giving

$$v = \tau\omega_u = \left[\left(k_p k_u\right) - 1\right]^{0.5} \tag{9.44}$$

From the phase angle criterion, we get

$$D = \left[\pi - 2\tan^{-1}(v)\right]/\omega_u \tag{9.45}$$

Eq.(9.41) becomes

$$\left[y(s_1)/u(s_1)\right]\left[(v/\omega_u)s_1 + 1\right]^2 - k_p \exp\left\{-(s_1/\omega_u)\left[\pi - 2\tan^{-1}(v)\right]\right\} = 0 \tag{9.46}$$

Solving numerically, Eq.(9.46), we get the value for k_p. Since

$$v = \left[\left(k_p k_u\right) - 1\right]^{0.5} \tag{9.47}$$

using $v = \tau\omega_u$ and Eq.(9.45), we get τ and D.

9.5 Simulation Study

Three simulation examples are considered here. For all the examples, a symmetric relay with relay height ±1 is used. The first example system is an SOPTD model with equal time constants. The second example is a third order plus time delay (TOPTD) system with equal time constants and the third example is of TOPTD system with unequal time constants.

9.5.1 Case study 1

Let us consider an SOPTD system with equal time constants as

$$G(s) = \exp(-0.5s)/(20s+1)^2 \tag{9.48}$$

Using the relay test, we get system output oscillations as shown in Fig. 9.3, from which we noted the amplitude $a_0 = 0.0177$ ($k_u = 71.934$) and $\omega_u = 0.4164$. It shows that the relay response consists of sinusoidal oscillations with exponentially increasing magnitude that reach steady-state oscillations after fewer cycles. This is a typical response of an SOPTD model with equal time constants for low D/τ ratios. Using the proposed method, the intermediate values are obtained as $s_1 = 0.0925$, $y(s_1) = -8.9212 \times 10^{-5}$ and $u(s_1) = -7.5929 \times 10^{-4}$. The proposed method gives an SOPTD model with equal time constants as $k_p = 1.058$, $D = 0.551$ and $\tau = 20.81$. Using the method of Srinivasan and Chidambaram (2004) [SC method], the FOPTD model parameters are calculated as $k_p = 0.3753$, $D = 3.8613$ and $\tau = 64.795$.

For the example considered, Thyagarajan and Yu method (TY method) is not applicable. If applied the method gives $k_p = 27.78$, $\tau = 107.32$ and $D = 0.1073$. The obtained gain is very high as compared to the actual model considered. Hence, as per the method reported by Thyagarajan and Yu (2003), ε is corrected as 0.0092 and the model parameters are obtained as $k_p = 3.03$, $\tau = 35.4277$ and $D = 0.3259$. Based on the method proposed by Li et al. (1991), the SOPTD model with equal time constant is calculated as $k_p = 1.1907$, $\tau = 22.095$ and $D = 0.52$ (Delay is measured from the response). Table 9.3 gives the identified model and the actual model considered for the simulation study. Figure 9.4 shows the open loop response of the actual and the identified models for a unit step change in the set point. The proposed method gives a response close to that of the actual system. Hence the proposed method is preferred. The model identified by Thyagarajan and Yu (2003) method deviates too much from the actual system hence it is not applicable for this case.

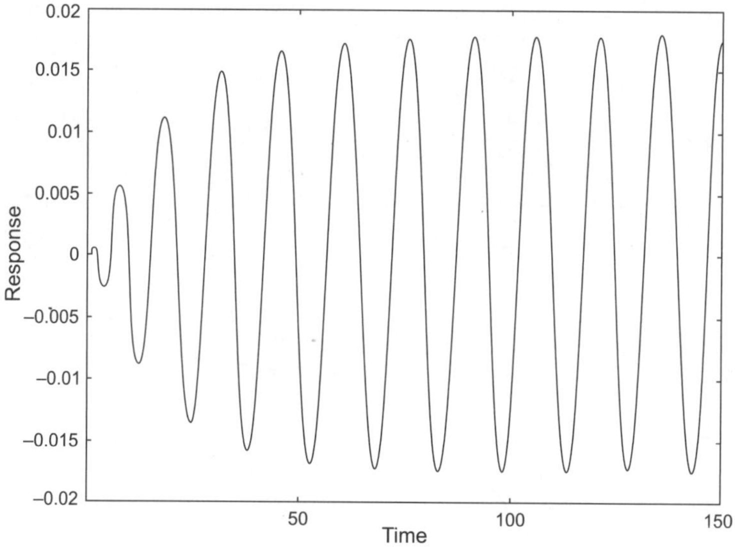

Fig. 9.3 *Relay feedback response of case study 1*

Table 9.3 *Identified model parameters and controller settings*

Example	Parameters	Actual	Proposed	SC method FOPTD	Li et al. method
Case study 1	k_p	1	1.0581	0.3753	1.1907
	D	0.5	0.5487	3.8613	0.52
	τ	20	20.8183	64.7945	20.0955
	k_c	120	102.96	22.3561	94.62
	τ_I	24	25.21	30.89	24.26
	τ_D	3.33	3.63	-	3.45
Case study 2	k_p	0.125	0.127	0.1319	0.2403
	D	1	1.4998	2.1484	1.05
	τ	1	1.3221	2.5556	2.0795
	k_c	-	6.9411	4.509	8.2417
	τ_I	-	2.6442	2.556	4.159
	τ_D	-	0.661	--	1.0398

Contd.

Table 9.3 *Contd.*

Case study 3	k_p	1	0.9693	1.8417	2.1504
	D	2	3.0075	7.6331	2.0
	τ_1	1	14.2083	90.3619	21.7902
	τ_2	10	-	-	-
	τ_3	20	-	-	-
	k_c	-	4.8739	3.2139	5.984
	τ_I	-	28.4166	61.065	37.79
	τ_D	-	7.1041	-	9.226

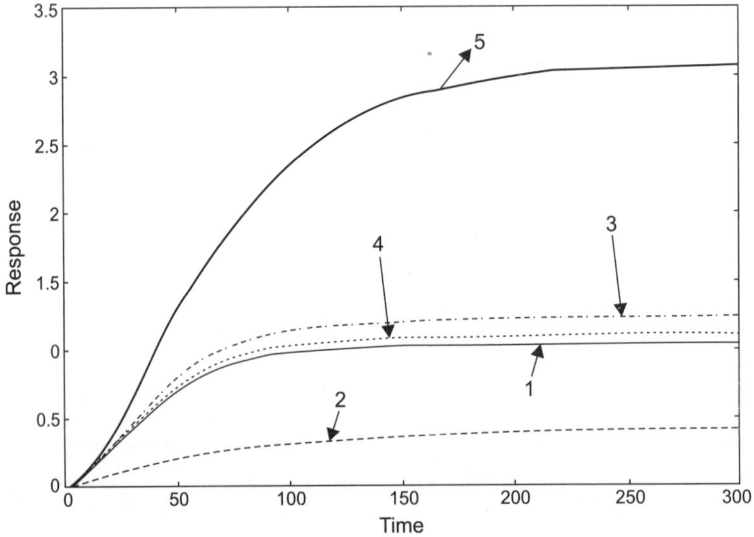

Fig. 9.4 *Open loop response comparison for case study* 1
Legend: 1–Actual system; 2–SC method for FOPTD system; 3–Li et al. method for SOPTD; 4–Proposed method; 5–T–Y method for SOPTD (Thick line)

Now let us design a controller based on the identified model parameters. Skogestad (2003) suggested simple IMC-based ideal PID settings to design a controller for good robustness and fast response with $\tau_c = D$ as

for $\tau_1 \le (8*D)$

$$k_c = \frac{0.5}{k_p}\frac{(\tau_1+\tau_2)}{D}; \qquad \tau_I = \tau_1+\tau_2; \qquad \tau_D = \frac{\tau_1}{\left(1+\left(\frac{\tau_1}{\tau_2}\right)\right)} \tag{9.49}$$

for $\tau_1 \ge (8*D)$

$$k_c = \frac{0.5}{k_p}\frac{\tau_1}{D}\left[1+\left(\frac{\tau_2}{(8*D)}\right)\right]; \qquad \tau_I = (8*D)+\tau_2; \qquad \tau_D = \frac{\tau_2}{\left[1+\left(\frac{\tau_2}{8*D}\right)\right]} \tag{9.50}$$

For an FOPTD model, Skogestad (2003) suggested to use the above equations with $\tau_2 = 0$. This will lead to a PI controller for the FOPTD systems. Now, based on the identified model, the controller is designed using Eq.(9.49) and Eq.(9.50) (refer to Table 9.3 for details) and the performance is evaluated on the actual system. Figure 9.5 shows the responses of the proposed method, Li et al. method and SC method. The controller designed on the model identified TY method gives a highly oscillatory response (not shown in Fig. 9.5). Because of the robustness of the IMC controller, the closed loop responses are almost same for the present method, Li et al. method and SC method. For the FOPTD model obtained from SC method, the controller is designed using Eq.(9.49) and Eq.(9.50) with $\tau_2 = 0$.

9.5.2 Case study 2

Let us consider a TOPTD with equal time constant system considered by Luyben (2001) as

$$G(s) = 0.125\exp(-s)/(s+1)^3 \tag{9.51}$$

The conventional relay analysis gives observed amplitude $a_0 = 0.0658$, $\omega_u = 0.9186$ and k_u as 19.3501. Figure 9.5 shows the system output response using the relay test. Figure 9.6 shows that there is initial dynamics in the output response. The intermediate values are obtained as $s_1 = 0.064$, $y(s_1) = -0.0129$ and $u(s_1) = -0.1306$. The proposed method gives the parameters of an SOPTD with equal time constants model as $k_p = 0.127$, $D = 1.4998$ and $\tau = 1.3221$. Using SC method, the FOPTD model is obtained with $k_p = 0.1319$, $D = 2.1484$ and $\tau = 2.5556$.

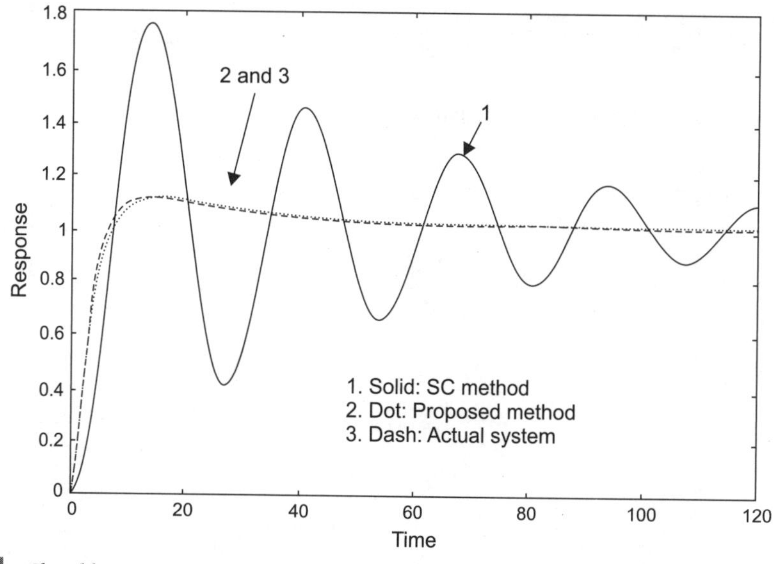

Fig. 9.5 *Closed loop response comparison for case study 1*

Since the relay output is an exponentially developing one and ε is greater than 0.01, TY method is not applicable. Based on the method proposed by Li et al. (1991), the SOPTD with equal time constant model is given by $k_p = 0.2403$, $\tau = 2.0795$ and $D = 1.05$ (delay is measured from the response). Table 9.3 gives the identified model and the actual model considered for simulation.

Figure 9.7 shows the open loop system response of the actual and identified models. The proposed method gives a response close to that of the actual system.

The PID controller is designed based on the identified models using Eq.(9.49) and Eq.(9.50) (refer to Table 9.3 for details) and the performance is evaluated on the actual system. Figure 9.8 shows the closed loop responses. ISE values for the proposed method, Li et al. method and SC method are respectively 3.85, 4.03 and 4.53. The ISE value indicates that the proposed method is preferred.

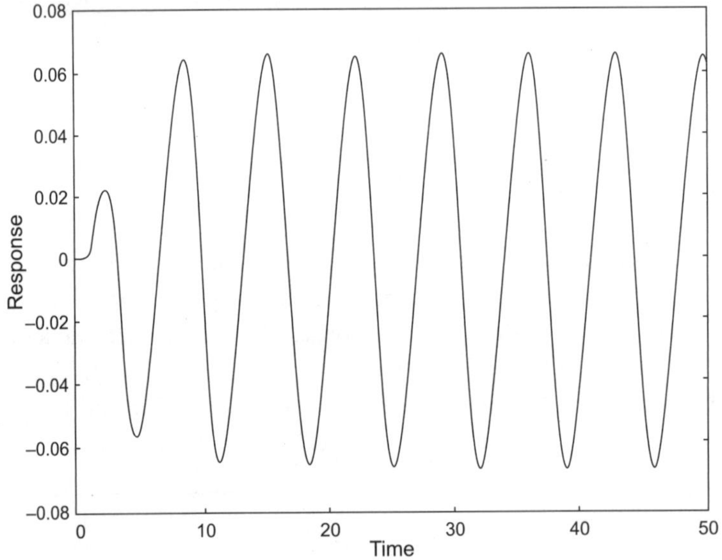

Fig. 9.6 *Relay feedback response of case study 2*

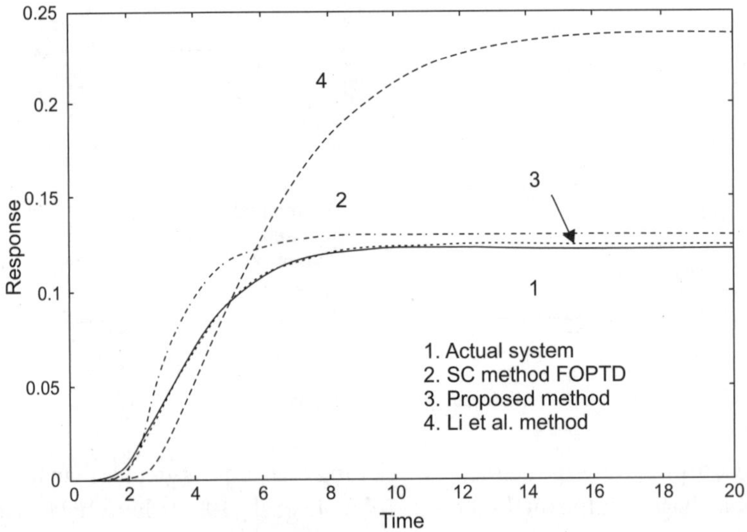

Fig. 9.7 *Open loop response of case study 2*

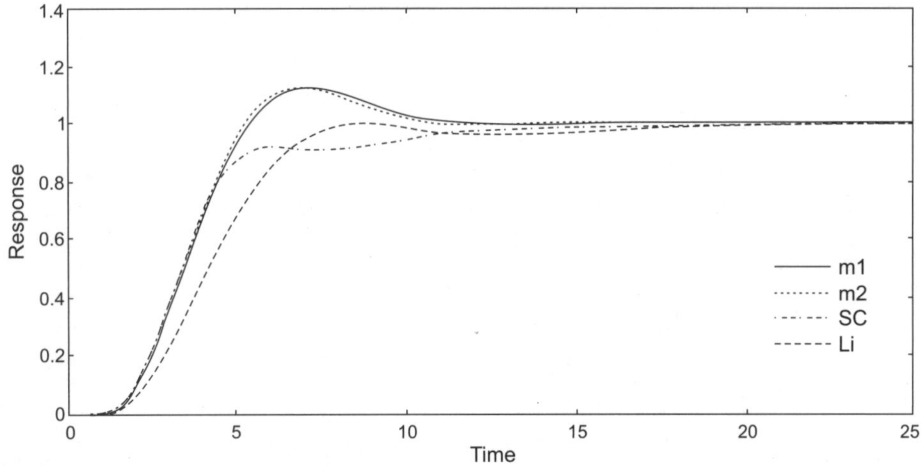

Fig. 9.8 *Closed loop response on the actual system for a step change in set point for case study 2 (controller settings as in Tables 9.3 and 9.6)*
Legend: m1–present method 1; m2–present method 2; SC-SC method; Li-Li et al. method

9.5.3 Case study 3

Let us consider a TOPTD with unequal time constant system considered by Li et al. (1991)

$$G(s) = \exp(-2s)/\left[(s+1)(10s+1)(20s+1)\right] \tag{9.52}$$

The symmetric relay test with relay height $h = \pm 1$ is conducted. Figure 9.9 shows the output response with initial dynamics. The observed amplitude $a_0 = 0.1219$ ($k_u = 10.445$) and $\omega_u = 0.2126$. The intermediate values are obtained as $s_1 = 0.0583$, $y(s_1) = -0.0429$ and $u(s_1) = -0.1763$. The proposed method gives an SOPTD (with equal time constant) model with parameters $k_p = 0.9693$, $D = 3.0075$ and $\tau = 14.2083$. Using Srinivasan and Chidambaram method, the identified FOPTD model parameters are $k_p = 1.8417$, $D = 7.6331$ and $\tau = 90.3619$. Since the relay output is not of exponential developing and ε is greater than 0.01, TY method is not applicable. Based on the method proposed by Li et al. (1991), the SOPTD model with equal time constants is calculated as $k_p = 2.15$, $\tau = 21.79$ and $D = 2$ (delay is measured from the response). Table 9.3 gives the parameters of the identified and actual models. Figure 9.10 shows the open loop performance for a unit step change in the set point. The proposed method gives a response close to that of the actual system. Hence the proposed method is preferred.

For closed loop performance, the controller is designed on the identified models and the performance is evaluated on the actual system (refer to Fig. 9.11) for a unit step change in the set point. Since error persists for long time, ITAE values are used to compare the closed loop performance. The ITAE for the proposed method is 339.45, for SC method it is 1007.83 and for Li et al. method it is 448.65. The ITAE value shows that the proposed method is preferred.

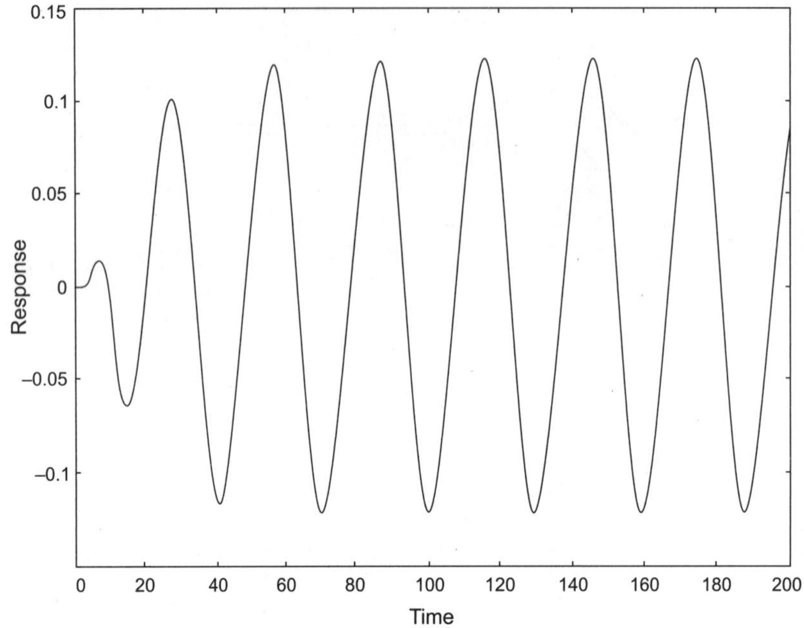

Fig. 9.9 *Relay response of case study 3*

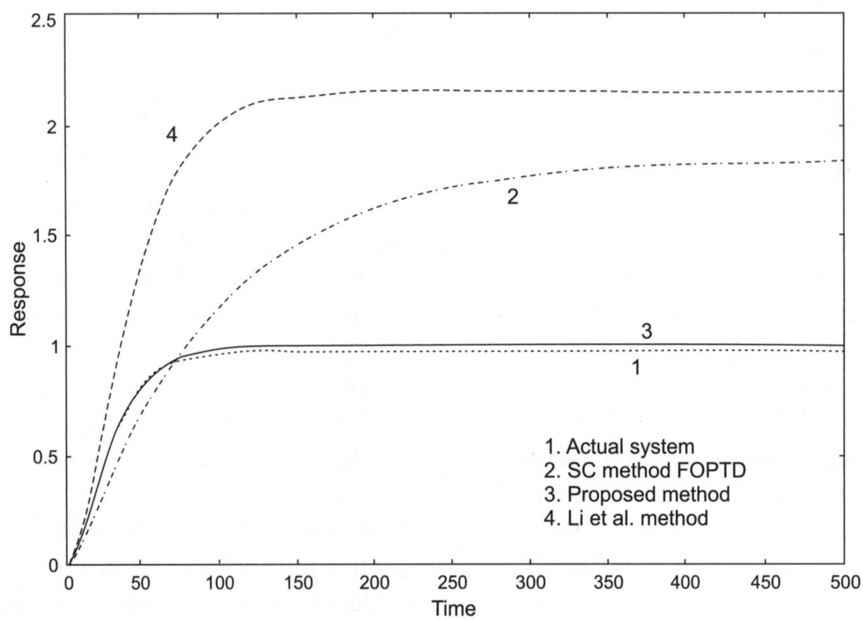

Fig. 9.10 *Open loop response of case study 3. SC method is for FOPTD model*

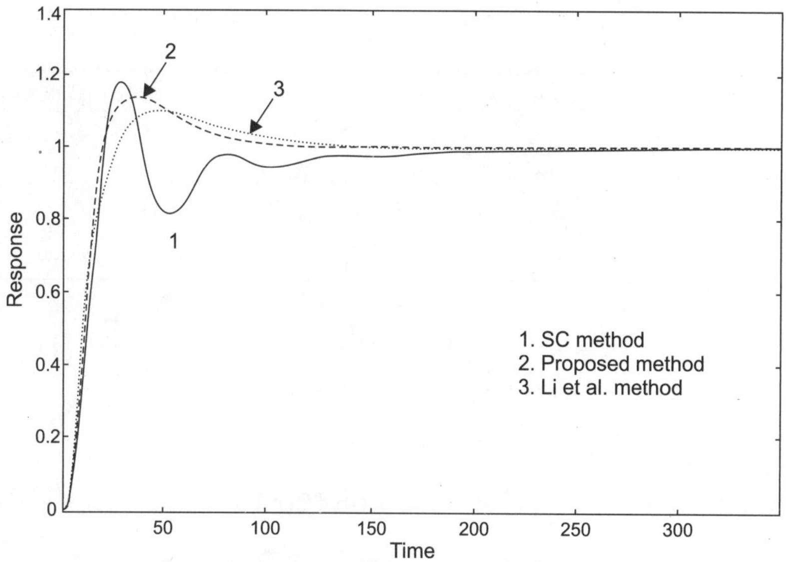

Fig. 9.11 *Closed loop response for unit step change in set point for case study 3*

9.5.4 Effect of measurement noise on identification

The effect of measurement noise on the accuracy of estimation of the parameters is considered. The example considered for the simulation study is an SOPTD model with equal time constants with a large D/τ ratio as

$$G(s) = \exp(-10s)/(s+1)^2 \tag{9.53}$$

The measurement noise with a zero mean Gaussian distribution and a standard deviation of 0.5%, 1% and 1.5% is added separately to the output of the system. In the identification test, the corrupted signal is used in the feedback control and for the system output. Once the initial dynamics are died out, the amplitude and period of oscillation are calculated by taking the average values at the various peak locations. Table 9.4 gives the identified parameters using the proposed method. As the noise level increases, the parameter estimation deteriorates. It is desirable to use a first-order filter to remove the noise.

9.5.5 Effect of load on model parameter identification

The effect of load on parameter identification is also considered [refer to Eq.(9.53)] for transfer function model. A load of 0.02 and separately of 0.04 is introduced. The transfer function for load is assumed as that of the process transfer function. The load-affected output signal is used for the feedback relay and as the system output for model identification purposes. The effect of load on model identification is reported in Table 9.4. It is observed that there is no significant effect of the load on the model identification. Hence the identification method is robust.

Table 9.4 *Effect of measurement noise and load on model identification using the proposed method*

Effect considered for		Identified parameters		
		k_p	τ	D
Measurement noise with a zero mean Gaussian distribution	$\sigma = 0.5\%$	0.9775	0.8166	9.994
	$\sigma = 1.0\%$	0.9995	0.9984	9.698
	$\sigma = 1.5\%$	1.002	0.9995	9.742
Load	0.02	0.9964	0.8159	10.072
	0.04	1.0278	0.9147	9.885

Note: Model used for simulation: $G(s)= \exp(-10s)/(s+1)^2$

9.6 Simulation Study of a Non-linear Bioreactor System

The proposed method of identification of open loop stable SOPTD system with equal time constants is applied to a non-linear continuous bioreactor. The dimensionless model equations are given by Agarwal and Lim (1984):

$$dX_1/dt = (\mu - D_1)X \tag{9.54}$$

$$dX_2/dt = (X_{2f} - X_2)D_1 - (\mu X_1/\gamma) \tag{9.55}$$

where

$$\mu = (\mu_m X_2)/(K_m + X_2) \tag{9.56}$$

Here X_1, X_2 and X_{2f} are dimensionless concentration of biomass cell, substrate and substrate feed, respectively. 'D_1', is the dilution rate and 'μ' is the specific generation rate. The model parameters are given by Agarwal and Lim (1984) as

$$\begin{array}{lll} \gamma = 0.4 \text{ g/g}; & X_{2f} = 1.0 \text{ g/l}; & \mu_m = 0.4h^{-1} \\ K_m = 0.05 \text{ g/l}; & k_f = 0.4545 \text{ g/g}, & D_1 = 0.36\ l/h^{-1} \end{array} \tag{9.57}$$

The solution of Eq.(9.61) to Eq.(9.62) gives the following stable steady-state:

$$[X_1, X_2] = [0.22, 0.45] \tag{9.58}$$

Initially, the system is assumed to be at stable steady-state condition. At time $t = 0$, $X_1 = 0.22$ and $X_2 = 0.45$. In this condition, dilution rate (D_1) is 0.36 l/hr. The substrate feed concentration is considered as the manipulated variable in order to control the cell mass concentration (X_1) at the stable steady-state at $X_1 = 0.22$. A delay of four hours is considered in the measurement of X_1. For the given operating point [refer Eq.(9.65)], the local linearized model is obtained as

$$G(s) = 0.401 \exp(-4s)/(63.2911s^2 + 25.5969s + 1) \tag{9.59}$$

A symmetrical relay with relay height (h) = 0.02 is conducted. The relay output response (deviation value from the steady-state point of X = 0.22) is shown in Fig. 9.12. For the identified SOPTD model with equal time constants, parameter values obtained by the proposed method are $k_p = 0.225$, $\tau = 6.52$ and $D = 3.843$. Using SC method, the identified FOPTD model parameters are $k_p = 1.1646$, $\tau = 78.572$ and $D = 6.0083$. Using Li et al. method, the SOPTD model parameters with equal time constants are calculated as $k_p = 0.2043$, $\tau = 6.1196$ and $D = 4.05$. Using TY method, the identified unstable model parameters are $k_p = 109.3862$, $\tau = 165.91$ and $D = 0.1659$. The relay response shows that TY method is not applicable for this case study. An open loop response of the non-linear model is evaluated for a step change in X_1 from 0.22 to 0.242. The identified model parameters and PID controller settings [using Eq.(9.49)] are given in Table 9.5. A PID controller is also designed for Eq.(9.59).

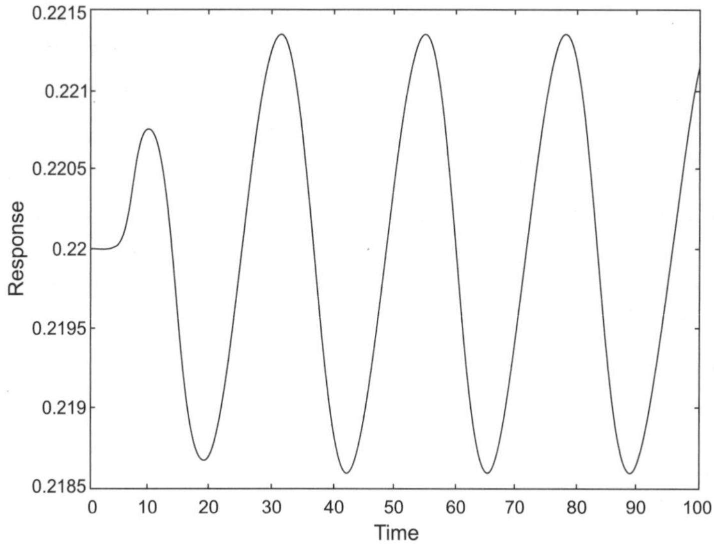

Fig. 9.12 *Relay feedback response of bioreactor system*

Figure 9.13 shows that the open loop response of the proposed method is closer to that of the actual system. Figure 9.14 shows the closed loop response for a unit step change in the set point. The ITAE values for the present method, Lee et al. method, locally linearized model method and SC method are respectively 6.52, 6.85, 4.4 and 60.23. The SOPTD model is much better than the FOPTD model. The proposed method of identifying an SOPTD model is preferred.

Table 9.5 *Controller performance based on non-linear identified models*

Parameter	Proposed	SC method	Li et al.	L-L model
k_c	7.54	5.6145	7.3881	7.97
τ_I	13.04	78.572	12.2022	25.59
τ_D	3.26		3.0505	2.4726
ITAE	6.529	60.2358	6.853	4.4

Note: L-L: Locally linearized model
PID controller settings by Eqs (9.49) and (9.50)

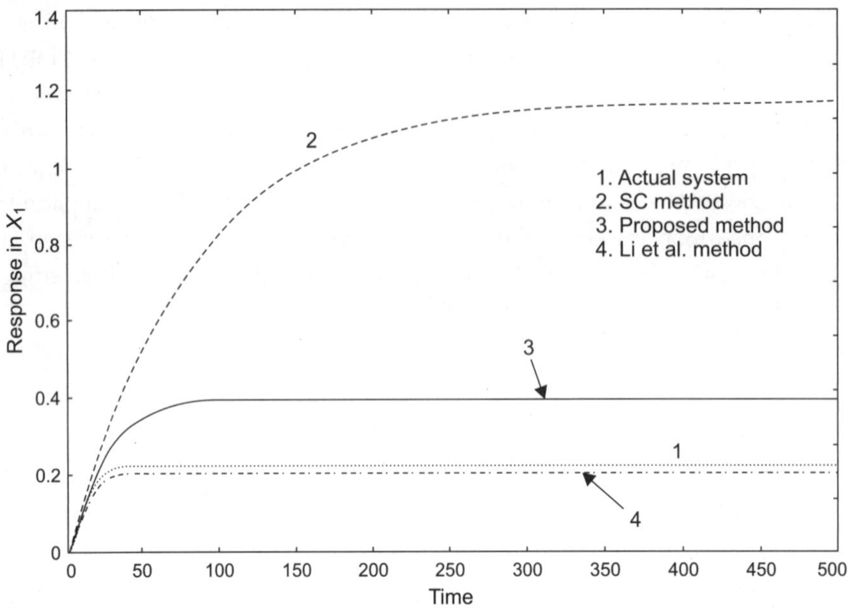

Fig. 9.13 *Open loop response of a bioreactor system (deviation from initial steady-state value is plotted for the actual system)*

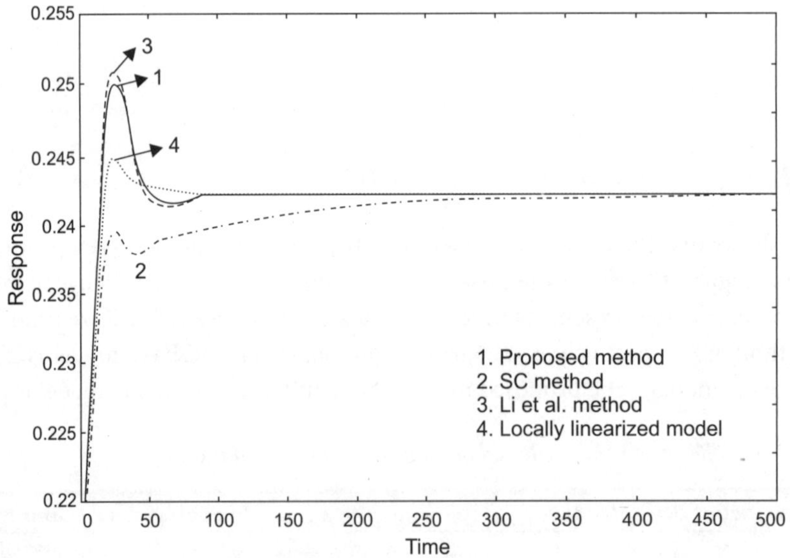

Fig. 9.14 *Closed loop response for unit step change in set point for a bioreactor system*

9.7 Estimation of k_p, D and τ without using k_u

Let us estimate the model parameters from the input/output data of the system under relay feedback without the knowledge of k_u.

From the definition of Laplace transform, we get

$$y(s) = \int_0^\infty y(t)\exp(-st)\,dt \tag{9.60}$$

The above integral can be evaluated for $s = 8/t_s$ where 't_s' is the time at which few (say 3) repeated cycles of oscillation appear in the output after the initial transient die out. The reason for taking $s = 8/t_s$ is that, for $t > t_s$, because of very small value of the term $\exp(-st)$, contribution by subsequent terms is negligible while evaluating the integral. Let the resulting integral value be denoted as $y(s_1)$.

Similarly,

$$u(s) = \int_0^\infty u(t)\exp(-st)\,dt \tag{9.61}$$

Equation (9.61) is also evaluated using $s = 8/t_s$. With the numerical values of $y(s_1)$ and $u(s_1)$, using the assumed model, an equation is formulated as given below:

$$y(s_1)/u(s_1) = k_p \exp(-D\,S_1)/(\tau s_1 + 1)^2 \tag{9.62}$$

Padmasree and Chidambaram (2001) first used this equation while analyzing the closed loop response of a PID controlled system. Here it is used for relay autotuning.

Since $y(t)$ and $u(t)$ are piecewise continuous and periodic, the Laplace transform of $y(t)$ and $u(t)$ can be written as (Kreyszig, 1996)

$$y(s) = \frac{1}{\left[1-\exp(-Ps)\right]}\int_0^P y(t)\exp(-st)\,dt \tag{9.63}$$

$$u(s) = \frac{1}{\left[1-\exp(-Ps)\right]}\int_0^P u(t)\exp(-st)\,dt \tag{9.64}$$

From Eq.(9.63) and Eq.(9.64), we get

$$G(s) = y(s)/u(s) = \left[\int_0^P y(t)\exp(-st)\,dt\right] \Big/ \left[\int_0^P u(t)\exp(-st)\,dt\right] \tag{9.65}$$

Where $P = 2\pi/\omega$, and ω is the frequency of oscillations observed in the output.

By substituting $s = j\omega$ and using Euler's expansion for $\exp(-j\omega t)$ in Eq.(9.65) we get

$$G(j\omega) = y(j\omega)/u(j\omega) = (c_1 - jd_1)/(c_2 - jd_2) \tag{9.66}$$

Where

$$c_1 = \int_0^P y(t)\cos(\omega t)\,dt \tag{9.67a}$$

$$d_1 = \int_0^P y(t)\sin(\omega t)\,dt \tag{9.67b}$$

$$c_2 = \int_0^P u(t)\cos(\omega t)\,dt \tag{9.67c}$$

$$d_2 = \int_0^P u(t)\sin(\omega t)\,dt \tag{9.67d}$$

Equations (9.67a) to (9.67d) can be evaluated numerically using $y(t)$ and $u(t)$ data of relay test. Hence Eq.(9.66) can be written as

$$G(j\omega) = p + jq \tag{9.68}$$

Where

$$p = (c_1 c_2 + d_1 d_2)/(c_2^2 + d_2^2) \tag{9.69a}$$

$$q = (d_2 c_1 - d_1 c_2)/(c_2^2 + d_2^2) \tag{9.69b}$$

By substituting $s = j\omega$ in Eq.(9.66), we get

$$G(j\omega) = k_P \exp(-Dj\omega)/(\tau j\omega + 1)^2 \tag{9.70}$$

On substituting $\exp(-Dj\omega) = \cos(D\omega) - j\sin(D\omega)$ in Eq.(9.70), we get

$$p + qj = k_p[\cos(D\omega) - j\sin(D\omega)]/(\tau j\omega + 1)^2 \tag{9.71}$$

On cross-multiplying and separating the real and imaginary terms we get

$$[p(1 - \tau_1^2\omega^2)] - (2\tau_1\omega q) - [k_p\cos(D\omega)] = 0 \tag{9.72a}$$

$$2\tau_1\omega p + [q(1 - \tau_1^2\omega^2)] + [k_p\sin(D\omega)] = 0 \tag{9.72b}$$

From the relay test, the three linear algebraic equations, Eqs (9.62), (9.72a) and (9.72b), can be solved to estimate the three unknown parameters: k_P, D and τ. In Eqs (9.72a) and (9.72b), the frequency of oscillation of the system output is used. It is to be noted that for the analysis of an asymmetric relay test, Srinivasan and Chidambaram (2003) used Eqs (2.48) and (2.49) and calculated the process gain from $u(t)$ and $y(t)$ data as

$$k_p = \left[\int y(t)\,dt\right]/\left[\int u(t)\,dt\right] \tag{9.73}$$

Integration is to be carried out for one period of oscillation. In the present method, because of a symmetric relay, Eq.(9.73) cannot be used. Hence, in symmetric relay autotuning, Eq.(9.62) is used. We also do not use the equation $k_u = 4h/(\pi a)$ and $\omega_u = 2\pi/P_u$. Thus using the single symmetrical relay test, all the three parameters of an SOPTD system are calculated.

9.8 Simulation Study

Let us consider case study 2 as reported earlier:

$$G(s) = 0.125\exp(-s)/(s + 1)^3 \tag{9.74}$$

The conventional relay analysis gives observed amplitude a_0 = 0.0658, ω_u = 0.9186 and k_u as 19.3501. The intermediate values are obtained as t_s = 125, s = 0.064, $y(s_1)$ = −0.0129 and $u(s_1)$ = −0.1306; p = −0.0499 and q = −2.415e−4. The proposed method gives the parameters of SOPTD with equal time constants model as k_p = 0.132, D = 1.62 and τ = 1.43. The parameters identified using the proposed methods are given in Table 9.6. The open loop response of the identified model is found close to that of the method given in Section 9.4 using the corrected k_u (the response is not shown in Fig. 9.7). PID controllers are designed and the settings are given in Table 9.6. The closed loop servo response is shown in Fig. 9.8. The response is close to that of the method given in Section 9.4. Similar analysis of higher order harmonics relay tuning of integrating plus FOPTD system is carried out and results are given in Appendix C.

Table 9.6 *Identified and controller parameters by the proposed methods for case study* 2

Parameters	Proposed method using corrected k_u	Proposed method Eqs (9.72a,b and 9.62)
k_p	0.127	0.128
D	1.50	1.461
τ	1.32	1.364
k_c	6.94	7.279
τ_I	2.64	2.728
τ_D	0.66	0.682

For simulation example: $G(s) = 0.125\ \exp(-s)/(s+1)^3$

Liu et al. (2008) proposed a method to identify the parameters of SOPTD systems. Let us compare the present two methods with that of Liu et al. (2008). Consider the system $G(s)$ = exp(−4s)/(0.5s+1)^3. This system has a large delay to time constant ratio (as 8) compared to the previous example, which has only the ratio as around 1. Liu et al. (2008) identified the system as $G(s)$ = exp(−4.2771s)/(0.6183s+1)^2. Consider proposed method 2. The intermediate values are t_s = 125, s_1 = 0.064, $y(s_1)$ = 1.0513, $u(s_1)$ = 1.4994, p = −0.8792, q = 0.0609 and ω_u = 0.5928. The proposed method 2 identifies the system as 0.9932 exp(−4.26s)/(0.6011s+1)^2. Let us consider proposed method 1. The system output shows an amplitude of 1 and has no initial dynamics. Hence N = 5 is considered. The intermediate values are s_1 = 0.064, t_s = 125, s_1 = 0.064, $y(s_1)$ = 1.0513, $u(s_1)$ = 1.4994 and ω_u = 0.5928. t^* = 6.65, $y(t^*)$ = 0.898, a(corrected) = 1.0754, k_u (corrected) = 1.184. The solution of a non-linear algebraic equation [Eq.(9.46)] gives k_p = 0.9868. Equation (9.44) gives τ = 0.6922 and Eq.(9.45) gives D = 3.9859. Hence the model is identified as $G(s)$ = 0.9868 exp(−3.9859)/(0.6922s+1)^2. Figure 9.15 shows the Nyquist plot for the actual system and the three identified models. All the three methods give a good match to the actual system. The proposed method 2 and Liu et al. method give similar performances. The proposed method 1 also gives a good performance. The proposed methods 1 and 2 are simple to formulate.

9.9 Conclusions

A method is proposed to consider higher order harmonics for calculating the controller settings of an SOPTD transfer function model with equal time constants using a single symmetric relay test. Two methods are used to identify all the three parameters of the SOPTD with equal time constant system. Simulation results show that for the three transfer function case studies and on a non-linear model of a bioreactor system, the method gives improved parameter estimates and hence an improved closed loop response. The method gives good results even in the presence of noise and separately a load is present in the input. The conventional method of analyzing the relay output for the calculation of ultimate controller gain gives a maximum error of 27% in k_u.

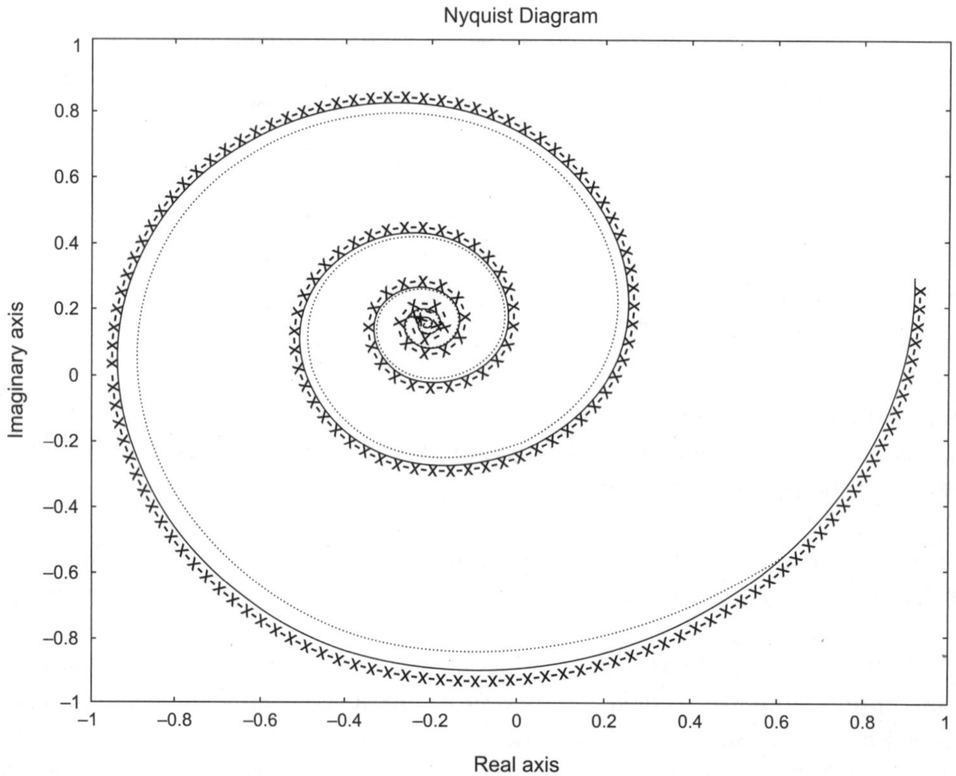

Fig. 9.15 *Nyquist plot of the transfer function*
Legend: solid–actual system; ++–Liu et al.; ...–proposed method 1; chain–proposed method 2

10 Estimation of SOPTD Transfer Function Model

In this chapter, using a single asymmetric relay feedback test, a method proposed by Ramakrishnan and Chidambaram (2003) is reviewed to identify four parameters of a transfer function model. The proposed method is used to identify all the parameters in a second order plus time delay model (SOPTD). The parameters estimated are of adequate accuracy for designing suitable controllers. The estimated SOPTD model has a step response behavior matching with that of the actual process. The method can also be used for identifying open loop unstable transfer function models. For unstable systems, the closed loop step responses are compared. Simulation results are given for four case studies.

10.1 Introduction

Identification of dynamic transfer function models from experimental data is essential for model-based controller design. Often derivation of rigorous models is difficult due to the complex nature of chemical processes. Hence, system identification is a valuable tool to identify low order models, based on input–output data, for controller design.

As stated earlier in Chapter 1, Åström and Hägglund (1984) suggested the relay feedback test to generate sustained oscillations of the controlled variable and get the ultimate gain (K_u) and ultimate frequency (ω_u) directly from the experiment. Since only process information K_u and ω_u are available, the additional information such as steady-state gain or time delay should be known *a priori* in order to fit a typical transfer function model such as first order plus time delay (FOPTD). Many methods have been proposed for obtaining this additional information. Li et al. (1991) suggested using two relay tests: one conventional and the other with a known dynamic element inserted in the loop. Friman and Waller (1997) proposed the use of a two-channel relay test. Scali et al. (1999) extended the technique suggested by Li et al. (1991), which assumed the delay of the process to be known, to estimate models with up to five parameters by means of a maximum of three relay tests with additional delay. But use of additional relay tests is tedious as well as time consuming. Hence, there is a need to propose a method by which models with more than two parameters can be identified accurately using a single relay feedback test without *a priori* knowledge of process gain and/or time delay.

SOPTD models can incorporate various processes such as under damped and higher order processes than FOPTD models and controllers designed based on SOPTD models give better closed loop response than those designed on FOPTD models. Certain higher order models when approximated to an FOPTD model give negative time constant; hence, a second order model is necessary. In addition, for unstable processes, a first order model is not sufficient for describing its dynamics because other than the unstable time constant, there are stable dynamic lags in the process. In general, an SOPTD transfer function model will be preferred to an FOPTD model. To identify the parameters of an SOPTD system, an improved relay feedback method is required.

Srinivasan and Chidambaram (2004) proposed a method to analyze the conventional relay autotune data for estimating the three parameters of an FOPTD model using only one relay experiment. An additional equation (proposed for process gain) along with the phase angle and amplitude criteria gives three parameters. Srinivasan and Chidambaram (2003) also proposed a modified asymmetrical relay feedback method to get improved estimates of the parameters of the FOPTD model. A single relay test is used and a method is suggested for formulating additional equations to evaluate all three parameters of the FOPTD model.

The objective here is to combine the methods suggested by Srinivasan and Chidambaram (2003; 2004) to estimate the four parameters of an SOPTD model using a single asymmetrical relay feedback method. Several simulation results are given to check the identification of the SOPTD model parameters.

10.2 Problem Description

An SOPTD model can be expressed as

$$G(s)=\frac{y(s)}{u(s)}=\frac{K_P \exp(-\theta s)}{(\tau_1 s+1)(\tau_2 s+1)} \tag{10.1}$$

where 'K_p' (process gain), 'θ' (time delay), 'τ_1' and 'τ_2' (time constants), are the parameters to be estimated.

10.3 Relay Test

Consider a relay feedback system (Fig. 10.1) where $G(s)$ is the process transfer function, y is the output, y_r is the set point, e is the error and u is the manipulated variable. An asymmetrical relay is placed in the feedback loop. Depending on the error and process, the relay switches between $+\gamma h$ and $-h$, where $\gamma \neq 1$.

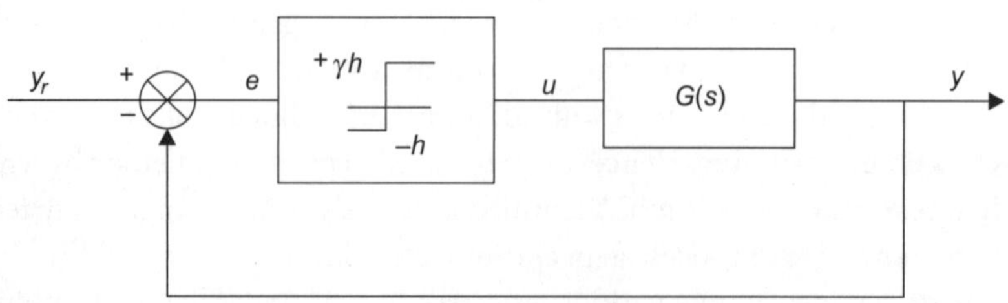

Fig. 10.1 *Block diagram for asymmetric relay feedback system*

10.4 Estimation of Process Gain

The steady-state gain of SOPTD model can be found by integrating the system input–output response.

$$k_P = \frac{\int_{t_o}^{t_o+P} y(t)dt}{\int_{t_o}^{t_o+P} u(t)dt} \tag{10.2}$$

where 't_o' is the starting time for integration and 'P' is the period of oscillations.

10.5 Estimation of θ, τ_1 and τ_2

Padmasree and Chidambaram (2001) proposed a method for formulation of an additional equation while analyzing the closed loop response of a PID controlled system. Srinivasan and Chidambaram (2004) applied this method to the conventional relay method to identify the three parameters of an FOPTD model. In the present work, this method is applied to an asymmetrical relay test to identify four parameters of an SOPTD model. From the definition of Laplace transform we get

$$y(s) = \int_0^\infty y(t)\exp(-st)dt \tag{10.3}$$

The above integral can be evaluated for a particular value of s (say s_1). It is suggested to use the value $s_1 = 8/t_s$, where 't_s' is the time at which three repeated cycles of oscillation appear in the output. The reason for taking $s_1 = 8/t_s$ is that, for $t > t_s$, because of very small value of the term $\exp(-s_1t)$, contributions by subsequent terms is negligible while evaluating the integral. Let the resulting integral value be denoted as $y(s_1)$. Similarly,

$$u(s) = \int_0^\infty u(t)\exp(-st)dt \tag{10.4}$$

Equation (10.4) is also evaluated using $s_1 = 8/t_s$. With the numerical values of $y(s_1)$ and $u(s_1)$, using the assumed model [Eq.(10.1)], an equation is formulated as given below:

$$\frac{y(s_1)}{u(s_1)} = \frac{K_P\exp(-\theta s_1)}{(\tau_1 s_1 + 1)(\tau_2 s_1 + 1)} \tag{10.5}$$

Since $y(t)$ and $u(t)$ are piece-wise continuous and periodic, the Laplace transform of $y(t)$ and $u(t)$ can be written as (Kreyszig, 1996)

$$y(s) = \frac{1}{[1-\exp(-Ps)]}\int_0^P y(t)\exp(-st)dt \tag{10.6}$$

$$u(s) = \frac{1}{[1-\exp(-Ps)]}\int_0^P u(t)\exp(-st)dt \tag{10.7}$$

From Eqs (10.6) and (10.7), we get

$$G(s)=\frac{y(s)}{u(s)}=\frac{\int_0^P y(t)\exp(-st)\,dt}{\int_0^P u(t)\exp(-st)\,dt} \tag{10.8}$$

where $P = 2\pi/\omega$ and ω is the frequency of oscillations observed in the output. By substituting $s = j\omega$ and using Euler's expansion for $\exp(-j\omega t)$ in Eq.(10.8) we get

$$G(j\omega)=\frac{y(j\omega)}{u(j\omega)}=\frac{c_1-jd_1}{c_2-jd_2} \tag{10.9}$$

where

$$c_1=\int_0^P y(t)\cos(\omega t)\,dt \tag{10.10a}$$

$$d_1=\int_0^P y(t)\sin(\omega t)\,dt \tag{10.10b}$$

$$c_2=\int_0^P u(t)\cos(\omega t)\,dt \tag{10.10c}$$

$$d_2=\int_0^P u(t)\sin(\omega t)\,dt \tag{10.10d}$$

Equations (10.10 a, b, c, d) can be evaluated numerically using $y(t)$ and $u(t)$ data. Hence, Eq.(10.9) can be written as

$$G(j\omega)=p+jq \tag{10.11}$$

where

$$p=\frac{(c_1c_2+d_1d_2)}{(c_2^2+d_2^2)} \tag{10.12a}$$

$$q=\frac{(d_2c_1-d_1c_2)}{(c_2^2+d_2^2)} \tag{10.12b}$$

By substituting $s = j\omega$ in Eq.(10.1), we get

$$G(j\omega)=\frac{K_P\exp(-\theta(j\omega))}{(\tau_1 j\omega+1)(\tau_2 j\omega+1)} \tag{10.13}$$

Equating the real and imaginary parts of Eqs (10.11) and (10.13) and after simplifications we get

$$p(1-\tau_1\tau_2\omega^2)-q\omega(\tau_1+\tau_2)-K_P\cos(\theta\omega)=0 \tag{10.14}$$

$$q(1-\tau_1\tau_2\omega^2)+p\omega(\tau_1+\tau_2)+K_P\sin(\theta\omega)=0 \tag{10.15}$$

From the relay test, values for ω_u and P are obtained. Using this, Eq.(10.2) can be solved to get K_p. Substituting ω_u for ω in Eq.10.10 (a, b, c, d), we can evaluate c_1, c_2, d_1 and d_2, and hence p and q from Eq.10.12 (a, b). With K_p, p and q known, Eqs (10.5), (10.14) and (10.15) can be solved to estimate the three unknown parameters: θ, τ_1 and τ_2.

10.6 Simulation Results

10.6.1 Case study 1

Consider the transfer function given by Li et al. (1991)

$$G(s) = \frac{\exp(-2s)}{(10s+1)(s+1)} \tag{10.16}$$

An asymmetrical relay with $h = 1$ and $\gamma = 2$ is used. The intermediate values, required for calculation of SOPTD model parameters, are given in Table 10.1. The proposed method gives the model parameters as $K_p = 1.05$, $\theta = 1.814$, $\tau_1 = 9.766$ and $\tau_2 = 1.271$. The parameters identified by Li et al. (1991) are $K_p = 0.8530$, $\theta = 2$, $\tau_1 = 7.416$ and $\tau_2 = 1.15$ (time delay is directly got from the response). The integral square error (ISE) of the proposed model is 0.1466 whereas that of Li et al. model is 0.8050. The proposed method gives better parameter estimates. The process output and step response are as shown in Fig. 10.2(a, b). The step response of the proposed model is closer to that of the actual system than of Li et al. model.

The effect of measurement noise is studied by adding a random noise (with mean = 0 and standard deviation = 0.5%) to the process output and the corrupted signal is used for relay feedback. The present method gives the model parameters as $K_p = 1.06$, $\theta = 2.04$, $\tau_1 = 10.73$ and $\tau_2 = 0.92$ (refer Table 10.1 for intermediate values). The values are close to the values obtained without any measurement noise. This indicates that the present method is robust for measurement noise.

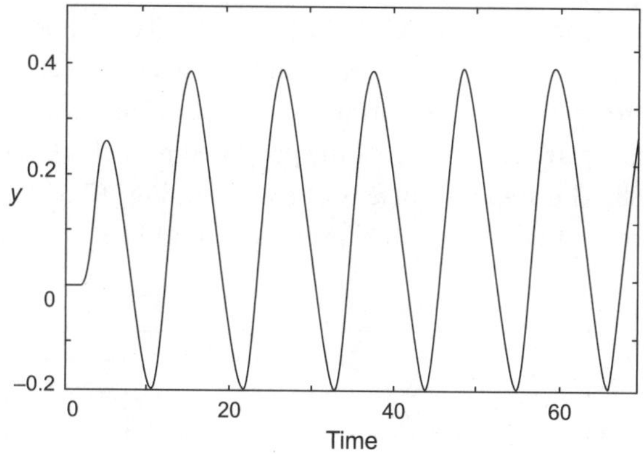

Fig. 10.2a *Output response for case study 1*

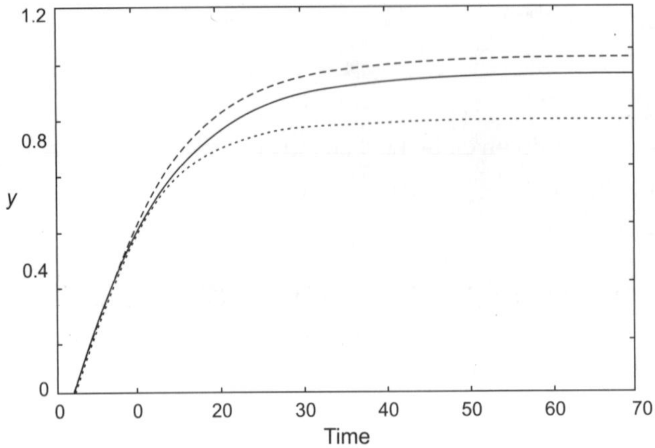

Fig. 10.2b Step response for case study 1
Legend: solid–actual; dash–proposed model; dash-dot–Li et al. model

10.6.2 Case study 2

Consider the higher order system given by Li et al. (1991)

$$G(s)=\frac{\exp(-4s)}{(0.5s+1)^3} \tag{10.17}$$

This is a third order system approximated to an SOPTD. An asymmetrical relay with $h = 1$ and $\gamma = 2$ is used. The intermediate values are given in Table 10.1. The proposed method gives model parameters as $K_p = 1.0046$, $\theta = 4.4239$, $\tau_1 = 0.2286$ and $\tau_2 = 0.8839$. Li et al. (1991) approximated this to an FOPTD. The parameters identified by Li et al. are $K_p = 1.0841$, $\theta = 4.04$ and $\tau = 1.6362$. The integral square error (ISE) of the proposed model is 0.0024 whereas that of the Li et al. model is 0.0988. The step response is as shown in Fig. 10.3. The step response of the proposed model is closer to that of the actual process than of Li et al. model.

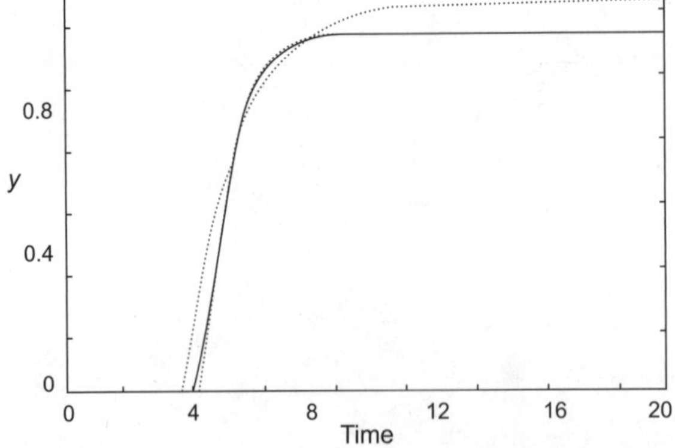

Fig. 10.3 Step response for case study 2
Legend: solid–actual; dash–proposed model; dash-dot–Li et al. model

10.6.3 Case study 3

Consider the non-minimum phase system

$$G(s)=\frac{(-3s+1)}{(5s+1)(s+1)} \tag{10.18}$$

Our aim here is to identify the four parameters, namely, K_p and the three time constants. An asymmetrical relay with $h = 1$ and $\gamma = 1.2$ is used. The intermediate values are given in Table 10.1. The proposed method gives model parameters as $K_p = 1.05$, $\tau_1 = -3.034$, $\tau_2 = 5.439$ and $\tau_3 = 0.9367$. The parameters identified are very close to the actual values. The step response is as shown in Fig. 10.4. The step response of the proposed model is very similar to that of the actual process.

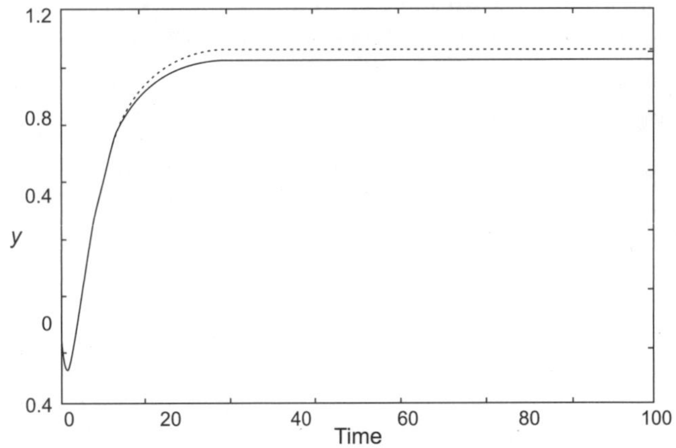

Fig. 10.4 *Step response for case study 3*
Legend: solid–actual; dash–proposed model

Table 10.1 *Intermediate values for the examples considered*

Case study	a	τ_u	P_u	t_s	$y(s_1)$	$u(s_1)$	p	q
1a	0.2906	0.5640	11.14	43.85	0.3755	1.8079	−0.1516	−0.0159
1b	0.2910	0.5661	11.10	41.80	0.1008	0.5064	−0.1514	−0.0156
2	1.4949	0.5850	10.74	45.97	1.9707	5.0819	−0.8830	0.0500
3	0.8499	0.5756	10.92	37.48	−0.1326	−0.9260	−0.5360	−0.1865
4	0.4519	1.1180	5.62	19.66	−0.0087	−0.0021	−0.3550	−0.0260

Note: 1a: without measurement noise; 1b: with measurement noise

10.6.4 Case study 4

Consider the open loop unstable system given by Huang and Chen (1999)

$$G(s)=\frac{\exp(-0.5s)}{(2s-1)(0.5s+1)} \tag{10.19}$$

An asymmetrical relay with $h = 1$ and $\gamma = 1.01$ is used. The intermediate values are given in Table 10.1. The proposed method gives the model parameters as $K_p = 1.00$, $\theta = 0.52$, $\tau_1 = 1.9999$ and $\tau_2 = 0.4837$. Model parameters identified by Huang and Chen are $K_p = 1.00$, $\theta = 0.509$, $\tau_1 = 2.015$ and $\tau_2 = 0.506$. But Huang and Chen use an asymmetrical relay with a lead lag element to evaluate the model parameters of an unstable system whereas the proposed method uses only an asymmetrical relay. PID controller is designed using the tuning rule suggested by Huang and Chen (1999). The values of the PID controller parameters for the proposed model are $K_c = 3.08$, $\tau_I = 6.08$ and $\tau_D = 0.48$, and for Huang and Chen model are $K_c = 3.13$, $\tau_I = 5.77$ and $\tau_D = 0.51$. The closed loop response to a step change in set point is as shown in Fig. 10.5. Both the methods give similar closed loop response.

The proposed method is also applied by simulation to a non-linear unstable CSTR. The model equations and model parameters are given by Huang and Chen (1999). The asymmetric relay test is carried out around the unstable operating point (0.4472, 2.7517). By the proposed method, the unstable SOPTD model identified is $0.23 \exp(-0.22s)/(1.91s - 1)(1.11s + 1)$. This is comparable with the model obtained from local linearization of the non-linear model around the unstable operating point $0.25 \exp(-0.2s)/(2.03s-1)(1.194s+1)$.

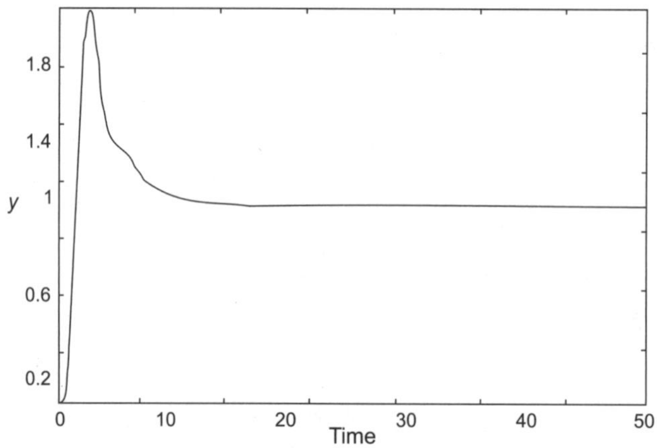

Fig. 10.5 *Closed loop response for case study 4*
Legend: solid–proposed; dash–Huang and Chen model (both lines coincide)

10.7 Conclusions

A method proposed by Ramakrishnan and Chidambaram (2003), for estimating the parameters of an SOPTD transfer function model using an asymmetric relay test is reviewed. The advantages of this method are single relay feedback test, four parameters can be identified, and the method can also be applied to open loop unstable processes without any change in the procedure. Simulation results show that, for three case studies, the method gives improved parameter estimates whereas for case study 4, it gives similar results as obtained from asymmetrical relay test with additional lead-lag element. The method takes lesser time for system identification as it requires only a single test. The method is also robust for measurement noise.

11 Estimation of Five Parameters of Unstable SOPTD Model with a Zero

Using a single relay feedback test, a method is proposed to identify all the five parameters of an unstable second order plus time delay model with a zero (SOPTDZ): $G_p = k_p(1\pm\tau_1 s)\exp(-Ds)/[(\tau_2 s\pm 1)(\tau_3 s-1)]$. In the present work, three simulation examples are given (delay system with one unstable pole, one stable pole and a negative zero; delay system with one unstable pole, one stable pole and a positive zero; delay system with two unstable poles and one negative zero). All the five parameters of the SOPTDZ model are estimated with adequate accuracy. Performance of the controller designed for the identified model is compared with that of the controller designed on the actual transfer function model. The method gives a closed loop performance closer to that of the actual system. The methods are given for an asymmetric relay test. A simulation example of a third order non-linear CSTR system is also given.

11.1 Introduction

Identification of transfer function models from experimental data is essential for model-based controller design. Often derivation of a mathematical model is difficult due to the complex nature of chemical processes. Hence system identification is a valuable tool to identify models based on input-output data.

Luyben (1987) suggested the use of relay testing for identifying a transfer function model. Using the values of k_u and ω_u in the phase angle and amplitude criteria for first order plus time delay (FOPTD) model, we get two equations relating the three parameters (k_p, τ and D). Since only k_u and ω_u are available, the additional information such as steady-state gain or time delay should be known *a priori* in order to fit a typical transfer function model such as an FOPTD model. Many methods are proposed for obtaining the additional information. Li et al. (1991) suggested the use of two relay tests. The first one is a relay test on the system and the second one on the system with a known dynamic element (additional dead time) inserted in the loop. However, the use of additional relay test is tedious and time consuming. Also use of additional delay to the system, particularly if the system is an unstable one, may not give oscillations if delay of the original system is large as compared to the time constant.

Friman and Waller (1997) proposed the use of a two-channel relay test (one operating on process output and other on the integral of the process output). An excellent review of relay feedback method is given by Yu (1999). Scali et al. (1999) extended the technique suggested by Li et al. (1991), which assumed the delay of the process to be known, to estimate models with up to five parameters by means of a maximum of three relay tests with addition of different time delays. Luyben (2001) proposed a method that included the effect of shape factor in identifying the FOPTD process model parameters. However, k_u calculation is based on the principal harmonics assumption that neglects higher order harmonics of the system, and the proposed method will not work for processes with strong inverse response. The method is not suitable for unstable systems. Thyagarajan and Yu (2003) proposed an identification method based on the shape of the response from the relay feedback test. They suggested a method to identify parameters of an FOPTD transfer function model, second order plus small dead time (SOPSDT) with equal time constants model, and higher order model without dead time. The proposed method works well

a) when the orders of the actual transfer function model and identified transfer function model are same

b) when the ratio of D/τ is less than 0.1, for the SOPSDT system

c) for higher order system without dead time (D)

The method will not identify more than three parameters.

Shen et al. (1996a) used an asymmetric relay test for getting the model parameters using a single relay test. Majhi and Atherton (1999) proposed a method to identify parameters of stable/unstable FOPTD or SOPTD systems. Sets of exact analytical expressions are derived from a single asymmetric relay test. The four parameters of unstable SOPTD system are identified using a single asymmetric relay test.

Marlin (1995) showed that a jacketed continuous stirred tank reactor (CSTR) carrying out a simple reaction gives an unstable transfer function model containing two complex conjugate poles and a negative zero when the model equations are linearized around an unstable operating point. Kaustubh and Chidambaram (2000) analyzed the behavior of a crystallizer and showed that linearizing the model equations around the steady-state condition results in an SOPTD transfer function with a negative zero and two complex conjugate poles. Bequette (2003) considered a CSTR carrying out a simple reaction. Linearization of the model equations around a steady-state operating point gives an unstable SOPTD system with an LHP zero transfer function model. Padmasree et al. (2004) used the model equations for an isothermal CSTR carrying out catalytic reaction. The model equations give a transfer function model with two unstable poles with one unstable zero. On considering the measurement delay term in the above transfer function, the delay term is included. Hence, there is a need to propose a method by which models with five parameters can be identified accurately using a relay feedback test without *a priori* knowledge of process gain and/or time delay.

For unstable FOPTD systems, the relay will give a sustained oscillation only for $D/\tau < 0.69$ and the value will be less for time delayed second order unstable system with two unstable poles and one negative zero. For such a situation, the time delay can be easily calculated by applying Majhi and Atherton (1999) method.

In the present work, a method is proposed to estimate all the five parameters of an unstable model with a zero (SOPTDZ), using a single asymmetric relay feedback method. Three simulation studies are given to check the proposed identification of unstable SOPTDZ model parameters. A simulation study on a non-linear CSTR problem to identify the five parameters of an unstable SOPTDZ is also given.

11.2 Problem Description

Consider an SOPTDZ model

$$G(s)=k_p\left(1\pm\tau_1 s\right)\exp\left(-Ds\right)/\left[\left(\tau_2 s\pm1\right)\left(\tau_3 s-1\right)\right] \tag{11.1}$$

where 'k_p' (process gain),'D' (time delay) and 'τ_1', 'τ_2' and 'τ_3' (time constants), are the five parameters to be estimated. To stabilize the system with two unstable poles by a proportional controller requires that the system should also have a negative zero and the same condition is required for the relay testing as well.

11.3 Relay Test

Consider a relay feedback system where $G(s)$ is the process transfer function, y is the output, y_r is the set point, e is the error and u is the manipulated variable. A feedback loop with an asymmetrical relay is used. The relay switches between $+\gamma h$ and $-h$ as shown in Fig. 10.1 for a positive gain process. For the symmetric relay testing, $\gamma = 1$.

11.4 Parameter Estimation

An asymmetric relay is used in the feedback loop (refer Fig. 11.1). The system input-output data are recorded and the open loop process gain (k_p) of the model is obtained by integrating the system input–output response as

$$k_p=\frac{\int_{t_o}^{t_o+p_u} y(t)dt}{\int_{t_o}^{t_o+p_u} u(t)dt} \tag{11.2}$$

Here y is the output variable, u is the input (manipulated) variable, t_o is the starting time of the integration after initial dynamics dies out, and p_u is the period of oscillations. Integration in Eq.(10.2) is to be carried out for one period of oscillation. Using the asymmetric relay, dead time (D) is obtained by aligning output response (y) to the input response (u) by shifting to the left. The dead time is noted as the time required to reach the peak amplitude (Majhi and Atherton, 1999). Since k_p and D are known, we can then formulate three more equations to get the remaining (τ_1, τ_2 and τ_3) parameters.

From the definition of Laplace transform, we get

$$y(s) = \int_0^\infty y(t)\exp(-st)\,dt \tag{11.3}$$

The above integral can be evaluated for $s = 8/t_s$, where 't_s' is the time at which few (say 3) repeated cycles of oscillation appear in the output after the initial transients die out. The reason for taking $s = 8/t_s$ is that, for $t > t_s$, because of a very small value of the term exp($-st$), the contribution by the subsequent terms is negligible while evaluating the integral. Let the resulting integral value be denoted by $y(s_1)$.

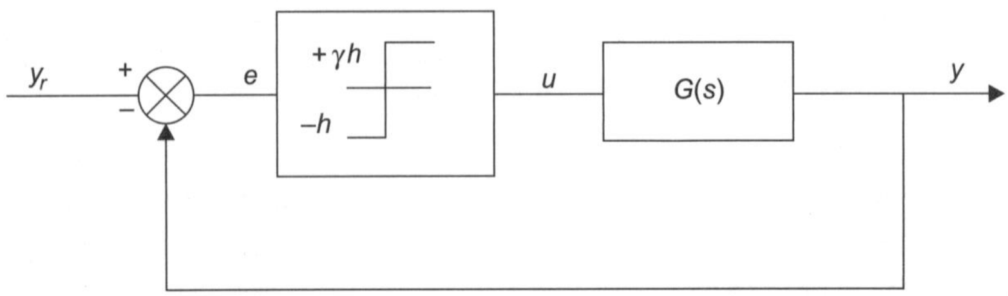

Fig. 11.1 *Block diagram for asymmetric relay feedback system*

Similarly,

$$u(s) = \int_0^\infty u(t)\exp(-st)\,dt \tag{11.4}$$

Equation (10.3) is also evaluated using $s = 8/t_s$. With the numerical values of $y(s_1)$ and $u(s_1)$ and using the assumed model, the following equation is formulated:

$$y(s_1)/u(s_1) = k_p\,(1 \pm \tau_1 s_1)\,\exp(-Ds_1)/\left[(\tau_2 s_1 \pm 1)(\tau_3 s_1 - 1)\right] \tag{11.5}$$

Since $y(t)$ and $u(t)$ are piece-wise continuous and periodic, the Laplace transform of $y(t)$ and $u(t)$ can be written as (Kreyszig, 1996)

$$y(s) = \frac{1}{\left[1 - \exp(-Ps)\right]}\int_0^P y(t)\exp(-st)\,dt \tag{11.6}$$

$$u(s) = \frac{1}{\left[1 - \exp(-Ps)\right]}\int_0^P u(t)\exp(-st)\,dt \tag{11.7}$$

From Eq.(11.6) and Eq.(11.7), we get

$$G(s) = y(s)/u(s) = \left[\int_0^P y(t)\exp(-st)\,dt\right] \Big/ \left[\int_0^P u(t)\exp(-st)\,dt\right] \tag{11.8}$$

where $P = 2\pi/\omega$ and ω is the frequency of oscillations observed in the output.
By substituting $s = j\omega$ and using the Euler's expansion for the exp($-j\omega t$) in Eq.(11.8), we get

$$G(j\omega) = y(j\omega)/u(j\omega) = (c_1 - jd_1)/(c_2 - jd_2) \tag{11.9}$$

where

$$c_1 = \int_0^P y(t)\cos(\omega t)\,dt \tag{11.10a}$$

$$d_1 = \int_0^P y(t)\sin(\omega t)dt \tag{11.10b}$$

$$c_2 = \int_0^P u(t)\cos(\omega t)dt \tag{11.10c}$$

$$d_2 = \int_0^P u(t)\sin(\omega t)dt \tag{11.10d}$$

Equations (11.10a) to (11.10d) can be evaluated numerically using $y(t)$ and $u(t)$ data of the relay test. Hence, Eq.(11.9) can be written as

$$G(j\omega) = p + jq \tag{11.11}$$

where

$$p = (c_1c_2 + d_1d_2)/(c_2^2 + d_2^2) \tag{11.12a}$$

$$q = (d_2c_1 - d_1c_2)/(c_2^2 + d_2^2) \tag{11.12b}$$

By substituting $s = j\omega$ and $\exp(-Dj\omega) = \cos(D\omega) - j\sin(D\omega)$ in Eq.(11.1), we get

$$p + jq = \{k_p(1 \pm \tau_1 j\omega)[\cos(-D\omega) - j\sin(-D\omega)]\}/[(\tau_2 sj\omega \pm 1)(\tau_3 j\omega - 1)] \tag{11.13}$$

In the above equation, + sign is used for the stable system and – sign is used for an unstable system. On cross-multiplying the right-hand side denominator of Eq.(11.13) to the left-hand side of Eq.(11.13) and separating the coefficients of the real and imaginary terms, we get:

$$\{k_p[\mp D_1 - BCD_1 \mp AE \mp ABCE \mp ABD_1 \pm ACD_1 + BE \mp CE]\}/ \\ \{p[(\mp 1 - BC)^2 + (B \mp C)^2]\} \tag{11.14}$$

$$\{k_p[BD_1 \mp CD_1 \pm ABE \mp ACE \mp AD_1 \mp ABCD_1 \pm E + BCE]\}/ \\ \{q[(\mp 1 - BC)^2 + (B \mp C)^2]\} \tag{11.15}$$

where $A = \tau_1\omega,\ B = \tau_2\omega,\ C = \tau_3\omega,\ D_1 = \cos(D\omega),\ E = \sin(D\omega)$ (11.16)

Using the relay test, the three linear algebraic Eqs (11.5), (11.14) and(11.15) can be solved to estimate the values for τ_1, τ_2 and τ_3.

For the process transfer function model,

$$G(s) = k_p(1 \pm \tau_1 s)\exp(-Ds)/[a_1 s^2 + a_2 s + 1] \tag{11.17}$$

the five equations to identify k_p, τ_1, $-\tau_d$, a_1 and a_2 are summarized as follows.

Consider Eq.(11.17) transfer function model for identification. Using asymmetric relay [refer to Eq.(11.2)] process gain k_p can be found. After aligning the output response to that of the input response, D is noted.

On substituting $s = j\omega$ and $\exp(-D\,j\omega) = \cos(D\,\omega) - j\sin(D\,\omega)$ in above equation we get

$$G(j\omega) = p + jq = \{k_p(1 \pm \tau_1 j\omega)[\cos(D\omega) - j\sin(D\omega)]\}/[a_1(j\omega)^2 + a_2(j\omega) + 1] \quad (11.18)$$

$$p + jq = \{k_p(1 \pm jA_1)[D_1 - jE_1]\}/[(1 - B_1) + jC_1] \quad (11.19)$$

where $A_1 = \tau_1\omega$, $B_1 = a_1\omega^2$, $C_1 = a_2\omega$, $D_1 = \cos(D\,\omega)$, $E_1 = \sin(D\,\omega)$.

On cross-multiplying, we get

$$[(p + jq)(1 - B_1 + jC_1)] = \{k_p(1 \pm jA_1)[D_1 - jE_1]\} \quad (11.20)$$

$$[p - (pB_1) + (jpC_1) + (jq) - (jqB_1) - (qC_1)] = k_p[D_1 - (jE_1) \pm (jA_1D_1) \pm (A_1E_1)] \quad (11.21)$$

Separating the coefficients of the real and imaginary terms, Eq.(11.21), we get

$$[p - (pB_1) - (qC_1)] = k_p[D_1 \pm (A_1E_1)] \quad (11.22)$$

$$[(pC_1) + q - (qB_1)] = k_p[-E_1 \pm (A_1D_1)] \quad (11.23)$$

$$[y(s_1)/u(s_1)] = k_p(1 \pm \tau_1 s_1)\exp(-Ds_1)/[a_1 s_1^2 + a_2 s_1 + 1] \quad (11.24)$$

Solving Eqs (11.22), (11.23) and (11.24), the remaining parameters (τ_1, a_1 and a_2) can be identified.

Thus, using a single asymmetrical relay test, all the five parameters of an SOPTDZ system can be calculated. For controlling the SOPTDZ model, a PID controller with lead-lag compensator proposed by Rao and Chidambaram (2007) is to be designed:

$$G_c(s) = k_c\left(1 + \frac{1}{\tau_I s} + \tau_D s\right)\frac{(\alpha s + 1)}{(\beta s + 1)} \quad (11.25)$$

where α and β are lead and lag time constants respectively. The design equations for the calculation of the controller parameters are given in Chapter 1.

11.5 Simulation Results

Three examples are considered for simulation using an asymmetric relay feedback test. For all the examples, $\gamma = 2$ and $h = \pm 1$ is considered.

11.5.1 Case study 1

Consider open loop unstable SOPTDZ model given by Yu (1999):

$$G(s) = [1.0(3s + 1)\exp(-s)]/[(10s + 1)(3s - 1)] \quad (11.26)$$

In Fig. 11.2, the process output response is aligned with the input response. Dead time is noted as the time required to the peak amplitude ($\tau_d = 1.0$). Using the proposed method, the intermediate values are calculated as $s_1 = 0.3271$, $y(s_1) = 0.41$, $u(s_1) = -0.0024$, $p = -0.1725$ and $q = -0.0213$ (refer Table 11.1). The identified model parameters are $k_p = 1.009$, $\tau_d = 1.0$, $\tau_1 = 2.9993$, $\tau_2 = 9.97$ and $\tau_3 = 3.05$. Table 11.2 gives the parameters of the identified model and the controller settings using Eq.(11.25) based on the identified model. Controller is also designed based on the actual system. The performances of both the controllers are evaluated on the actual system (refer to Fig. 11.3). Table 11.2 and Fig. 11.3 shows that the model parameters of the identified model are closer to that of the actual system. The controller designed on the identified model works very well on the actual system. There is no difference in the closed loop performances among the two set of controllers.

The effect of load (entering along with u) or noise corrupted during the identification step is studied and the identified model parameters are given in Table 11.3. PID settings for the identified model estimated under load condition are calculated as $k_c = 5.806$, $\tau_I = 20.79$, $\tau_D = 4.99$, $\alpha = 0.5$ and $\beta = 2.50$. The closed loop performance on the actual system is evaluated and the results are given in Fig. 11.4. PID settings for the identified model estimated under measurement noise are calculated as $k_c = 6.13$, $\tau_I = 18.92$, $\tau_D = 5.32$, $\alpha = 0.5$ and $\beta = 2.783$. The closed loop performance on the actual system is evaluated and performance similar to Fig. 11.4 is obtained. The performances are very close to each other.

Table 11.1 *Intermediate values for the case studies considered*

Case study	s_1	$y(s_1)$	$u(s_1)$	P	q
1	0.3271	0.41	–0.0024	–0.1725	–0.0213
2	0.0267	5.0349	–3.1146	–0.3437	–0.0492
3	0.2502	0.4176	0.1091	–0.1141	–0.0160

Table 11.2 *Controller parameters based on identified model and actual process parameters*

Parameters	Actual system			Identified system		
	Case study			Case study		
	1	2	3	1	2	3
k_p	1.0	1.0	1.0	1.009	1.01	0.9651
τ_d	1.0	1.0	1.0	1.0	1.0	1.0
τ_1	3.0	–1.0	5.0	2.9993	–0.9601	4.97
τ_2	10.0	10.0	10.0	9.97	10.1399	8.8786
τ_3	3.0	20.0	5.0	3.05	20.1566	5.42
λ	6	8	6.5	6	8	6.5
k_c	5.64	6.82	9.49	5.66	6.92	9.40
τ_I	19.45	30.47	14.02	19.40	30.29	13.99
τ_D	5.28	6.74	6.47	5.26	6.77	6.50
α	0.5	0.5	0.5	0.5	0.5	0.5
β	2.534	0.172	4.931	2.528	0.178	4.898

Note: For case study 3, the denominator is of the form $(\tau_2 s - 1)(\tau_3 s - 1)$

Table 11.3 *Effect of noise and load disturbance on identification for case study 1*

Parameters	Load	Noise
	0.02	σ
k_p	0.748	0.989
τ_d	1.0	1.0
τ_1	2.997	3.198
τ_2	7.987	10.72
τ_3	2.75	2.89

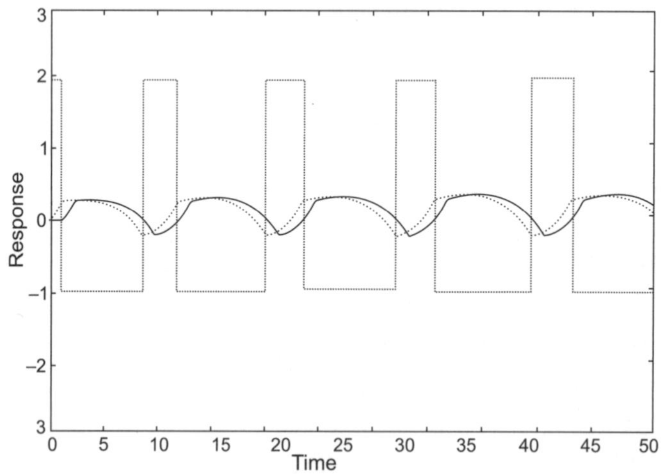

Fig. 11.2 *Process input-output relay response (case study 1)*
Legend: solid–process output response; dash–shifted process output response; dash-dot–relay output response

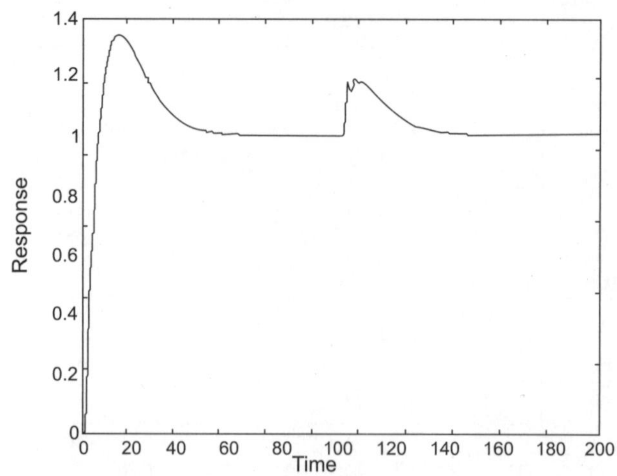

Fig. 11.3 *Controllers performance evaluated on the actual system (case study 1)*
Legend: solid–controller designed on actual system; dash–controller designed on identified model (solid and dash lines overlap)

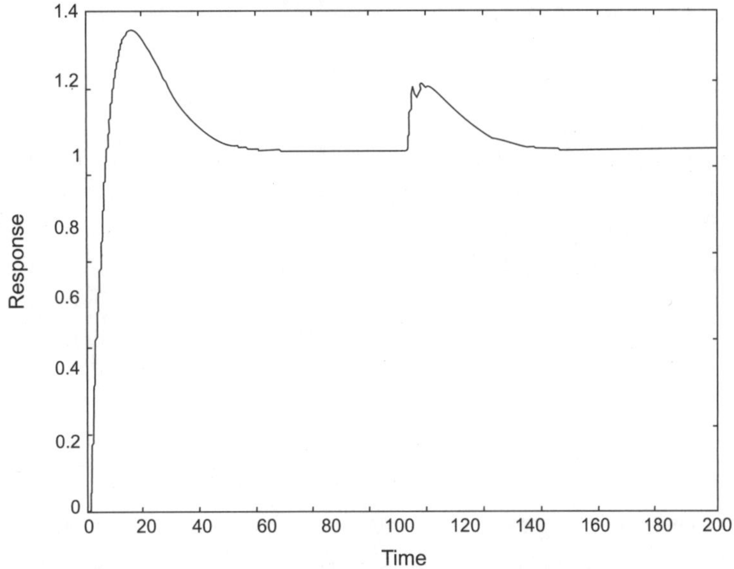

Fig. 11.4 *Controllers performance evaluated on the actual system (case study 1, effect of load on model identification)*
Legend: solid–controller designed on actual system; dash–controller designed on identified model (with load effect)

11.5.2 Case study 2

Consider open loop unstable SOPTDZ model with a stable pole and an unstable pole and a positive zero:

$$G(s)=\left[1.0(1-s)\exp(-s)\right]/\left[(10s+1)(20s-1)\right] \tag{11.27}$$

Figure 11.5 shows the process output response obtained using an asymmetric relay. After aligning the output response to the input response, value of τ_d is noted as 1. Using the proposed method, intermediate values are obtained as at t = 299.49, s_1 = 0.0267, $y(s_1)$ = 5.0349, $u(s_1)$ = −3.1146, p = −0.3437 and q = −0.0492. The identified model parameters are k_p= 1.01, τ_d = 1.0, τ_1 = −0.9601, τ_2= 10.1399 and τ_3= 20.1566. Table 11.2 gives the identified model parameters and the controller settings based on the identified model. Controller is also designed for the actual system. The parameters of the identified model are closer to that of the actual system. Figure 11.6 shows the closed loop performance on the actual system with the controller designed on the identified model and also designed on the actual system. Both the controllers are simulated on the actual system. The controller designed on the identified model works very well on the actual system. There is no difference in the closed loop performances among the two sets of controllers.

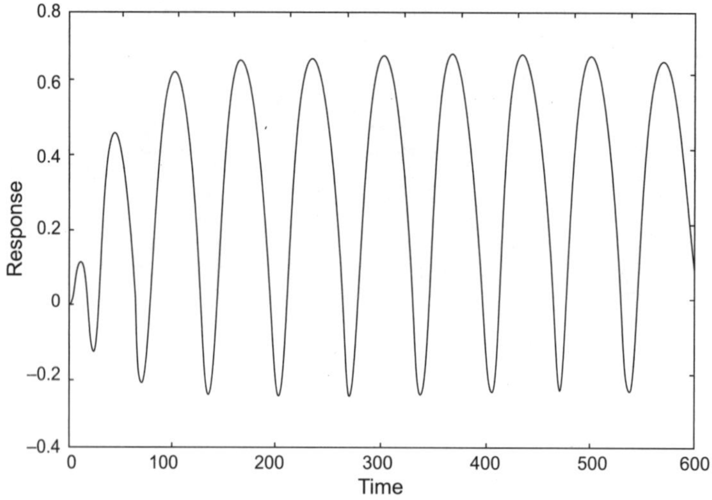

Fig. 11.5 *Process output relay response (case study 2)*

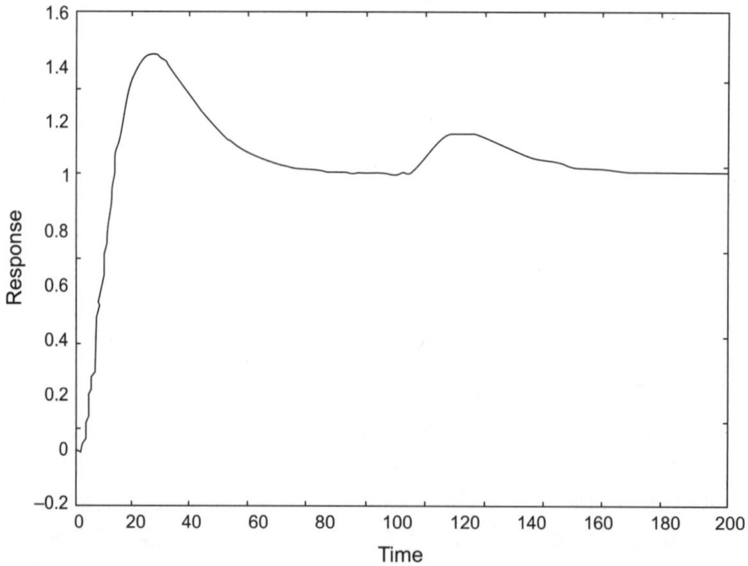

Fig. 11.6 *Controllers performance evaluated on the actual system (case study 2)*
Legend: solid–controller designed on actual system; dash–controller designed on identified model (solid and dash lines overlap)

11.5.3 Case study 3

Consider open loop unstable SOPTDZ model with two unstable poles and one negative zero as

$$G(s) = \left[1.0(5s+1)\exp(-s)\right] / \left[(10s-1)(5s-1)\right] \tag{11.28}$$

Figure 11.7 shows the process output response for the relay testing. After aligning the output response to the input response, value of τ_d is noted as 1. Using the proposed method, intermediate values are obtained as at $s_1 = 0.2502$, $y(s_1) = 0.4176$, $u(s_1) = -0.1091$, $p = -0.1141$ and $q = -0.016$. The identified model parameters are $k_p = 0.9651$, $\tau_d = 1.0$, $\tau_1 = 4.97$, $\tau_2 = 8.8786$ and $\tau_3 = 5.42$. Table 11.2 gives the model parameters of the identified model and the controller settings based on the identified model. Controller is also designed on the actual system. Figure 11.8 shows the closed loop performance of the controllers on the actual system. Table 11.2 and Fig. 11.8 show that the model parameters of the identified model are closer to that of the actual system. The controller designed on the identified model works very well on the actual system. There is no difference in the closed loop performances among the two sets of controllers.

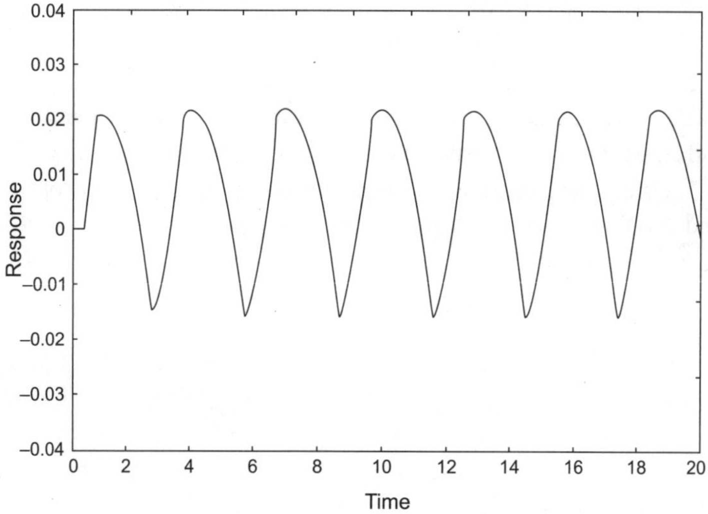

Fig. 11.7 *Process output relay response (case study 3)*

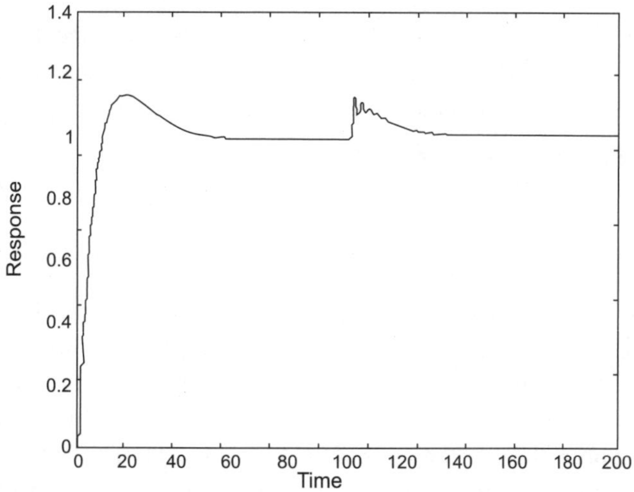

Fig. 11.8 *Controllers performance evaluated on actual system (case study 3)*
Legend: solid–controller designed on actual system; dash–controller designed on identified model (solid and dash lines overlap)

11.6 Simulation Application to a Non-linear Continuous Stirred Tank Reactor (CSTR)

The proposed method of identification is applied to a non-linear CSTR process considered by Russo and Bequette (1995). The reaction is an irreversible first order exothermic reaction. The dynamic material and energy balance and additional energy balance around cooling jacket is expressed by the following dimensionless differential equations:

$$dx_1/dt = \left[-D_a x_1 \exp\left(x_2/\left(1+\left(x_2/E_a\right)\right)\right)\right] + (1 - x_1) \tag{11.29a}$$

$$dx_2/dt = \left[Q D_a x_1 \exp\left(x_2/\left(1+\left(x_2/E_a\right)\right)\right)\right] - \left[(1+h_c)x_2\right] + h_c x_3 \tag{11.29b}$$

$$dx_3/dt = 10\left[q_c(-1 - x_3) + h_c(x_2 - x_3)\right] \tag{11.29c}$$

Here, x_1, x_2, x_3 and q_c are the dimensionless concentration, reactor temperature, cooling jacket temperature and coolant flow rate respectively. The parameter values are $Q = 11$, $D_a = 0.135$, $E_a = 20$ and $h_c = 1.5$. Using these parameter values and taking x_2 as the controlled variable and q_c as a manipulated variable, the steady-state solution of Eq.(11.29a) to Eq.(11.29c) for $q_c = 3.2$ is given by Wang and Hwang (2005) as

$$\left[x_1, x_2, x_3\right] = \left[0.3195, 3.2, 0.3429\right] \tag{11.30}$$

Let us consider initially, the system is at this steady-state condition at $t = 0$, $x_1 = 0.3195$, $x_2 = 3.2$ and $x_3 = 0.3429$ [refer to Eq.(11.30)]. Coolant flow rate (q_c) is considered as the manipulated variable in order to control the reactor temperature (x_2). As suggested by Wang and Hwang (2005), delay of 0.15 is considered in the measurement of x_2. For the given condition of the operating point, the locally linearized model is obtained as

$$G(s) = -1.77(0.3186s+1)\exp(-0.15s)/\left(0.028s^3 + 1.324s^2 - 0.479s + 1\right) \tag{11.31}$$

An asymmetrical relay with relay height (h) = 0.16 and $\gamma = 2$ is conducted. Relay output response (deviation value from the steady-state point) is shown in Fig. 11.9. The model parameter values are identified by the proposed method and the corresponding model is given by

$$G(s) = -1.6569(0.4598s+1)\exp(-0.22s)/\left(1.59s^2 - 0.32s + 1\right) \tag{11.32}$$

Wang and Hwang (2005) identified the model as

$$G(s) = -1.80(0.3172s+1)\exp(-0.15s)/\left(0.0311s^3 + 1.34s^2 - 0.491s + 1\right) \tag{11.33}$$

The controller (PID with lead-lag filter) is designed based on the identified model using the method of Rao and Chidambaram (2007). The controller parameters based on identified model parameters are given in Table 11.4. The performance is evaluated on the non-linear model equations for a change in the set point (step change in x_2 from 3.20 to 3.52). The proposed method gives a response close to the method of Wang and Hwang (refer to Fig. 11.10). The performance in terms of ISE and IAE values are reported in Table 11.4, which shows that the proposed method and the Wang and Hwang method give good performances. However, the model identified by Wang and Hwang gives a slightly better performance. As stated earlier, the method of Wang and Hwang (2005) requires more computations.

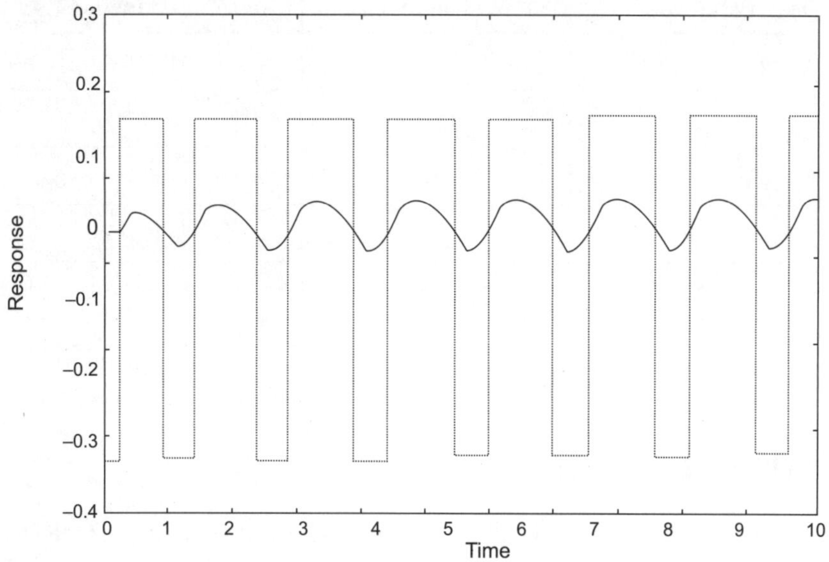

Fig. 11.9 *Process asymmetric relay response of the CSTR*

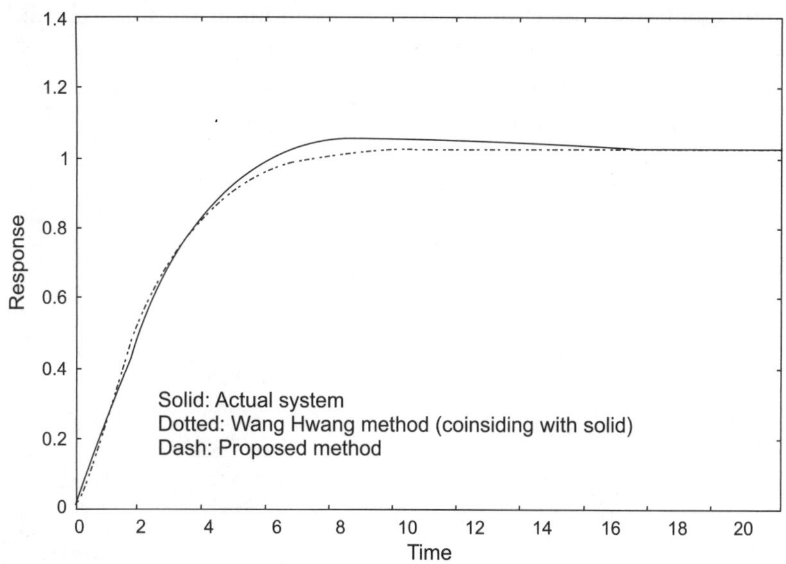

Fig. 11.10 *Closed loop response of actual system for step point in the set point*
Legend: solid-controlled designed on actual model; dash-controller designed on proposed identified model; dotted-controller based on identified model of Wang Hwang method (dotted line coincide with solid line)

Table 11.4 *Controller parameters and ISE and IAE values for the CSTR control problem*

Parameter	WH method	Proposed method
k_c	−1.33	-1.016
τ_I	3.16	2.86
τ_D	2.32	2.55
ISE	2.46	2.203
IAE	1.31	1.28

Note: $\alpha = 0.11$, $\beta = 0.303$ (present method); $\alpha = 0.075$, $\beta = 0.171$ (WH method)

11.7 Conclusions

A method is proposed for estimating all the five parameters of an unstable SOPTDZ transfer function model using a single asymmetric relay test. The gain and time delay are calculated straight forward. Solution of three linear algebraic equations gives the other three parameters (system zero and two poles). The method is simple to use. Simulation results show that for three case studies on transfer function models and a non-linear reactor problem, the method estimates the model (with five parameters) with adequate accuracy. The controller designed on the identified model works well on the actual system. The closed loop performance of the identified system is found to be closer to that of the actual system.

12 Identification of FOPTD Multivariable Systems

In this chapter, a technique for identification of multivariable transfer function matrix proposed by Ganesh and Chidambaram (2003) is reviewed. They extended the method of Srinivasan and Chidambaram (2003) meant for asymmetric relay feedback for a scalar system to a multivariable system. The method identifies the FOPTD model for each element of the transfer function matrix model without any *priori* knowledge of the value of time delay or gain of the system. In this method, a decentralized relay feedback method suggested by Wang et al. (1997) is applied to the $m \times m$ system and 'm' relay feedback tests are required for identifying the entire transfer function matrix. The identification method is a closed loop relay feedback method and hence this method is less susceptible to disturbance and measurement noise and can be used for identifying an unstable system. The method is applied for various 2×2 transfer function matrices as Wood and Berry's methanol–water distillation column process (1973) and a multivariable system with higher order individual elements. IMC controller (Tanttu and Lieslehto, 1991) is designed for the identified model, and using this controller the closed loop performances of actual and identified models are compared.

12.1 Introduction

The transfer function model of the system is required for designing a suitable control system. The dynamic models of process are determined from the fundamental physical and chemical laws. But for the process that is already in operation, there is an alternative approach based on experimental dynamic data obtained from plant test. This approach is also used when the process is too complex to model. A number of techniques have been proposed for process identification. Identification of transfer function can be carried out either by open loop or by closed loop method. Closed loop identification method is preferable, as it is less susceptible to disturbance and measurement noise. For the design of controllers, identification of a first order plus time delay (FOPTD) model is adequate and also most of the chemical processes can be best approximated by the FOPTD models.

Extension of the autotuning technique to multivariable systems is non-trivial and it has attracted much attention in literature. When relay technique is extended to multivariable system there are three possible relay feedback schemes (Wang et al. 1997):

1. Independent single relay feedback:

   Only one loop is subjected to relay feedback while the other loop is manual.

2. Sequential relay feedback:

   This method tunes the multivariable system loop by loop closing each loop once it is tuned, until all loops are tuned.

3. Decentralized relay feedback:

   All the loops are simultaneously subjected to relay feedback.

Independent relay feedback and sequential relay feedback are only partial closed loop tests while decentralized relay feedback is a completely closed loop test.

Luyben (1987) used independent relay feedback method to identify the multivariable transfer function matrix. To identify each element of the transfer function matrix, two equations were formulated using phase angle and amplitude criteria. The steady-state gains were found by step response tests. Wu et al. (1994) also discussed the identification procedure by independent relay feedback. They found the steady gains directly from the information from limit cycle while the other parameters as time delay and time constant were obtained from phase angle and amplitude criteria. By keeping one loop open, the system is susceptible for measurement noise and disturbance. Also it cannot be used for identifying unstable systems.

Wang et al. (1997) obtained the process steady-state gain and frequency response for multivariable systems and used the results for designing the controller. The objective of this chapter is to use Wang et al. (1997) method to obtain the steady-state gain matrix and process frequency response matrix, and use Srinivasan and Chidambaram (2003) method to get the model parameters for each element of the transfer function matrix. The chapter is organized as follows. In section 12.2, relay test for multivariable system and estimation of steady-state gain matrix and process frequency response are discussed. In section 12.3, identification of time constant and delay for each element of transfer function matrix is presented. In section 12.4, comparison of closed loop performance for actual and identified model by designing IMC controller is presented. Simulation examples are given in section 12.5, followed by conclusions in section 12.6.

12.2 Relay Identification

The multivariable system is subjected to a decentralized relay feedback. In decentralized relay feedback, all the loops of multivariable system are subjected to relay feedback simultaneously. The decentralized relay feedback scheme for TITO case is shown in Fig. 12.1. Under decentralized relay feedback, the outputs of multivariable system will oscillate with limit cycle after initial transients. It was found by Atherton (1975) for a typically coupled multivariable system under decentralized relay feedback, all the outputs normally have the same frequency.

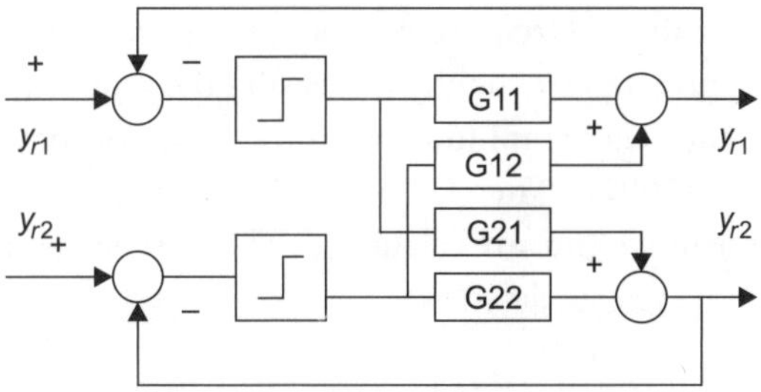

Fig. 12.1 *Decentralized relay feedback*

The procedure to obtain the steady-state gain matrix and frequency response matrix given by Wang et al. (1997) is as follows:

1. Multivariable system is subjected to the decentralized relay feedback as shown in Fig. 12.1. In order to get asymmetric limit cycles oscillations in the outputs and in the dominant loop, a bias is induced. If the system is having significant interactions, oscillation frequency in both the loops will be same. Then, direct current components and the first harmonic of these periodic waves (for 2 × 2 system) are extracted as

2. $$U^1(0) = \begin{bmatrix} \int_0^{Tc} u_1(t)dt \\ \int_0^{Tc} u_2(t)dt \end{bmatrix} \tag{12.1}$$

$$Y^1(0) = \begin{bmatrix} \int_0^{Tc} y_1(t)dt \\ \int_0^{Tc} y_2(t)dt \end{bmatrix} \tag{12.2}$$

$$U^1(j\omega) = \begin{bmatrix} \int_0^{Tc} u_1(t)e^{-j\omega t}dt \\ \int_0^{Tc} u_2(t)e^{-j\omega t}dt \end{bmatrix} \tag{12.3}$$

$$Y^1(j\omega) = \begin{bmatrix} \int_0^{Tc} y_1(t)e^{-j\omega t}dt \\ \int_0^{Tc} y_2(t)e^{-j\omega t}dt \end{bmatrix} \tag{12.4}$$

where T_c is the period of oscillation and ω is the frequency of oscillation. Also, we have

$$Y^1(0) = G(0)U^1(0) \tag{12.5}$$

$$Y^1(j\omega) = G(j\omega)U^1(j\omega) \tag{12.6}$$

3. To get the steady-state gain and frequency response matrix, one more relay test (for $m \times m$ system, 'm' tests) is conducted by slightly increasing the amplitude of the dominant loop. By slightly changing the relay amplitude, the process is still having the same oscillation frequency [Wang et al. (1997)]. Then from Eqs (12.1), (12.2), (12.3) and (12.4), the values of $U^2(0)$, $Y^2(0)$, $U^2(j\omega)$ and $Y^2(j\omega)$ are calculated. This is repeated for all 'm' tests. $G(0)$ and $G(j\omega)$ for 2×2 system are calculated as

$$\left[Y^1(0)\ Y^2(0)\right] = G(0)\left[U^1(0)\ U^2(0)\right] \tag{12.7}$$

$$\left[Y^1(j\omega)\ Y^2(j\omega)\right] = G(j\omega)\left[U^1(j\omega)\ U^2(j\omega)\right] \tag{12.8}$$

12.3 Estimation of Model Parameters (τ and θ)

FOPTD model is written as

$$G(s) = k_p \exp(-\theta s)/(\tau s + 1) \tag{12.9}$$

For this model, frequency response can be written by substituting $s = j\omega$ as

$$G(j\omega) = k_p\left[\cos(\theta\omega) - j\sin(\theta\omega)\right]/(1 + j\tau\omega) \tag{12.10}$$

From the frequency response matrix, $G(j\omega)$ for each element of transfer function matrix is known and is written as

$$G(j\omega) = a + jb \tag{12.11}$$

On equating the real and imaginary parts of (12.10) and (12.11), we get

$$a - b\tau\omega - k_p \cos(\theta\omega) = 0 \tag{12.12}$$

$$a\tau\omega + b + k_p \sin(\theta\omega) = 0 \tag{12.13}$$

Value of K_p for each element is taken from the steady-state gain matrix given by Eq.(12.7). Then Eq.(12.12) and Eq.(12.13) are solved analytically to get the values of time constant and delay for each element of transfer function matrix.

12.4 Comparision of Closed Loop Performance

Here the IMC controller is designed by the method suggested by Tanttu and Lieslehto (1991) for the identified model to compare the closed loop performance. Controller ($k_{c,ij}$) for each scalar transfer function ($g_{p,ij}$) is first designed.

For an FOPTD model, PI controller settings are given by

$$\left(k_p k_c\right)_{ij} = \frac{2\tau + \theta}{2\lambda}; \tag{12.14}$$

$$\text{and } \tau_{I,ij} = \tau + 0.5\theta; \tag{12.15}$$

where, λ $(\lambda \geq 0.2\tau)$ is the tuning the parameter.
The multivariable centralized controller is given by

$$G_c(s) = K_c + \left(\frac{K_I}{s}\right) \tag{12.16}$$

where, K_c, K_I are matrices of size $n \times n$ given by

$$K_c = \begin{bmatrix} \frac{1}{k_{c,11}} & \frac{1}{k_{c,12}} & .. & \frac{1}{k_{c,1n}} \\ \frac{1}{k_{c,21}} & \frac{1}{k_{c,22}} & .. & \frac{1}{k_{c,2n}} \\ .. & .. & .. & .. \\ \frac{1}{k_{c,n1}} & \frac{1}{k_{c,n2}} & .. & \frac{1}{k_{c,nn}} \end{bmatrix}^{-1} \tag{12.17}$$

and

$$K_I = \begin{bmatrix} \frac{1}{k_{I,11}} & \frac{1}{k_{I,12}} & .. & \frac{1}{k_{I,1n}} \\ \frac{1}{k_{I,21}} & \frac{1}{k_{I,22}} & .. & \frac{1}{k_{I,2n}} \\ .. & .. & .. & .. \\ \frac{1}{k_{I,n1}} & \frac{1}{k_{I,n2}} & .. & \frac{1}{k_{I,nn}} \end{bmatrix}^{-1} \tag{12.18}$$

where,

$$k_{I,ij} = k_{c,ij} / \tau_{I,ij} \tag{12.19}$$

Using the above controller design formulae, a controller is designed for the identified model. In the above controller design formulae, a same filter constant is used for each of the scalar systems. Closed loop performance for both the actual and the identified models and ISE values are compared.

12.5 Examples

12.5.1 Example 1

Wood and Berry (1973) binary distillation column plant

$$G(s) = \begin{bmatrix} \frac{12.8e^{-s}}{1+16.7s} & \frac{-18.9e^{-3s}}{1+21s} \\ \frac{6.6e^{-7s}}{1+10.9s} & \frac{-19.4e^{-3s}}{1+14.4s} \end{bmatrix} \tag{12.20}$$

For this process, in the first relay test, relay in loop 1 is ideal with switching levels 1 and −1 while, relay in loop 2 is a bias relay with switching levels 1.5 and −1. The oscillation frequency in both the outputs is same ($\omega = 0.485$). In the second relay test, the switching levels in loop 2 are changed to 1.8 and –1.2. The oscillation frequency in test 2 is nearly same as in test 1. The steady-state gain matrix and frequency response matrix is obtained from Eq.(12.7) and Eq.(12.8). The steady-state gain matrix is obtained as

$$G(0)=\begin{bmatrix} 12.0529 & -18.27 \\ 6.708 & -19.530 \end{bmatrix} \tag{12.21}$$

The process frequency response matrix at $\omega = 0.485$ is obtained as

$$G(j\omega)=\begin{bmatrix} -0.5542-1.4536j & 1.799+0.4009j \\ 0.1304+1.1926j & 2.699+0.7169j \end{bmatrix} \tag{12.22}$$

Then for each individual transfer function matrix, Eq.(12.12) and Eq.(12.13) are solved to get the time constant and time delay. The identified transfer function matrix is

$$G(s)=\begin{bmatrix} \dfrac{12.05e^{-1.018s}}{1+15.84s} & \dfrac{-18.27e^{-2.98s}}{1+20.297s} \\ \dfrac{6.708e^{-7.028s}}{1+11.34s} & \dfrac{-19.5301e^{-2.99s}}{1+14.23s} \end{bmatrix} \tag{12.23}$$

The centralized controller based on identified model is designed by the procedure given in section 12.4 as

$$K_c=\frac{1}{\lambda}\begin{bmatrix} 2.3236 & -1.5699 \\ 0.8456 & -1.3768 \end{bmatrix} \tag{12.24}$$

and

$$K_I=\frac{1}{\lambda}\begin{bmatrix} 0.1731 & -0.162 \\ 0.0595 & -0.1069 \end{bmatrix} \tag{12.25}$$

The tuning parameter, $\lambda = 6$ (used here), is selected by trial and error.

Performance of the actual and the identified models for step change in inputs is given in Fig. 12.2. A very close matching of the performance shows that the identified model is very close to the actual model. The ISE value comparison is given in Table 12.1. These ISE values also show a close match between the actual and the identified models.

12.5.2 Example 2

Multivariable system with individual higher order elements

$$G(s)=\begin{bmatrix} \dfrac{(2s+1)e^{-10s}}{s^2+2s+1} & \dfrac{(s+1)e^{-0.5s}}{s^2+3s+4} \\ \dfrac{e^{-s}}{s^2+4s+5} & \dfrac{(s+2)e^{-8s}}{s^2+2s+2} \end{bmatrix} \tag{12.26}$$

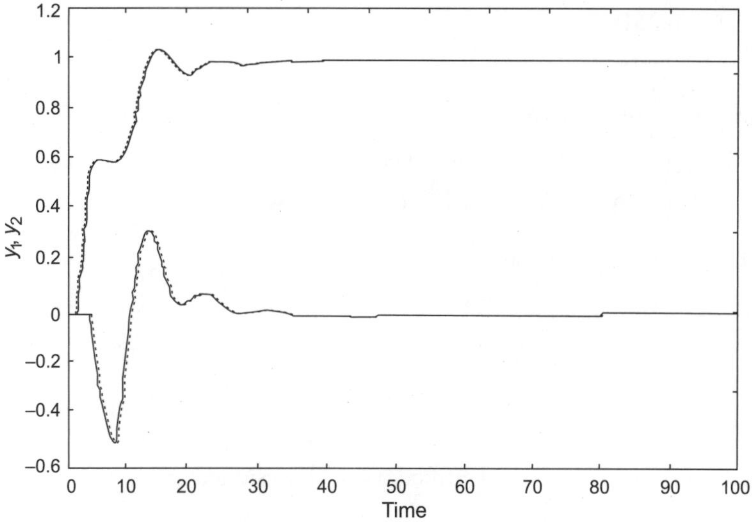

Fig. 12.2a *Comparison of closed loop performance for step change in Y_1*
Legend: solid–actual model; dot–identified model for example 1

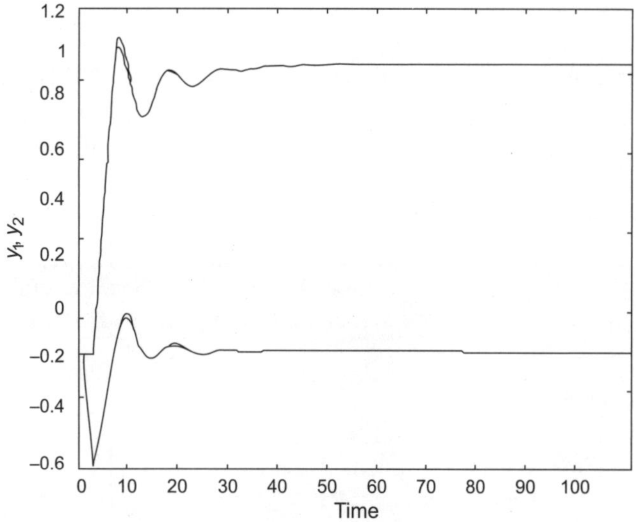

Fig. 12.2b *Comparison of closed loop performance for step change in Y_2*
Legend: solid–actual model; dot–identified model for example 1

For this process, in the first relay test, the switching levels of relay in loop 1 are 1.5 and –1 while relay in loop 2 is an ideal relay with switching levels 0.2 and –0.2. The oscillation frequency in both the outputs is same ($\omega = 0.304$).

In the second relay test, the switching levels in loop 1 are changed to 1.8 and –1.2. The steady-state gain matrix is calculated from Eq.(12.7).

$$G(0)=\begin{bmatrix} 1 & 0.2518 \\ 0.2 & 0.9989 \end{bmatrix} \tag{12.27}$$

The process frequency response matrix at $\omega = 0.304$ is given by Eq.(12.8)

$$G(j\omega)=\begin{bmatrix} -1.067-0.0583j & 0.2639+0.00888j \\ 0.1686-0.1034j & -0.8615-0.5348j \end{bmatrix} \tag{12.28}$$

The identified transfer function matrix is

$$G(s)=\begin{bmatrix} \dfrac{1e^{-10.15s}}{3.882s+1} & \dfrac{0.2519e^{-0.4965s}}{5.5818s+1} \\ \dfrac{0.2e^{-0.987s}}{0.715s+1} & \dfrac{1.e^{-8.5s}}{0.079s+1} \end{bmatrix} \tag{12.29}$$

The centralized controller based on identified model is designed as

$$K_c = \frac{1}{\lambda}\begin{bmatrix} 12.39 & -2.3181 \\ -8.8769 & 5.9898 \end{bmatrix} \tag{12.30}$$

and

$$K_I = \frac{1}{\lambda}\begin{bmatrix} 0.1731 & -0.162 \\ 0.0595 & -0.1069 \end{bmatrix} \tag{12.31}$$

The tuning parameter obtained by trial and error procedure is $\lambda = 25$. The comparison of ISE values of closed loop performance of the actual and identified models is given in Table 12.1. This comparison shows a close match between the actual and identified models.

Table 12.1 *ISE value comparison for actual and identified models for the above examples*

Example	Step in	ISE values for actual model		ISE values for identified model	
		Y1	Y2	Y1	Y2
Example 1	Y1	3.10	1.067	3.24	1.13
	Y2	4.28	0.36	4.26	0.344
Example 2	Y1	16.02	2.36	17.13	1.17
	Y2	1.479	14.19	0.097	14.94

12.6 Conclusions

In this chapter, the asymmetric relay feedback method for a scalar system is extended to identify multivariable FOPTD systems. The method identifies each element of transfer function matrix without any *a priori* knowledge of the model parameters as time delay or steady-state gain. This identification method being a closed loop one is able to cope with disturbance and measurement noise. This method can also be used for unstable multivariable processes. The examples illustrated show that this method gives an accurate process model. Comparison of the closed loop performance and the ISE values show that the identified model is a good representation of the actual model.

13 Identification of SOPTD Multivariable Systems

In this chapter, the method discussed in Chapter 10 on identifying second order plus time delay (SOPTD) transfer function models by asymmetric relay tuning method is extended to identify a multivariable system. The decentralized relay feedback method suggested by Wang et al. (1997) is applied for the $m \times m$ system and m relay tests are required for identifying the entire transfer function matrix. Although most of the processes can be adequately approximated by a first order plus time delay (FOPTD) model, some of the processes are under-damped and higher order processes can be better incorporated by an SOPTD than an FOPTD model. Certain higher order stable models when approximated to an FOPTD model give a negative time constant, hence, identifying a second order model is necessary. The proposed method (Ganesh and Chidambaram, 2005) is applied to a 2×2 transfer function matrix. Multivariable IMC controllers are designed for the identified model and the closed loop performances of the actual and identified models are compared.

13.1 Introduction

As stated earlier in Chapter 2, Srinivasan and Chidambaram (2004) proposed a method to analyze the conventional relay autotune data for estimating the three parameters of the FOPTD model using only one relay experiment. They proposed an additional equation, which along with the phase angle and amplitude criteria gives three parameters. Srinivasan and Chidambaram (2003) also proposed a modified asymmetrical relay feedback method to get improved estimates of the parameters of the FOPTD model. A single relay test is used to evaluate all three parameters of the FOPTD model without any *a priori* information about model parameters as time delay or steady-state gain. Ganesh and Chidambaram (2003) extended the method to identify multivariable systems. However, they identified each subsystem as an FOPTD system. However, for higher order system, identification of the FOPTD system may not be adequate and sometimes identifying the FOPTD model may not be possible.

Ramakrishnan and Chidambaram (2003) proposed a method to identify an SOPTD transfer function using the biased relay tuning method. In the present method, the biased relay feedback method proposed by Ramakrishnan and Chidambaram (2003) for identifying the SOPTD model for a scalar system is extended to multivariable systems. The extension of the autotuning technique to multivariable systems is non-trivial, and it has attracted much attention in the literature. When the relay technique is extended to multivariable systems using the decentralized relay feedback wherein all the loops are simultaneously subjected to relay feedback. The decentralized relay feedback method is a completely closed loop test. Luyben (1987) and Wu et al. (1997) proposed SOPTD model identification using an independent relay feedback. In addition, they used additional relay feedback tests to get the extra parameters. Wang et al. (1997) obtained the process steady-state gain and frequency response for multivariable system and used it for designing the controller. The objective is to use Wang et al. (1997) method to obtain the steady-state gain matrix and process frequency response matrix, and use Srinivasan and Chidambaram (2003) method and Ramakrishnan and Chidambaram (2003) method to get the model parameters for each element of transfer function matrix.

The chapter is organized as follows. In section 13.2, relay test for multivariable system and estimation of steady-state gain matrix and process frequency response is discussed. In section 13.3, identification of time constant and delay for each element of transfer function matrix is presented. In section 13.4, comparison of closed loop performance for the actual and identified models by designing IMC controllers is presented. Simulation examples are given in section 13.5, followed by conclusions in section 13.6.

13.2 Relay Identification

The multivariable system is subjected to a decentralized relay feedback. Under decentralized relay feedback, the outputs of multivariable system will oscillate with limit cycle after initial transient. The relay feedback procedure as described by Ganesh and Chidambaram (2003) is performed and the limit cycle data vectors $U^1(0)$, $Y^1(0)$, $U^1(j\omega)$ and $Y^1(j\omega)$ are found by the equations given by Ganesh and Chidambaram (2003). Here ω is the frequency of oscillation of the output of the process. Srinivasan and Chidambaram (2003) proposed a method for formulating additional equation for a scalar system. This method can be extended to multivariable systems to get an additional equation.

The vectors $U^1(s)$ and $Y^1(s)$ are given by

$$U^1(s) = \begin{bmatrix} \int_0^{\infty} u_1(t)e^{-st}dt \\ \int_0^{\infty} u_2(t)e^{-st}dt \end{bmatrix} \tag{13.1}$$

$$Y^1(s) = \begin{bmatrix} \int_0^{\infty} y_1(t)e^{-st}dt \\ \int_0^{\infty} y_2(t)e^{-st}dt \end{bmatrix} \tag{13.2}$$

To evaluate the integrals

$$u(s) = \int_0^{\infty} u(t)\exp(-st)dt \quad \text{and} \quad y(s) = \int_0^{\infty} y(t)\exp(-st)dt,$$

Padmasree and Chidambaram (2001) suggested using $s_1 = 8/t_s$ where 't_s' is the time at which few (say 3) repeated cycles of oscillation appear in the output. The reason for taking $s_1 = 8/t_s$ is that, for $t > t_s$, because of very small value of the term exp $(-st)$, contribution by the subsequent terms is negligible while evaluating the integral value. Also, we have

$$Y^1(0) = G(0)U^1(0) \tag{13.3}$$

$$Y^1(j\omega) = G(j\omega)U^1(j\omega) \tag{13.4}$$

$$Y^1(s_1) = G(s_1)U^1(s_1) \tag{13.5}$$

Since these equations are vector equations, they can not be solved to get $G(0)$, $G(j\omega)$ and $G(s)$. Hence, one more ('m' tests for $m \times m$ system) relay feedback test, by slightly varying the relay parameter is performed as described by Ganesh and Chidambaram (2003). From the first test, vectors of $U^1(0)$, $Y^1(0)$, $U^1(j\omega)$ and $Y^1(j\omega)$ are calculated. Also vectors $U^1(s_1)$ and $Y^1(s_1)$ are calculated from Eqs (13.1) and (13.2). A MATLAB program is written to calculate value of $U(s_1)$ and $Y(s_1)$. This is repeated for all 'm' tests. $G(0)$, $G(j\omega)$ and $G(s_1)$ for 2×2 system are calculated as

$$\left[Y^1(0)\ Y^2(0)\right] = G(0)\left[U^1(0)\ U^2(0)\right] \tag{13.6}$$

$$\left[Y^1(j\omega)\ Y^2(j\omega)\right] = G(j\omega)\left[U^1(j\omega)\ U^2(j\omega)\right] \tag{13.7}$$

$$\left[Y^1(s_1)\ Y^2(s_1)\right] = G(s_1)\left[U^1(s_1)\ U^2(s_1)\right] \tag{13.8}$$

13.3 Estimation of Model Parameters

SOPTD model can be expressed as

$$g(s) = \frac{y(s)}{u(s)} = \frac{k_P \exp(-\theta s)}{(\tau_1 s+1)(\tau_2 s+1)} \tag{13.9}$$

where 'k_p' (process gain), 'θ' (time delay), 'τ_1' and 'τ_2' (time constants), are the parameters to be estimated. For this model, the frequency response can be written by substituting $s = j\omega$ as

$$g(j\omega) = \frac{k_P \exp\left(-\theta(j\omega)\right)}{\left(\tau_1 j\omega+1\right)\left(\tau_2 j\omega+1\right)} \tag{13.10}$$

From the frequency response matrix $g(j\omega)$, for each element of transfer function matrix is known and is written as

$$g(j\omega) = a + jb \tag{13.11}$$

Equating the real and imaginary parts of Eqs (13.10) and (13.11) and after simplifications, we get

$$a\left(1-\tau_1\tau_2\omega^2\right) - b\omega\left(\tau_1+\tau_2\right) - k_p\cos\left(\theta\omega\right) = 0 \tag{13.12}$$

$$b\left(1-\tau_1\tau_2\omega^2\right)+a\omega\left(\tau_1+\tau_2\right)+k_p\cos\left(\theta\omega\right)=0 \tag{13.13}$$

Also we have,

$$g\left(s_1\right)=\frac{k_P\exp\left(-\theta s_1\right)}{\left(\tau_1 s_1+1\right)\left(\tau_2 s_1+1\right)} \tag{13.14}$$

The value of k_p for each element is taken from the steady-state gain matrix given by Eq. (13.3). Equations (13.12), (13.13) and (13.14) can be solved to get the model parameters τ_1, τ_2 and θ. MATLAB routine is used to solve Eqs (13.12), (13.13) and (13.14) to get the model parameters.

Measurement noise is an important issue in the identification problem. Filtering the noise from the corrupted signal by using a first order filter is a recommended one (Åström and Hägglund, 1984). The measurement noise is usually of high frequency while for the controller design the process frequency of interest is usually in the low frequency region. Hence, a low pass filter can be employed to reduce the measurement noise. Shen et al. (1996a) recommended adopting more oscillation time (periods) in calculating the static gain and critical points of the process.

13.4 Comparison of Closed Loop Performance

The IMC controller is designed by the method suggested by Tanttu and Lieslehto (1991) for the identified model to compare the closed loop performance as described by Ganesh and Chidambaram (2003). Controller ($k_{c,ij}$) for each scalar transfer function ($g_{p,ji}$) is first designed. For the SOPTD model, PI controller settings are given by

$$\left(k_p k_c\right)_{ij}=\left[2\left(\tau_1+\tau_2\right)+\theta\right]/2\lambda \tag{13.15a}$$

$$\tau_{I,ij}=\left(\tau_1+\tau_2\right)+0.5\theta \tag{13.15b}$$

where $\lambda\left[\lambda\geq 0.2\left(\tau_1+\tau_2\right)\right]$ is the tuning parameter. Multivariable PI controllers are designed for the identified model by the equations given by Ganesh and Chidambaram (2003). Closed loop performance of the actual and identified models for above designed controller is compared. ISE values are also compared.

13.5 Simulation Examples

Example 1: The transfer function of a process is given by

$$G\left(s\right)=\begin{bmatrix}\dfrac{2e^{-3s}}{1+7s} & \dfrac{e^{-s}}{1+3s}\\ \dfrac{\left(s+1\right)e^{-s}}{\left(s^2+4s+5\right)} & \dfrac{e^{-7s}}{1+1.5s}\end{bmatrix} \tag{13.16}$$

For this process, in the first relay test, relay in loop 1 is a bias with switching levels 1.2 and –1 while relay in loop 2 is an ideal relay with switching levels 0.2 and –0.2. The oscillation frequency in both the outputs is found to be same ($\omega = 0.57$). From the process input and output data, the values of $U^1(0)$, $Y^1(0)$, $U^1(j\omega)$ and $Y^1(j\omega)$ are determined. In the second relay test, the switching levels in loop 1 are changed to 1.5 and –1.2. The oscillation frequency in test 2 is nearly same as in test 1. From the process input and output data, $U^2(0)$, $Y^2(0)$, $U^2(j\omega)$ and $Y^2(j\omega)$ are determined. From the Eqs (13.6) and (13.7), the steady-state gain matrix and frequency response matrix are calculated. In both the tests, at $t_s = 60$, all the outputs have few repeated cycles of oscillation. So at $s_1 = 8/t_s$, $U(s_1)$ and $Y(s_1)$ are also calculated. $G(0)$ and $G(j\omega)$ are calculated as

$$G(0) = \begin{bmatrix} 1.9733 & 1.0739 \\ 0.2026 & 0.9979 \end{bmatrix} \tag{13.17}$$

$$G(j\omega) = \begin{bmatrix} -0.479 - 0.0371j & -0.0567 - 0.4258j \\ 0.1836 - 0.1085j & -0.0696 + 0.7365j \end{bmatrix} \tag{13.18}$$

In both the tests, at $t_s = 60$, all the outputs have few repeated cycles of oscillation. So at $s_1 = 8/t_s = 0.1333$, $U(s_1)$ and $Y(s_1)$ are also calculated from Eqs (13.1), (13.2) and (13.8). MATLAB package is used to calculate $U(s_1)$ and $Y(s_1)$

$$U(0.1333) = \begin{bmatrix} 5.5109 & 9.4652 \\ -1.3624 & -1.0793 \end{bmatrix} \tag{13.19}$$

$$Y(0.1333) = \begin{bmatrix} 2.9732 & 5.8964 \\ 0.5387 & 1.3406 \end{bmatrix} \tag{13.20}$$

$$G(0.1333) = \begin{bmatrix} 0.6944 & 0.6265 \\ 0.1792 & 0.3295 \end{bmatrix} \tag{13.21}$$

Now, if we try to model each element of transfer function matrix as an FOPTD by the procedure given in Chapter 12, then for $g_{21,}$ we get a negative time constant ($\tau = -0.1484$ and $\theta = 0.9364$). Hence, this transfer function has to be identified by the SOPTD model. By using Eq.(13.8), identified transfer function matrix is obtained as

$$G(s) = \begin{bmatrix} \dfrac{1.9733e^{-3.05s}}{1+6.98s} & \dfrac{1.07e^{-s}}{1+4.02s} \\ \dfrac{0.6338e^{-0.385s}}{(s^2+1.7411s+3.375)} & \dfrac{0.9979e^{-6.8s}}{1+1.595s} \end{bmatrix} \tag{13.22}$$

The multivariable PI controller is designed by the Tanttu and Lieslehto method. For the identified model, the resulting multivariable PI controller is given by

$$K_c = \frac{1}{\lambda}\begin{bmatrix} -15.9051 & 16.3986 \\ 21.1771 & -17.179 \end{bmatrix} \tag{13.23}$$

$$K_I = \frac{1}{\lambda}\begin{bmatrix} 0.5638 & -0.5239 \\ -0.1129 & 1.0361 \end{bmatrix} \tag{13.24}$$

The tuning parameter λ is selected by trial and error method as 60. The performance of the actual and identified models is compared. Figure 13.1 shows performance of the actual and the identified models to be very close to each other. Table 13.1 gives the ISE values comparison. The ISE value comparison also shows that the identified model is very close to the actual system.

Table 13.1 *Closed loop ISE values comparison between the identified model and actual system for example 1*

Step change in	ISE values for actual model		ISE values for identified model	
	Y1	Y2	Y1	Y2
Y1	47.74	4.36	46.68	5.34
Y2	5	54.38	5.72	54.07

The method proposed in this chapter identifies individual element of transfer function matrix as SOPTD models without any additional tests. Also it is illustrated that, some higher order processes, when identified as an FOPTD model, gives negative time constant. So higher order modeling is necessary for these processes. The proposed method also does not require any *a priori* knowledge of process.

Example 2: The transfer function matrix for the actual system is considered as

$$G(s) = \begin{bmatrix} \dfrac{(2s+1)e^{-10s}}{s^2+2s+1} & \dfrac{e^{-0.5s}}{s^2+3s+4} \\ \dfrac{(s+1)e^{-s}}{s^2+4s+5} & \dfrac{(s+2)e^{-8s}}{s^2+2s+2} \end{bmatrix} \tag{13.25}$$

For this process, in the first relay test, relay in loop 1 is bias one with switching levels 1.5 and −1 while relay in the loop 2 is an ideal relay with switching levels 0.2 and −0.2. The oscillation frequency in both the outputs is found to be same ($\omega = 0.304$). From the process input and output data, $U^1(0)$, $Y^1(0)$, $U^1(j\omega)$ and $Y^1(j\omega)$ are determined. In the second relay test, the switching levels in loop 1 are changed to 1.8 and −1.2. The oscillation frequency in test 2 is nearly the same as in test 1. From the process input and output data, $U^2(0)$, $Y^2(0)$, $U^2(j\omega)$ and $Y^2(j\omega)$ are determined. From Eqs (13.6) and (13.7), steady-state gain matrix and frequency response matrix are calculated. In both the tests, at $t_s = 60$, all the outputs have few repeated cycles of oscillation. Using $s_1 = 8/t_s$, $U(s_1)$ and $Y(s_1)$ are calculated. We get $G(0)$ and $G(j\omega)$ as

$$G(0) = \begin{bmatrix} 0.9993 & 0.2383 \\ 0.2 & 1.0015 \end{bmatrix} \tag{13.26}$$

$$G(j\omega) = \begin{bmatrix} -1.0673 - 0.0587j & 0.2404 - 0.0658j \\ 0.2 - 0.052j & 0.8608 - 0.5336j \end{bmatrix} \tag{13.27}$$

In both the tests, at $t_s = 60$, all the outputs have few repeated cycles of oscillations. So at $s_1 = 8/t_s = 0.1333$, $U(s_1)$ and $Y(s_1)$ are also calculated from Eqs (13.1), (13.2) and (13.8). We get the values of $G(0.1333)$ as

$$G(0.1333)=\begin{bmatrix} 0.2598 & 0.21183 \\ 0.1786 & 0.3211 \end{bmatrix} \tag{13.28}$$

The identified transfer function matrix is given by

$$G(s)=\begin{bmatrix} \dfrac{e^{-8.76s}}{1.589\,s^2+1.26s+1} & \dfrac{(0.2386)e^{-0.08s}}{0.766\,s^2+0.7572\,s+1} \\ \dfrac{(0.2)e^{-0.086s}}{0.6199\,s^2+0.7205\,s+1} & \dfrac{e^{-8.1s}}{0.2154\,s^2+0.394\,s+1} \end{bmatrix} \tag{13.29}$$

The multivariable PI controller is designed based on the identified model as

$$K_c=(1/\lambda)\begin{bmatrix} -5.8396 & 7.7671 \\ 6.7989 & -4.5989 \end{bmatrix} \tag{13.30}$$

$$K_I=\begin{bmatrix} 1.0504 & -0.2506 \\ -0.2111 & 1.0504 \end{bmatrix} \tag{13.31}$$

The tuning parameter (λ) is selected by trial and error as 30. The closed loop responses are obtained by using the SIMULINK package for a set point change in the set point of y_1 first and separately in y_2. Both the model and actual system responses match very well (refer to Fig. 13.2). The ISE values are calculated and are given in Table 13.2. A very closed match is obtained between the performance of the same controller on the process and on the model.

Table 13.2 *Closed loop ISE values comparison for the actual and identified models for example* 2

Step change in	ISE values for actual model		ISE values for identified model	
	Y1	Y2	Y1	Y2
Y1	27.107	1.8417	27.18	1.86
Y2	2.77	24.004	2.81	23.99

13.6 Conclusions

Asymmetric relay feedback method for the scalar system is extended to a multivariable system. Decentralized relay feedback tests are carried out. The method identifies each element of the transfer function matrix as either an FOPTD or SOPTD model without any *a priori* knowledge of the model parameters such as time delay or steady-state gain. It is shown that for some higher order processes when identified as FOPTD model gives a negative time constant. Modeling the system as an SOPTD gives a better performance. Simulation results on two examples show that the method gives identified model giving a similar closed performance as that of the actual system with the same controllers.

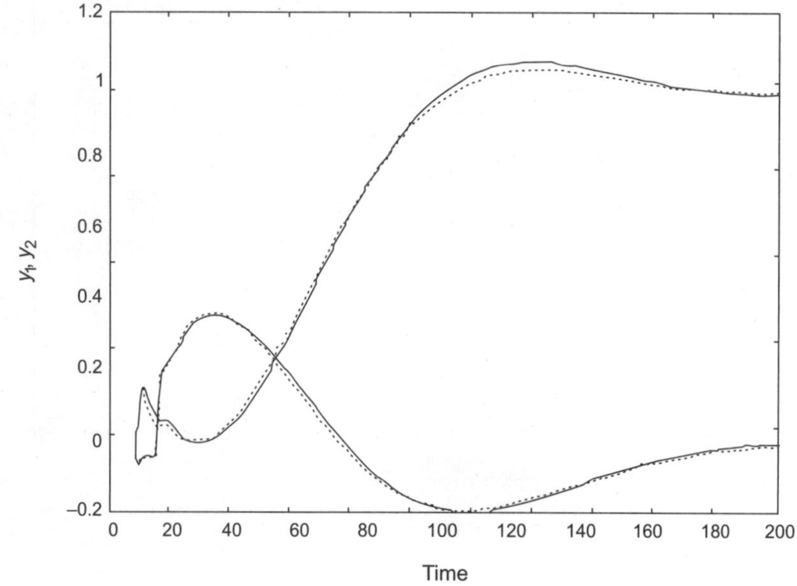

(A)

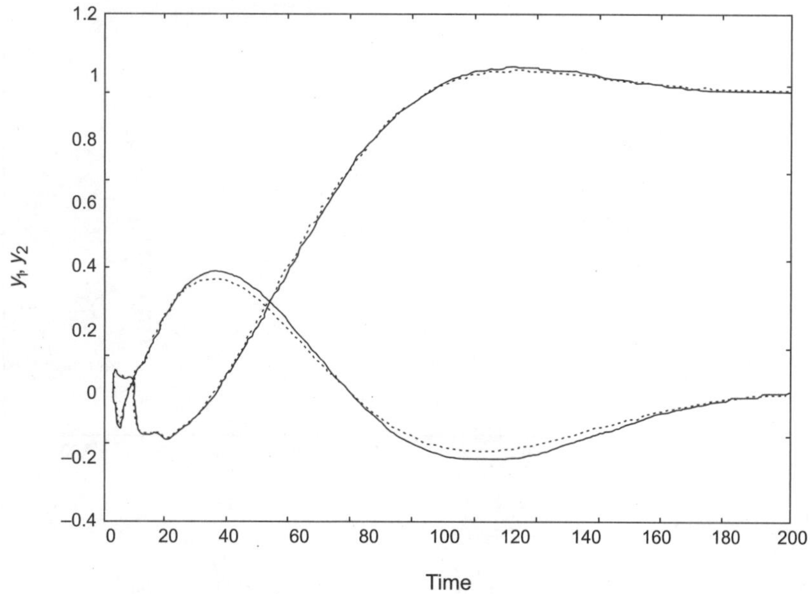

(B)

Fig. 13.1 *Comparison of performance of actual and identified models for example* 1
Legend: upper curves–response; lower curves–interaction solid-actual model; dot–identified model (A-step change in Y1, B-step change in Y2)

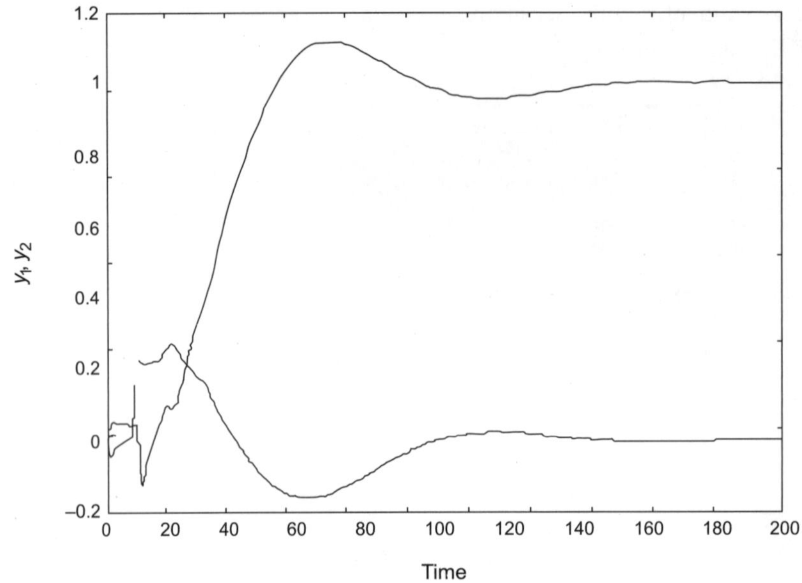

(A)

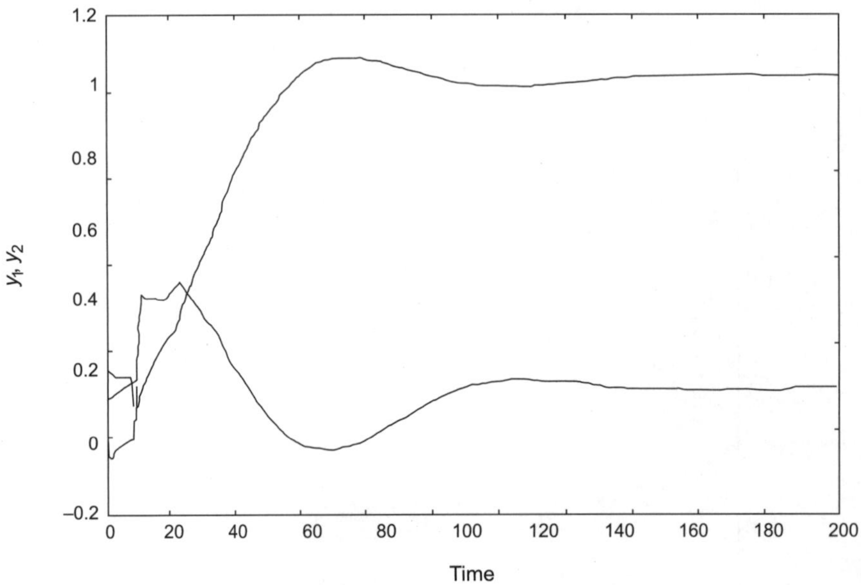

(B)

Fig. 13.2 *Comparison of performance of actual and identified models for example 2*
Legend: upper curves–response; lower curves–interaction solid-actual model; dot–identified model (A-step change in Y1, B-step change in Y2)

14 Tuning of Multivariable Controllers for Non-Minimum Phase Systems

The system with one or more right half plane transmission (RHPT) zero is generally called as a non-minimum phase system. In this chapter, two methods of designing multivariable controllers for the systems with RHPT zeros are compared (Reddy et al., 2006). The methods used are decoupled internal model controller (IMC) (Wang et al., 2002) and a simple tuning method (Davison, 1976). In decoupled IMC method (Wang et al., 2002), controller design procedure is developed with the help of a model reduction theorem. Davison (1976) proposed a simple tuning method, which is based on the inverse of the steady-state gain matrix $[G(0)]$. Previously the method was applied to multivariable minimum phase systems only. In the present work, Davison method is applied to systems with multivariable right half plane zero. The decoupled IMC method (Wang et al., 2002) involves some complex calculations in designing the controllers whereas the Davison method (1976) is simple to apply. Simulation results are given for the following 2×2 transfer function matrix of the systems: (i) four-tank system, a benchmark example of multivariable system (Johansson, 2000) (ii) system (Morari et al., 1987) with individual and multivariable transmission zero at $s = 1$ (iii) binary distillation column (Wang et al., 2002). The performance comparisons are given by the sum of IAE values based on response and interaction both for servo and regulatory problems. The simple tuning method gives an improved performance.

14.1 Introduction

Most of the industrial processes are multivariable in nature. In a multi-input and multi-output (MIMO) system, interaction and location of transmission zero are also important in addition to responses. Depending on the location of transmission zero, the system exhibits minimum or non-minimum phase characteristics that arise completely from the multivariable nature of the system. Multivariable zeros play an important role in deciding the controllability of such systems. According to Skogestad and Postlethwaite (1996), 'Multivariable zeros are those points at which process transfer function losses rank below its normal rank, i.e., det $G(s) = 0$'. Thus, if the system does not have any individual transfer function zero, even then it is possible for the system to have a multivariable zero. If one or more zeros of the system should lie in the right half plane, the system is known as a 'non-minimum phase' system. Stability of any open loop system depends on the pole location, whereas the open loop zero location limits the value of the controller gain that could be used. The RHP zeros close to the origin of the complex plane affect dramatically the closed loop performance. The RHP zeros exhibit properties like, some sort of inverse response, some sort of transmission blocking and most of all they must be invariant under feedback control and become the poles of inverse. Multivariable zeros that exhibit these properties are called RHP transmission zeros. Owing to the presence of RHPT zeros, it is difficult to control the non-minimum phase systems.

In the internal model control (IMC) framework (Morari et al., 1987), it was shown that perfect control depends on the inevitability of the plant transfer function matrix. The presence of non-minimum phase elements, i.e., RHPT zeros, limits the ability for perfect control since the inverse of the process is not realizable. Holt and Morari (1985) showed that in MIMO case, the RHPT zeros are characteristic only of square systems. They were identified in chemical engineering systems such as simple distillation columns (Wang et al., 2002), binary distillation column (Rosenbrock, 1962), boilers (Szita and Sanathanan, 2000), grinding circuit system (Pomerleau et al., 2000), etc.

Three types of control strategies can be applied to non-minimum phase MIMO systems: decoupling, decentralized and centralized. Few reported methods are available for non-minimum phase systems to design multivariable controllers such as frequency-domain method (Szita and Sanathanan, 2000), simple tuning method (Pomerleau and Pomerleau, 2001), decoupling internal model control method (Wang et al., 2002), etc. Some of these methods involve complicated control strategies and lengthy calculations. Some simple methods are reported in literature such as simple tuning method proposed by Davison (1976) for minimum phase systems. Control of such a system needs the implementation of simple approach to design the multivariable controllers. Davison (1976) suggested the method but no simulation application was given. These methods were applied to the physical examples. In this work, Davison method is tested for multivariable non-minimum phase systems. A method of selecting the tuning parameters is also given.

In this chapter, a more rigorous decoupled internal model control method proposed by Wang et al., (2002) and the simple tuning method proposed by Davison (1976) are compared on three examples. The transfer function matrix model considered are (i) four tank control system, (ii) example considered by Morari et al. (1987) and (iii) Wood and Berry distillation column model. Section 14.2 gives a brief view of these methods. In section 14.3, a comparison criterion for controller performance is discussed. Simulation results of the examples considered are presented along with ISE values in section 14.4.

14.2 Controller Design Methods

14.2.1 Simple tuning method

In the case of systems with multivariable zero(s), some portion of the process model cannot be inverted that include RHP zeros. To overcome this problem, controller design method, which is not based on the inversion principle or which takes in to account only the invertible portion of the steady-state model, can be used. Davison (1976) proposed a simple tuning method to design the multivariable centralized PI controllers, which is based on inverse of the steady-state gain matrix [G(0)]. Previously the method was applied to multivariable minimum phase systems only. Here Davison method is applied to systems with multivariable right half plane zero. In this method the tuning parameters are selected as

$$K_c = \delta * \left[G^{-1}(0)\right] \tag{14.1}$$

$$K_I = \varepsilon * \left[G^{-1}(0)\right] \tag{14.2}$$

K_c is the proportional gain matrix and K_I is the integral gain matrix. $G^{-1}(0)$ is known as rough tuning matrix. Here δ and ε are the fine-tuning parameters that range in the interval (0,1). Davison (1976) did not give any method to select these tuning parameters. In the range of values for the tuning parameters (0 to 1), beyond a certain value, the closed loop system will not be stable. In the present work, the tuning parameters are selected from the characteristic equation $|I + KG|$, by applying Routh test for stability. (Pade's approximation is to be used for the time delay terms.) The values obtained are divided by 2 and around these values the parameters are tuned by simulation so as to get a better closed loop response. Another approach is based on the detuning method. In any multivariable system, after designing individual controllers, they must be detuned by using a common detuning factor (F). This is to take care of the loop interactions caused by multiple inputs by compromising on performances. The individual controller gains should be divided by F and the corresponding integral time (τ_I) is to be multiplied by F. Since K_I is given by K_c/τ_I, the values of the tuning parameters are given by ($1/F$) and ($1/F^2$). The factor F varies from 1.5 to 5 depending on the interaction of the given multivariable system. By extending this approach to the Davison method, the values of δ and ε can be considered as ($1/F$) and ($1/F^2$). Around these values, these parameters can be tuned by simulation so as to get a better performance.

14.2.2 Decoupled internal model controller method

Wang et al., (2002) proposed a method to design multivariable centralized controllers. This controller design method is a new approach to the IMC analysis and the controller design is proposed for decoupling and stabilizing of multivariable stable process. The characteristics of all the stabilizing IMC controllers that solve this decoupling problem and resulting closed-loop systems are given in terms of the open-loop system time delays and non-minimum phase zeros.

Decoupling IMC system design

For multivariable decoupling IMC control, the closed loop transfer function $H(s)$ in case of no process-model mismatch is H = diag $\{h_{ii}\}$ = diag $\{\tilde{g}_{ii}, k_{ii}\}$ and thus the key point in design lies in specifying the m decoupled loop transfer function $h_{ii}(s)$. Unlike the SISO case, $h_{ii}(s)$ cannot be chosen based solely on $\tilde{g}_{ii}(s)$ as this may cause an unrealizable and/or unstable controller. For best performance of the IMC system and simplicity of the controller it is undesirable to include any time delays and non-minimum phase zeros in $h_{ii}(s)$ more than necessary, that is, only the minimum time delay and minimum multiplicity of unavoidable non-minimum phase zeros is to be specified. This leads to

$$h_{ii} = e^{-(\tau|G|-\tau_i)s} \prod_{z \in Z^{+}_{|G|}} \left(\frac{z-s}{z+s} \right)^{\eta_z(|G|)-\eta_i(z)} \tag{14.3}$$

where $f_i(s)$ is the ith loop IMC filter,

$$\tau_i \cong \min \tau(G_{ii}) \tag{14.4}$$

$$\eta_i(z) \cong \min \eta_z(G_{ii}) \tag{14.5}$$

The form of IMC filter $f_i(s)$ is usually chosen as, but not restricted to, $f_i(s) = 1/(T_i s + 1)^{ri}$, where T_i and r_i are the filter time constant and order, respectively. The filter time constant determines the bandwidth of the closed loop system and thus serves as a tuning parameter for performance and robustness. The filter order of this low pass filter should be chosen to provide sufficient high frequency roll-off rate required by the properness of the controller as well as to ensure the robustness of the system.

The design of $h_{ii}(s)$ will determine $k_{ii}(s)$ as

$$k_{ii}(s) = h_{ii}(s) * \tilde{g}_{ii}^{-1}(s) \tag{14.6}$$

where $\tilde{g}_{ii}(s) = |G|/G_{ii}$

Which in turn yield off-diagonal elements $k_{ji}(s)$ uniquely by

$$k_{ji}(s) = -(G_{ij}/G_{ii}) * k_{ii}(s), \ \forall i, j \in m, j \neq i \tag{14.7}$$

For a multivariable process, IMC controller designed by the above procedure usually results in a complicated higher order form, which is difficult to implement. So a model reduction method is used to find a second order rational transfer function plus dead time model (Rangaiah and Krishnaswamy, 1994), which is much simpler yet good approximation to the theoretical controller.

$$|\hat{G}| = \frac{[b_2 s^2 + b_1 s + G(0)]}{[a_2 s^2 + a_1 s + 1]} e^{-Ls} \tag{14.8}$$

By the above method, $\hat{k}_{ii}(s)$ and $\hat{k}_{ji}(s)$ are calculated. According to Wang et al. (2002) the reduced order of the model Eq.(14.8) has been chosen as to cover a wide range of dynamics and yet to be low enough for economic implementation. The filter time constant should be greater than 0.2 times the dominant time constant of the open loop system. In the present work, the filter time constants are obtained by tuning around 0.4 to 0.8 times of the open loop time constant so as to get a better performance of the closed system.

14.3 Comparison Criterion of Controllers Performance

There are three ways to decide the controller criterion: integral of absolute error (IAE), integral square error (ISE) and integral of time weighted absolute error (ITAE). In multivariable systems, different criteria have been considered to compare the simulation results. Stephanopoulos (1984) discussed that ISE should be used to suppress large errors, while IAE should be used to suppress small errors and to suppress errors that persist for long times, ITAE criterion should be used:

$$\text{IAE} = \int_0^\infty \left| y_s(t) - y(t) \right| dt \tag{14.9}$$

$$\text{ISE} = \int_0^\infty \left[y_s(t) - y(t) \right]^2 dt \tag{14.10}$$

In the multivariable system, most of the controller settings give comparable responses. To distinguish the responses, IAE or ISE values should be compared. One important thing is that while comparing two controller settings, the ISE values in response may be lower but at the same time it may be higher in interaction. In this case, the criterion used by Lin et al. (1999) is better, which asks for minimization of sum of the IAE or ISE values in response and interaction. Frank (1974) showed that the relationship between the location of the zero and loss of performance is consistent. From the theorem proposed by him the value of ISE to a unit step in the set point is given by

$$\text{ISE} = \sum \frac{2}{Z_i}, \quad \text{for RHP zeros at } Z_i \tag{14.11}$$

This important result quantifies the impact of an RHP zero on the dynamic resilience. It indicates that the deterioration in the quality of control (as measured by ISE) is inversely proportional to the location of the RHP zero. Zeros close to the origin lead to a greater loss than those far away.

14.4 Examples

14.4.1 Example 1

To compare the multivariable controllers designed by using the methods discussed, following four-tank liquid level system (Johansson, 2000), which is a benchmark example of non-minimum phase system is considered.

$$G(s) = \begin{bmatrix} \dfrac{1.5}{(63s+1)} & \dfrac{2.5}{(63s+1)(39s+1)} \\ \dfrac{2.5}{(91s+1)(56s+1)} & \dfrac{1.6}{(91s+1)} \end{bmatrix} \tag{14.12}$$

This system does not have any individual positive zero, but it has one multivariable RHP zero located at 0.013.

Decoupling IMC method

Step (i): The determinant calculated from the above system gives the location of zeros. The application of reduced order method to $|G|$ gives $|\hat{G}|$ = $(5241.6s^2 + 228s - 3.85)^*\ \exp(-50s)/(11886s^2 + 200.98s + 1)$, it has a non-minimum phase zero at $z = 0.013$. From Eqs (14.4) and (14.5), we get $\tau_1 = 0$ and $\eta_1(z) = 0$.

Step (ii): From Eq.(14.3), with filter chosen as $1/(54s+1)$ we have $h_{11} = [(s-0.013)/(s+0.013)(54s+1)]e^{-50s}$. k_{11} is obtained from Eq.(14.6). The application of reduced order theorem to k_{11} yields $\hat{k}_{11}$. Making use of k_{11}, the value k_{21} is evaluated from Eq.(14.7) and then $\hat{k}_{21}$.

Step (iii): Loop two is treated similarly, where h_{22} is chosen as $h_{22} = [(s-0.013)/(s-0.013)(36s+1)]e^{-50s}$ with filter as $1/(36s+1)$ and controllers $\hat{k}_{22}$ and $\hat{k}_{12}$ are evaluated from Eqs (14.6) and (14.7). This results in overall controller as

$$K = \begin{bmatrix} \dfrac{-1021.13s^2 - 42.39s - 0.4186}{1361s^2 + 94.6s + 1}e^{-50s} & \dfrac{3309s^2 + 95.46s + 0.6494}{2917s^2 + 119.83s + 1}e^{-63.7s} \\ \dfrac{1595s^2 + 66.23s + 0.6494}{5398s^2 + 133.44s + 1}e^{-60.6s} & \dfrac{-1985s^2 - 57.27s - 0.3896}{1361.3s^2 + 94.6s + 1}e^{-50s} \end{bmatrix} \quad (14.13)$$

Three values for the filter time constants are considered (36, 54, 72) for step (ii) and three values for the filter time constants (24, 36, 48) are considered in step (i). For these filter time constants, the controller transfer function matrix is calculated. From the closed loop simulation, the sum of ISE values for the servo problem is found lowest for the filter time constants (54, 36). The responses with these filter constant values are given in Figs 14.1 to 14.4.

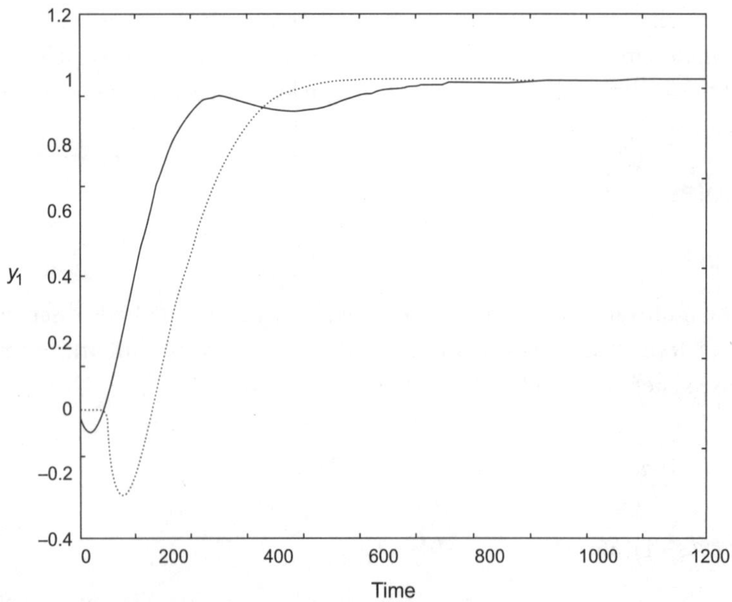

Fig. 14.1 *Control performance of four-tank system for unit step change in y1*
Legend: solid–simple tuning method; dot–decoupled IMC

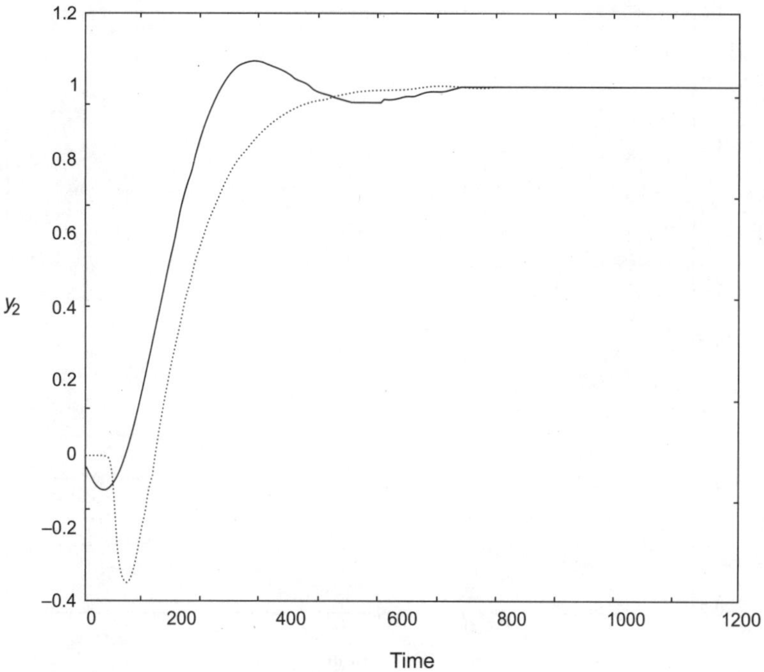

Fig. 14.2 *Control performance of four-tank system for unit step change in y2*
Legend: solid–simple tuning method; dot–decoupled IMC

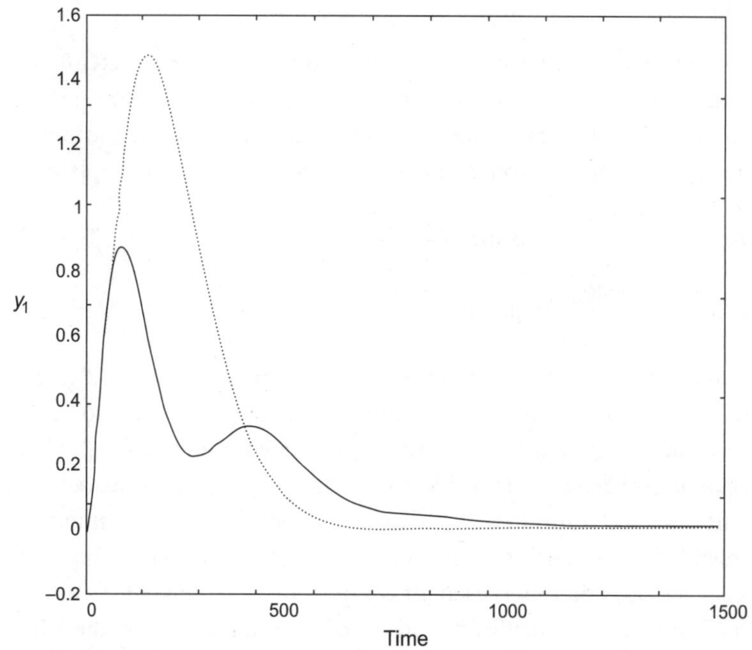

Fig. 14.3 *Control performance of four-tank system for unit step change in V1*
Legend: solid–simple tuning method; dot–decoupled IMC

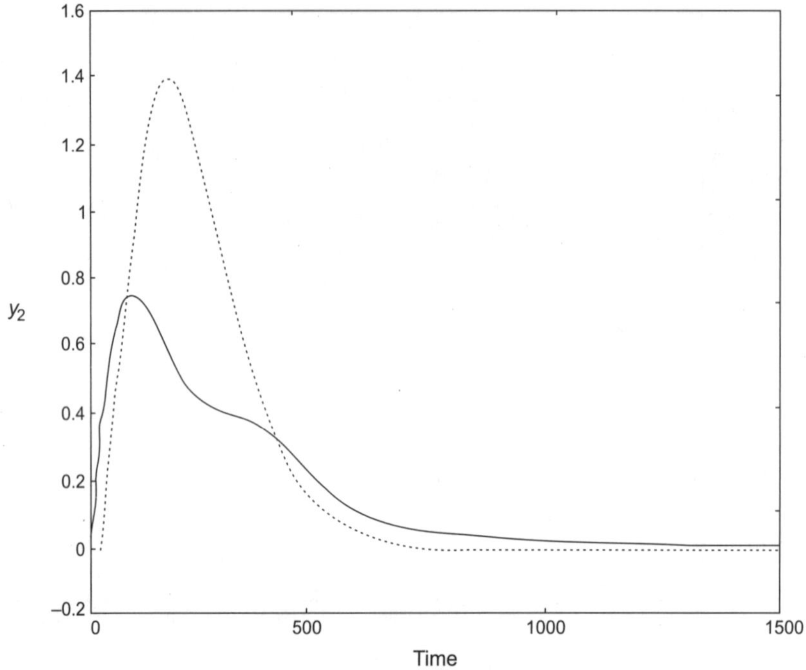

Fig. 14.4 *Control performance of four-tank system for unit step change in V2*
Legend: solid–simple tuning method; dot–decoupled IMC

Simple tuning method

In this rough tuning matrix $G^{-1}(0)$ is obtained from steady-state gain matrix $G(0)$. From the Routh array test, the values of δ and ε for stability are less than 1.0 and 0.025, respectively. By tuning around half of these values we get by simulation of the closed loop system the tuning parameter as $\delta = 0.72$ and $\varepsilon = 0.0062$. The controller gain matrix is then given by a simple PI form as

$$K = \begin{bmatrix} -0.2992 - \dfrac{0.00258}{s} & 0.4675 + \dfrac{0.00403}{s} \\ 0.4675 + \dfrac{0.00403}{s} & -0.2805 - \dfrac{0.00241}{s} \end{bmatrix} \tag{14.14}$$

The servo and regulatory responses of the controlled variables of the resulting controllers are shown in Figs 14.1 to 14.4. The load transfer function matrix is assumed to be the same as that of the process transfer function matrix. ISE values calculated from Eq.(14.10) for both servo and regulatory problems are reported in Table 14.1. These values are compared for performance evaluation of the designed controllers. It is clear from Table 14.1 that the interaction is lower in case of decoupled IMC method but the response is better in case of the simple tuning method. Here the Lin et al., (1999) criterion is used for comparison. The sum of ISE values for the simple tuning method is smaller than that of the proposed decoupled IMC method. This shows that simple tuning method is giving better performance than that of decoupled IMC method.

In order to evaluate the robustness of these two methods, we increased the steady-state gains of the process transfer function separately by 10%, 15% and 20%. The resulting ISE values calculated are given in Table 14.2. These values show that the controller designed by the simple tuning method is still giving a better robust performance.

Table 14.1 *ISE values of servo and regulatory problem for four-tank liquid level system*

Method	Servo problem			
	Step in	ISE values		Sum of ISE values
		*Y*1	*Y*2	
Decoupled IMC	*Y*1	227.84	0.09	227.93
	*Y*2	0.23	219.61	219.84
Simple tuning	*Y*1	106.02	12.08	118.1
	*Y*2	9.38	140.41	149.79
	Regulatory problem			
Method	Step in	ISE values		Sum of ISE values
		*Y*1	*Y*2	
Decoupled IMC	*V*1	319.92	795.13	1187.05
	*V*2	1019.9	364.04	1383.95
Simple tuning	*V*1	110.63	374.09	484.12
	*V*2	329.59	117.63	447.21

Table 14.2 *ISE values for robustness analysis (perturbation in all the steady-state gains)*

Method	Step in	ISE values		
		10%	15%	20%
Decoupled IMC	*Y*1	224.86	224.62	225.1
	*Y*2	217.02	216.9	217.65
Simple tuning	*Y*1	117.48	118.3	119.94
	*Y*2	151.42	153.5	156.91

14.4.2 Example 2

The system considered by Morari et al. (1987) had the following transfer function matrix with both individual and multivariable RHP zeros located at 1:

$$G(s)=\begin{bmatrix}\frac{(-s+1)}{(s+2)} & \frac{(-s+1)}{(s+2)}\\ \frac{1}{(s+2)} & \frac{2}{(s+2)}\end{bmatrix} \tag{14.15}$$

It follows from the decoupled IMC design procedure that with filters chosen as $f_1(s) = f_2 = 1/(0.125s+1)$, the overall controller is obtained as

$$K = \begin{bmatrix} \dfrac{(2s+4)}{(0.125s^2+1.125s+1)} & \dfrac{(s^2+s-2)}{(0.125s^2+1.125s+1)} \\ \dfrac{(-s-2)}{(0.125s^2+1.125s+1)} & \dfrac{(-s^2-s+2)}{(0.125s^2+1.125s+1)} \end{bmatrix} \tag{14.16}$$

The three sets of filter constants considered are (0.125, 0.125), (0.25, 0.25) and (0.375, 0.375). From the closed loop servo response it is found that the set (0.125, 0.125) gives the lowest sum of ISE values. Hence, the results with these filter time constants are considered for comparison with the simple tuning method. From the Routh array test, the values of δ and ε for stability are found to be less than 0.5 and 0.72. As discussed earlier, by tuning around 1/2 of these values, we select the values of the tuning parameters as $\delta = 0.3$ and $\varepsilon = 0.7$. The resultant controller is obtained as

$$K = \begin{bmatrix} 1.2 + \dfrac{2.8}{s} & -0.6 - \dfrac{1.4}{s} \\ -0.6 - \dfrac{1.4}{s} & 0.6 + \dfrac{1.4}{s} \end{bmatrix} \tag{14.17}$$

14.4.3 Example 3

The Wood and Berry process for binary distillation column studied by Wang et al. (2002) had the following transfer function matrix:

$$K = \begin{bmatrix} \dfrac{12.8e^{-s}}{(16.7s+1)} & \dfrac{-18.9e^{-9s}}{(21s+1)} \\ \dfrac{6.6e^{-2s}}{(10.9s+1)} & \dfrac{-19.4e^{-15s}}{(14.4s+1)} \end{bmatrix} \tag{14.18}$$

This system has a multivariable positive zero located at 0.136. From decoupled IMC system with filters chosen as $f_1(s) = f_2(s) = 1/(s+1)$ the overall controller is obtained as

$$K = \begin{bmatrix} \dfrac{0.3656s^2+0.0524s+0.002}{s^2+0.2265s+0.0124}e^{-13.1s} & \dfrac{-0.2375s^2-0.0294s-0.0011}{s^2+0.179s+0.007}e^{-8s} \\ \dfrac{0.1551s^2+0.0291s+0.0012}{s^2+0.2726s+0.0226} & \dfrac{-0.2003s^2-0.0317s-0.0014}{s^2+0.2215s+0.0136} \end{bmatrix} \tag{14.19}$$

From the Routh array test, the values of δ and ε for stability are found to be less than 1.0 and 0.067. As discussed earlier, by tuning around 1/2 of these values, we select the values of the tuning parameters as $\delta = 0.5$ and $\varepsilon = 0.03$. The resultant controller is obtained as

$$K = \begin{bmatrix} 0.0785 + \dfrac{0.0047}{s} & -0.0765 - \dfrac{0.00459}{s} \\ 0.0266 + \dfrac{0.00159}{s} & -0.0518 - \dfrac{0.00311}{s} \end{bmatrix} \tag{14.20}$$

For examples 2 and 3, the performance of the resultant controllers by simple tuning method and decoupled IMC method are compared. The calculated ISE values are reported in Table 14.3a and Table 14.3b for servo and regulatory problems respectively. These values also show that the simple tuning method gives a better performance over that of the decoupled IMC method. Similar results as that of example 1 are observed in examples 2 and 3 for robustness study also.

In IMC method, the resulting structure of the controller is not of PI form. In simple method, we assume the structure of the controller is of PI form. The controller gains (K_c and K_I) are assumed to be proportional to the inverse of the steady-state gain of the process. The controller system has two tuning constants. The controller designed by the simple method gives an improved performance than that of the IMC method.

14.5 Conclusions

In this chapter, a simple tuning method proposed by Reddy et al. (2006) as applied to multivariable non-minimum phase systems is reviewed. Performance of the controller designed is compared using ISE values with that of the decoupled IMC method proposed by Wang et al. (2002). Simulation examples have illustrated that simple tuning method gives an improved performance than that of the decoupled IMC method. Yet, the resulting controller obtained by simple tuning method is in the form of simple PI and easy to implement. The controller obtained by decoupled IMC method is of second order and difficult to implement. Perturbation studies are also made separately for 10%, 15% and 20% change in steady-state gains. In all the cases, the controller designed by simple tuning method is giving improved performances over those of the decoupled IMC controller. The controller designed by the decoupled IMC method becomes complicated for the systems other than 2 × 2 or systems having individual higher order transfer functions. By the simple tuning method, it is very easy to design the controllers and the method can be applied to any multivariable stable systems irrespective of their order.

Table 14.3a *ISE values of servo problem for the examples 2 and 3*

Example	Servo problem		
	Step in	Sum of ISE values*	
		Decoupled IMC	Simple tuning
Considered by Morari et al.	Y1	2.13	1.564
	Y2	2.13	1.433
Wood and Berry distillation column	Y1	26.14	18.25
	Y2	26.77	31.68

Table 14.3b *ISE values of regulatory problem for the examples 2 and 3*

Example	Regulatory problem		
	Step in	Sum of ISE values*	
		Decoupled IMC	Simple tuning
Considered by Holt and Morari	$V1$	2.802	2.272
	$V2$	3.106	2.986
Wood and Berry distillation column	$V1$	2890.9	2386.81
	$V2$	12388.6	9624.4

Note: *ISE values of response + ISE values of interaction

15 Tuning of Multivariable Controllers by Genetic Algorithms

Use of genetic algorithm for the tuning of decentralized and centralized controllers reported by Sadasivarao and Chidambaram (2003) is reviewed. The knowledge of the system is used to improve the search of genetic algorithms. All the loops are first tuned independently using the Ziegler–Nichols method or synthesis method. The detuning factors of 1.5 and 2 are used to get the range of the controller parameters. In this way, the unstable region (i.e., those settings that make the closed loop system unstable) is reduced from the region of search. The method is applied to a two input and two output distillation column model, given by Wood and Berry (1973). The knowledge-based genetic algorithm is converged faster than the method proposed by Vlachos et al. (1999).

15.1 Introduction

Many industrial control problems are of multi-input and multi-output (MIMO) in nature. The conventional multi-loop single-input and single-output (SISO) control systems are still used in industry because the formulation of MIMO control systems is not as straightforward. Centralized control scheme and decentralized control scheme are the two methods available for the control of such multivariable systems. In decentralized control system, single loop controllers are designed for each of the loops. Many methods are proposed for the design of decentralized controllers in the literature. Balachandran and Chidambaram (1996) compared BLT method, sequential design method and the method of inequalities for the design of decentralized controllers, and proved that the method of inequalities gave better performances than the other two methods. In the method of inequalities, the performance criterion is expressed as a set of inequalities and random search is used for the designing of decentralized controllers. But the random search method does not guarantee a global solution. For some systems where the interaction is significant, the decentralized controllers may not give satisfactory responses. For such systems, full multivariable controllers (also known as centralized controllers) should be used. However, centralized control system requires $n \times m$ controllers (n is the number of manipulated variables and m is the number of controlled variables) whereas the decentralized control system requires only n controllers. Maciesowski method, Davidson method and Tanttu–Lieslehto method are used for the designing of centralized controllers (Chidambaram, 1998). In all these methods, tuning is carried out in a trial and error manner. To get an improved performance, the controller design can be carried out by a suitable optimization method.

Vlachos et al. (1999) proposed a simple genetic algorithm for the designing of decentralized PI controllers. Genetic algorithms are search and optimization procedures that are motivated by the principles of natural genetics and natural selection (Goldberg, 1989). The powerful capability of genetic algorithms in locating the global optimal solution is used in the design of PI controllers. In their method, the different performance criteria were combined into a single objective function. The range of values for the proportional and integral term of the controller was given randomly. The disadvantage of this method was that a large number of computations were involved and it was time consuming. In the present method, the knowledge of the system is used to improve the search of genetic algorithms. The method is applied to the distillation column model (Wood and Berry, 1973) considered by Vlachos et al. (1999).

15.2 Genetic Algorithms

A simple genetic algorithm is used in this work. The binary alphabet and gray coding is used for the encoding of controller parameters. A generation gap of 0.9, linear rank-based fitness assignment with a selective pressure of two, proportionate selection with stochastic universal sampling, simple (single-point) crossover and fitness-based re-insertion is used as chosen by Vlachos et al. The size of the population is chosen to be 80 in accordance with the experimental studies of Grefenstette (1986), so that search space is attacked at many points simultaneously, thus resulting in faster convergence to the global optimum (which is a minimum in this case). The initially randomly selected population is left to evolve for 100 generations. A 12-bit string element is used for the encoding of each of the controller parameters. The crossover and mutation probabilities are chosen to be 0.45 and 0.02, respectively. The genetic algorithm MATLAB tool box developed by the Department of Automatic Control and Systems Engineering, University of Sheffield is used (Chipperfield et al., 1986).

In the present method, all loops are first tuned independently using the Ziegler–Nichols reaction curve method. The detuning factors of 1.5 and 2 are used. The lower limit of proportional gain is taken as 0 and the upper limit as $K_{c,\max}/1.5$. The lower limit of integral time is taken as $T_{I,\min}/2$. The parameters are encoded as $K_{c,ij}$ and $K_{c,ij}/T_{I,ij}$ in the algorithm. No upper limit is imposed on the integral time. Given the range of the controller parameters, the genetic algorithm will generate initial guesses randomly in that range and evaluate the objective function. In this way, the unstable region (i.e., those settings that make the closed loop system unstable) is eliminated from the region of search. In the next section, it will be shown that the present improved genetic algorithm is able to reduce the number of computations and converge faster than the method proposed by Vlachos et al. (1999).

15.3 Objective Function

Objective function (sum of deviation from performance criteria) is a mathematical way of representing the desired response. The method of Vlachos et al. (1999) is followed to represent the objective function. The objective function for the ith output and the jth set point pattern can be expressed as

$$J_{ij}\left(K_c, T_I\right) = \int_0^{t\,\max} \left[\max\left\{f_{ij}^{(l)}(t) - y_i(t), 0\right\} + \max\left\{y_i(t) - f_{ij}^{(u)}(t), 0\right\}\right] dt \tag{15.1}$$

where K_c and T_I are parameters of the n PI controllers associated with n loops of the multivariable system. $f_{ij}^{(l)}$ and $f_{ij}^{(u)}$ are user-defined lower and upper boundaries of the region representing the performance objectives for the ith output and jth set point pattern respectively. A unit step change is given to one of the set points (jth set point) by keeping the remaining set points constants and the performances of all outputs are evaluated using the above equation. The results are weighted and added together to form a single performance. The procedure is repeated for the other set points (j varies from 1 to n) and the final objective function is represented as

$$J_0\left(K_c, T_I\right) = \max_{1\le j\le n}\left\{\sum_{i=1}^{n} W_{ij}\, J_{ij}\left(K_c, T_I\right)\right\} \tag{15.2}$$

J_{ij} denotes the objective function for the ith output and jth set point. W_{ij} denotes the weighing factor of the objective function for the ith output and jth set point. Higher the weighing factor, the more important the corresponding objective function becomes. In case of centralized control systems, K_c and T_I are parameters of the $n \times n$ PI controller matrix. In the present case, w_{ij} is taken as 1 when $i = j$ and is taken as 0.25 when $i \neq j$ to give more importance to set point tracking objectives.

15.4 Design Example

The simulation example used by Vlachos et al. (1999) is considered here. It is a transfer function model of a pilot-scale distillation column, developed by Wood and Berry (1973). It is a two-input and two-output model with strong interactions between the input-output pairs.

$$G(s) = \begin{bmatrix} \dfrac{12.8e^{-s}}{16.7s+1} & \dfrac{18.9e^{-3s}}{21s+1} \\ \dfrac{6.6e^{-7s}}{10.9s+1} & \dfrac{19.4e^{-3s}}{14.4s+1} \end{bmatrix} \tag{15.3}$$

The centralized controller for the above example will be of the form:

$$C = \begin{bmatrix} C_{11} & C_{12} \\ C_{21} & C_{22} \end{bmatrix} \tag{15.4}$$

Where C_{ij} represents the controller for the loop corresponding to ith output and jth input. The response of the controller C_{ij} will be as follows:

$$U_{ij} = K_{c,ij}\, e_j + \left(K_{c,ij}/T_{I,ij}\right)\left(\int e_i\, dt\right) \tag{15.5}$$

Using the Ziegler–Nichols reaction curve method, the individual controller settings $(K_{c,\max}, T_{I,\min})_{ij}$ for the centralized controller are calculated as

$$(1.1742, 3.33)_{11}, (0.3333, 9.99)_{12}, (0.2132, 23.31)_{21}, \ (0.2227, 9.99)_{22} \tag{15.6}$$

For the decentralized controller, only diagonal controllers C_{11} and C_{22} will be used and C_{21} and C_{12} will be equal to zero.

The performance objectives are chosen (Vlachos et al., 1999) as follows:

$$\begin{aligned} &\text{Peak overshoot} &&: \leq 20\% \\ &\text{Settling time} &&: \leq 50 \text{ min (with 5\% settling error)} \\ &\text{Loop coupling} &&: \leq 50\%, 0 \leq t \leq 25 \text{ min} \\ & && \ \ \leq 5\%, 25 \leq t \leq 100 \text{ min} \end{aligned} \tag{15.7}$$

15.5 Decentralized Controllers

The decentralized controllers for the above example are designed by tuning both the loops independently using the Ziegler–Nichols reaction curve method. The detuning factors of 1.5 and 2 are used to get the range of the controller parameters.

$$\begin{aligned} &K_{c,11} = (0, 0.7828) && K_{c,22} = (0, 0.1485) \\ &K_{c,11}/T_{I,11} = (0, 0.4702) && K_{c,22}/T_{I,22} = (0, 0.0297) \end{aligned} \tag{15.8}$$

The improved genetic algorithm is able to converge within 35 generations. In Vlachos et al. (1999) where the range of controller parameters is chosen randomly (proportional gain 0–10 and integral time 0.1–100), the algorithm takes nearly 50 generations to converge. The convergence results are shown in Fig. 15.1, which shows the better convergence of the improved genetic algorithm over that of the Vlachos et al. (1999) method. The closed loop servo responses of the system are shown in Figs 15.2 and 15.3. These figures illustrate that both the methods give similar performance. The optimal PI controller parameters and the objective function values are tabulated in Table 15.1, which shows that both the methods satisfy the performance criteria. The generations taken by the Vlachos et al. method is nearly twice than that of the present method. The same MATLAB tool box is used in both the methods. The only difference between the two methods is that in the selection of the ranges of controller parameters. The regulatory response and servo response are also evaluated for both the methods and they show similar performances. The robustness of the controller settings are studied with +20% perturbations in time delay, time constant and process gain, considering one at a time. Both the methods give robust settings. Method used

Table 15.1 *Optimal PI controller parameters for decentralized control system*

Method used	**Controllers**	**Proportional gain**	**Integral gain**	**No of generations to converge**	**Objective function value**
Improved GA	C11	0.1598	0.0282	35	0 (optimal)
	C22	0.0756	0.0106		
Vlachos et al. method	C11	0.1514	0.0265	50	0 (optimal)
	C22	0.0805	0.0115		

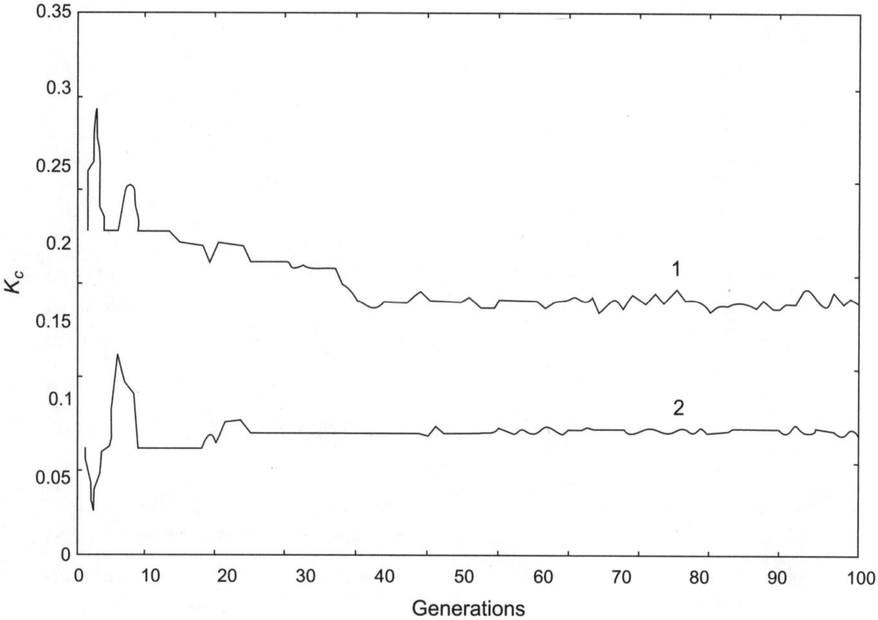
0.35
0.3
0.25
0.2
0.15
0.1
0.05
0
K_c
1
2
0
10
20
30
40
50
60
70
80
90
100
Generations

(a)

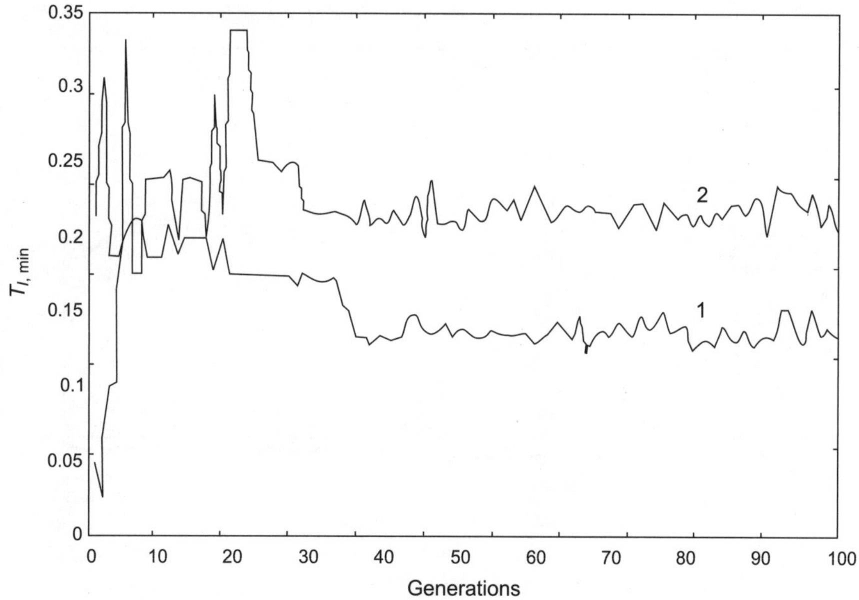
0.35
0.3
0.25
0.2
0.15
0.1
0.05
0
$T_{I,\,min}$
2
1
0
10
20
30
40
50
60
70
80
90
100
Generations

(b)

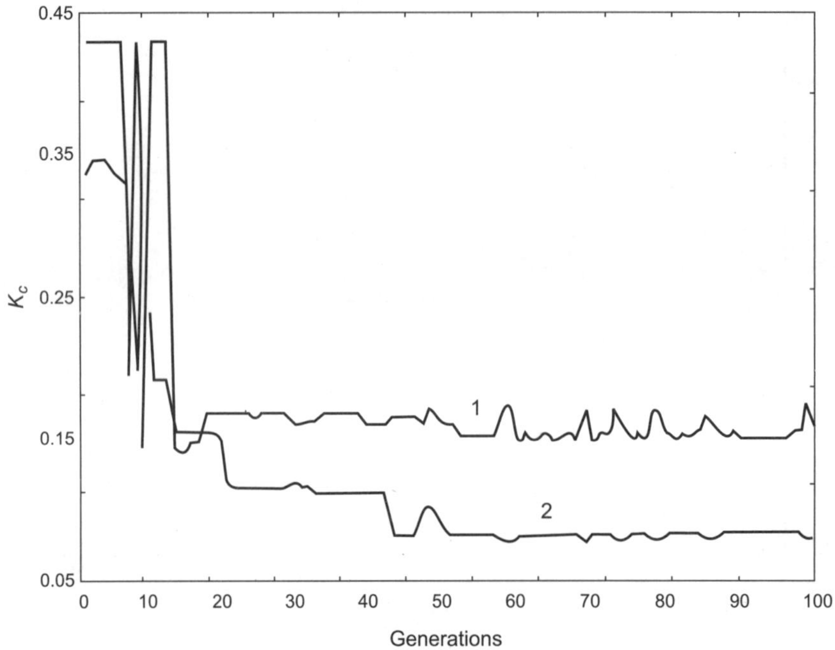

(c)

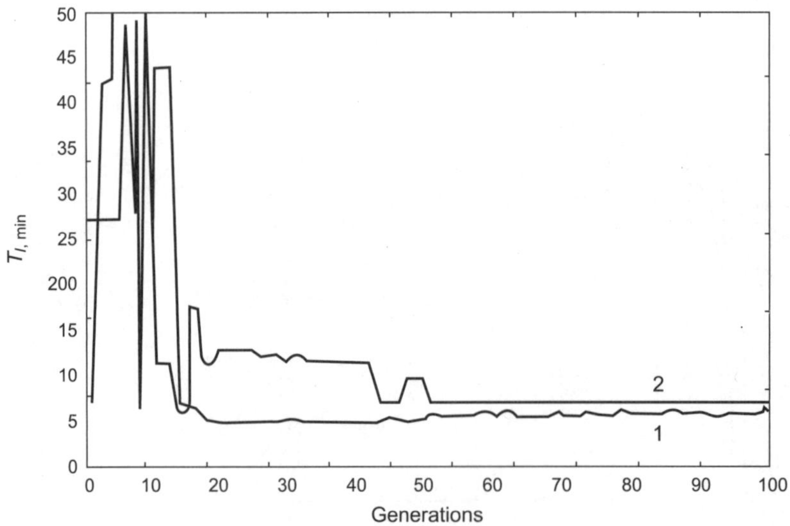

(d)

Fig. 15.1 *Convergence of PI controller parameters (1 and 2 represent the controllers C11 and C22, respectively)*
a. Proportional gain convergence by improved genetic algorithm
b. Integral time convergence by improved genetic algorithm
c. Proportional gain convergence by Vlachos et al. (1999)
d. Integral time convergence by Vlachos et al. (1999)

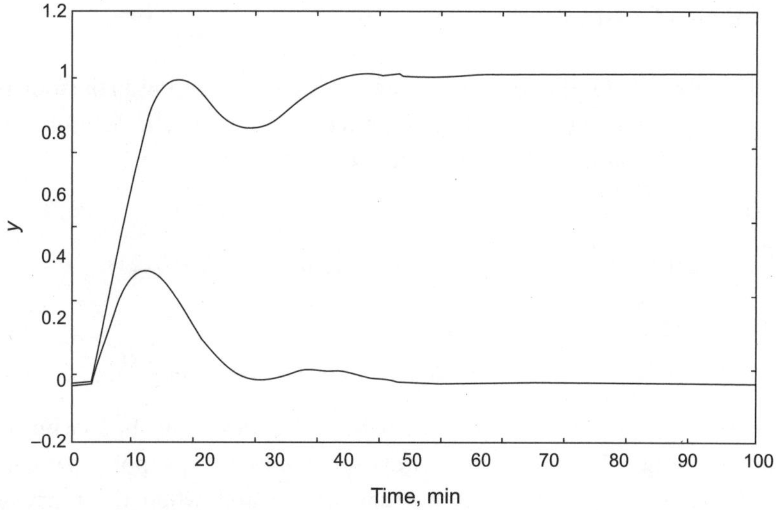

Fig. 15.2 *Closed loop system response of decentralized controllers for a step change in set point Y1 (the upper curves show the response and the lower curves show the interaction)*
Legend: thick line–improved GA; dotted line–Vlachos et al. method

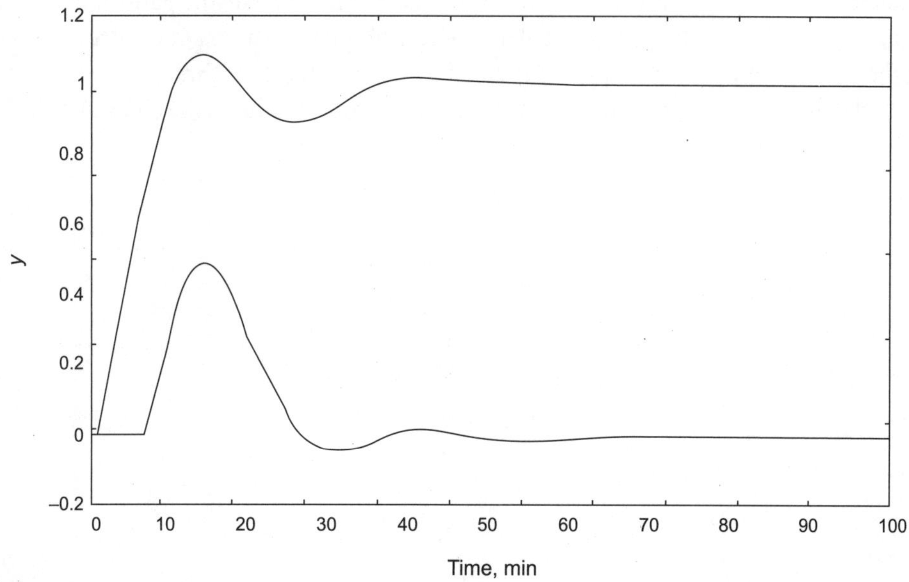

Fig. 15.3 *Closed loop system response of decentralized controllers for a step change in set point Y2 (the upper curves show the response and the lower curves show the interaction)*
Legend: same as in Fig. 15.2

15.6 Centralized Controllers

The centralized controllers for the same example are designed by tuning all the transfer functions independently using the Ziegler–Nichols reaction curve method. The detuning factors of 1.5 and 2 are used to get the range of the controller parameters.

$$
\begin{aligned}
K_{c,11} &= (0, 0.7828) & K_{c,21} &= (0, 0.1416) \\
K_{c,11}/T_{I,11} &= (0, 0.4702) & K_{c,21}/T_{I,21} &= (0, 0.0121) \\
K_{c,12} &= (0, 0.2222) & K_{c,22} &= (0, 0.1485) \\
K_{c,12}/T_{I,12} &= (0, 0.0445) & K_{c,22}/T_{I,22} &= (0, 0.0297)
\end{aligned} \tag{15.9}
$$

The improved genetic algorithm is converged within 25 generations. In Vlachos et al. (1999), where the range of controller parameters is chosen randomly (proportional gain 0–10 and integral time 0.1–100) the algorithm is not converging and when the range is increased (proportional gain 0-10 and integral time 0.1-1000), the convergence is very slow and the optimal solution is reached after 900 generations. The closed loop servo responses of the system are shown in Fig. 15.4 and Fig. 15.5. These figures illustrate that both the methods give similar performance. The optimal PI controller parameters and the objective function values are tabulated in Table 15.2, which shows that both the methods satisfy the performance criteria. The generations taken by the Vlachos et al. method is 36 times than that of the present method. When the number of parameters to be estimated by the genetic algorithm is large (as in case of centralized controllers), the proposed method is very useful. It reduces the computational time drastically as shown in this example.

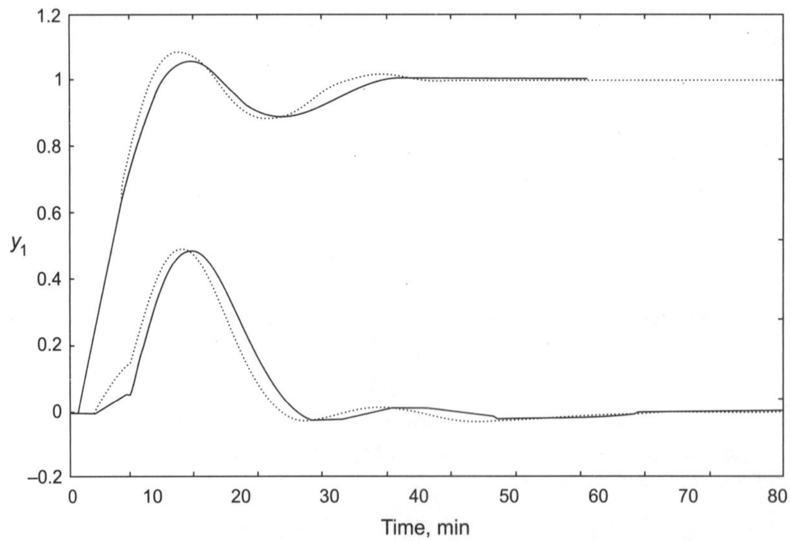

Fig. 15.4 *Closed loop system response of centralized controllers for a step change in set point Y1 (the upper curves show the response and the lower curves show the interaction)*
Legend: same as in Fig. 15.2

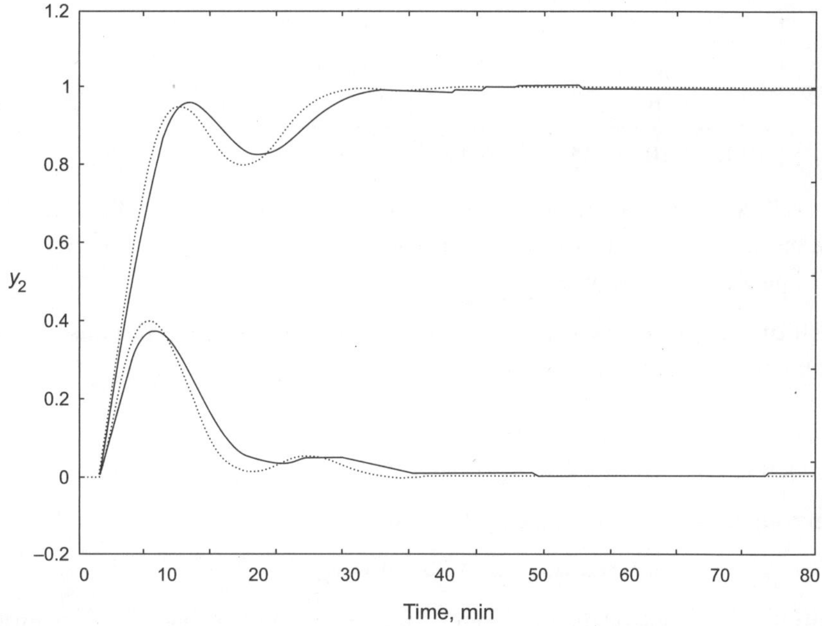

Fig. 15.5 *Closed loop system response of centralized controllers for a step change in set point Y2 (the upper curves show the response and the lower curves show the interaction)*
Legend: same as in Fig. 15.2

Table 15.2 *Optimal PI controller parameters for centralized control system*

Method used	Controllers	Proportional gain	Integral gain	No of generations to converge	Objective function value
Improved GA	C11	0.077618	0.030428	25	0 (optimal)
	C12	0.0094415	0.00041294		
	C21	0.051834	0.00027775		
	C22	0.094286	0.012308		
Vlachos et al. method	C11	0.16606	0.025751	900	0 (optimal)
	C12	0.017094	1.7628e–005		
	C21	0	0		
	C22	0.065934	0.011062		

15.7 Example 2: Niederlinski Model

A two-input and two-output transfer function model used by Niederlinski (1971) is considered in this example. It is a higher order system as given here.

$$G(s)=\begin{bmatrix} \dfrac{0.5}{(0.1s+1)^2(0.2s+1)^2} & \dfrac{-1.0}{(0.1s+1)(0.2s+1)^2} \\ \dfrac{1.0}{(0.1s+1)(0.2s+1)^2} & \dfrac{2.4}{(0.1s+1)(0.2s+1)^2(0.5s+1)} \end{bmatrix} \tag{15.10}$$

The decentralized control structure is chosen for this example. Using Relative Gain Array (RGA) the input-output parings are selected. The inputs $U1$ and $U2$ are chosen to control the outputs $Y1$ and $Y2$, respectively.

The performance objectives are chosen by studying the open loop response of the system and are as follows:

$$\begin{aligned} &\text{Peak overshoot} &:\quad &\leq 20\% \\ &\text{Settling time} &:\quad &\leq 10 \text{ min (with 5\% settling error)} \\ &\text{Loop coupling} &:\quad &\leq 50\,\%,\ 0 \leq t \leq 10 \text{ min} \\ & & &\leq 5\,\%,\ 10 \leq t \leq 30 \text{ min} \end{aligned} \tag{15.11}$$

The decentralized PI controllers are designed for this example using genetic algorithm. The SISO controller settings $(K_{c,\,des}, T_{I,\,des})_{ij}$ for the diagonal transfer functions in the process $G(s)$ are calculated using the Ziegler–Nicholas tuning method and are as follows.

$(4.05, 0.75)_{11}$, $(1.026, 1.1667)_{22}$

The ranges of the controller parameters are calculated as explained in the previous section and are as follows.

$$\begin{aligned} &K_{c,11} = (0, 2.7) & &K_{c,22} = (0, 0.6840) \\ &K_{c,11}/T_{I,11} = (0, 7.2) & &K_{c,22}/T_{I,22} = (0, 1.1724) \end{aligned} \tag{15.12}$$

In Vlachos et al. method, the ranges of controller parameters are chosen in a trial and error manner and are as follows.

$$\begin{aligned} &K_{c,11} = (0.01, 50) & &K_{c,22} = (0.01, 50) \\ &K_{c,11}/T_{I,11} = (0.01, 50) & &K_{c,22}/T_{I,22} = (0.01, 50) \end{aligned} \tag{15.13}$$

In the improved GA where the search region is chosen according to the previous section, the optimum (i.e., the objective function value of zero) is reached within three generations. In Vlachos et al. method where the search region is chosen in a trial and error manner, the optimum is reached after 25 generations. The optimal PI controller parameters and the objective function values are given in Table 15.3, which shows that both the methods satisfy the performance criteria. But the generations taken by the Vlachos et al. method is nearly eight times that of the present method. The servo response of the system is evaluated and it is found that both the methods give similar performance. The response of the system for a unit step change in the load variable is evaluated. The load transfer function matrix is assumed as a process transfer function matrix. It is found by simulation studies that both the methods give similar load rejection.

The variation of the manipulated for a unit step change in the set point is evaluated. Both the methods give a similar response in which the manipulated variable is not varying randomly. The robustness of the controller settings is also verified with +20% perturbations in the steady-state gain. It is found by simulation that both the methods give stable servo responses.

Table 15.3 *Optimal PI controller parameters for example 2 using the improved GA and the Vlachos et al. method*

Method used	Controllers	Proportional gain, K_c	Integral gain, K_c / T_I	No of generations to converge	Objective function value
Improved GA	C11	0.0428	0.4132	3	0 (optimal)
	C22	0.2903	0.2774		
Vlachos et al. method	C11	0.3785	0.4161	25	0 (optimal)
	C22	0.1221	0.2696		

15.8 Conclusions

The inclusion of knowledge of the problem for the ranges of the PI control parameters into the genetic algorithm improves the convergence of the algorithm significantly. If the ranges of the controller parameters are given randomly, the genetic algorithm is taking more time to converge. This is shown by designing the decentralized and centralized controllers for a two-input and two-output transfer function model. As the number of parameters to be estimated increases (as in case of centralized controllers), the improved genetic algorithm shows significant improvement in the convergence.

16 Summary and Conclusions

This work focuses attention on (i) incorporation of higher order harmonics in the analysis of relay tuning of controllers for a single loop controller, cascade control systems, single loop saturation relay test, single loop unstable FOPTD system and single loop stable SOPTD system; (ii) providing a simple method of designing P/PI controllers for cascade control schemes; (iii) estimation of model parameters of unstable FOPTD, stable SOPTD and unstable SOPTDZ systems using a single relay feedback test and (iv) application to multivariable systems.

16.1 Improved Autotune Identification Method

A method is suggested to formulate an additional equation so that the process gain can also be estimated using the conventional relay autotune method. This method avoids getting a negative time constant of an FOPTD model. For systems showing higher order harmonics in the response, a modification of the calculation for the model parameters of FOPTD model using the conventional relay feedback method is also proposed. This method does not assume the complete filtering of higher order harmonics. The method of calculation is also simple. The present method gives an improved value for the controller ultimate gain. The method gives a more accurate result [on k_u and on the identified FOPTD model parameters] than that proposed by Luyben (1987) and Li et al. (1991). Simulation results show that the present method gives improved open loop as well as closed loop performances.

The proposed modification in the asymmetrical relay test gives improved values of the parameters of the FOPTD model. Analytical solutions are given for the evaluation of the model parameters. The method needs only one relay test to identify the parameters of an FOPTD model. The method is also applicable for identifying an unstable FOPTD model. Simulation results are given for stable and unstable transfer function models. The proposed method gives a better closed loop response than that of Huang et al. (1996a, b) and Shen et al. (1996a).

16.2 Series Cascade Controller Tuning

The sequential autotune procedure is used for tuning cascade control systems. Hang et al. (1994) proposed a sequential relay autotuning of cascade control loops. The conventional relay autotune method gives an error in k_u calculations. In the present work, two methods are proposed to improve the performance of the autotuning of controllers.

16.2.1 Symmetric relay method

Srinivasan and Chidambaram (2003) method, proposed to get an accurate value of k_u for a single feedback loop, is extended to tune cascade control systems. The relay is used in the inner loop and the outer loop is kept under manual mode. From the relay oscillations, frequency ω_u is noted. From the output oscillations, it is possible to calculate $y(t^*)$ at time $t^* = 0.5\ \pi/\omega_u$. If the observed output oscillations are close to a rectangular waveform, then amplitude (a) is calculated as

$$y(t^*) = a[1-(1/3)+(1/5)-(1/7)+(1/9)-\cdots] \tag{16.1}$$

Let the number of terms to be considered in the above equation be denoted as N. If the oscillation deviates from pure sine wave and depending on the extent of deviation, N = 3 or 5 or 7 can be considered. Similarly, if the observed output oscillations are close to a triangular waveform then the amplitude is calculated as

$$a = y(t^*)\big/\left[1+(1/9)+(1/25)+(1/49)+(1/81)+\cdots\right] \tag{16.2}$$

After getting the above-corrected amplitude, the value of k_u is calculated as $k_u = 4h/(a\pi)$.

Using the values of k_u and ω_u, PI controller settings are calculated by using the Ziegler and Nichols tuning method. Using the PI controller settings in the inner loop instead of the relay and introducing a relay in the outer loop, the resulting oscillations in the outer loop are noted. Based on the improved relay test results, the outer loop PID controller is designed using the Ziegler and Nichols tuning method. The closed loop servo response is evaluated for a unit step change in the set point. A highly oscillatory response is obtained for conventional relay analysis whereas the present method gives a good response. Similar response is obtained for a regulatory problem as well [with $(k_L G_L)_2 = 1$].

16.2.2 Asymmetric relay method

The asymmetric relay autotune method is applied with a relay height of +2 and −1. Shen et al. (1996a) used a biased relay for getting the model parameters of an FOPTD model. Here, the method proposed by Srinivasan and Chidambaram (2003) is extended to series cascade systems. The relay is used in the inner loop and the outer loop is kept under manual mode. Here, only the final equations are given for the application to a cascade control system. Let us denote $G(s)$ as the FOPTD model to be identified. With asymmetric relay, the responses in output (y) and input (u) are recorded. The process gain k_p is calculated as

$$k_p = \int_0^{2\pi} y(t)\,d(\omega t)\Big/\int_0^{2\pi} u(t)\,d(\omega t) \tag{16.3}$$

$G(j\omega)$ can be written by substituting $s = j\omega$ in the above equation

$$G(j\omega) = a + jb = (c_1 - j\,f_1)/(c_2 - j\,f_2) \tag{16.4}$$

Where a is the real part and b is the imaginary part, and

$$c_1 = \int_0^P u(t)\cos(\omega t)dt;\; f_1 = \int_0^P u(t)\sin(\omega t)dt \tag{16.5}$$

$$c_2 = \int_0^P y(t)\cos(\omega t)dt;\; f_2 = \int_0^P y(t)\sin(\omega t)dt \tag{16.6}$$

where $P = 2\pi/\omega_u$ and ω (=ω_u) is the frequency of oscillation observed in the output response. Equations (15.5) and (15.6) are to be evaluated numerically. Here a and b are given by

$$a = (c_1\, c_2 + f_1\, f_2)/(c_2^2 + f_2^2) \tag{16.7}$$

$$b = (f_2\, c_1 - f_1\, c_2)/(c_2^2 + f_2^2) \tag{16.8}$$

We can write $G(j\omega)$ of an FOPTD model as

$$a + jb = k_p\left[\cos(D\omega) - j\sin(D\omega)\right]/(1 + j\tau\omega) \tag{16.9}$$

On cross-multiplying and equating the resulting real and imaginary parts to zero, we get

$$a - b\,\tau\omega - k_p\cos(D\omega) = 0 \tag{16.10}$$

$$a\tau\omega + b + k_p\sin(D\omega) = 0 \tag{16.11}$$

Substituting ω_u for ω, an analytical solution of Eq.(15.10), Eq.(15.11) gives the values for τ and D. The value of k_u for the inner loop based on the assumed FOPTD model is obtained. Based on the FOPTD model, a PI controller is designed by the Ziegler–Nichols method. The inner loop is kept under a PI controller and an asymmetric relay is then used in the outer loop. Based on the oscillation obtained in y_1 and hence based on the identified FOPTD model the value of k_u is obtained for the outer loop. PID controller is designed for the outer loop based on the values of k_u and ω_u. The servo response in y_1 for a unit step change in the set point is evaluated and the performance is found as good as the performance of the proposed symmetric relay method.

In the modified symmetrical relay method proposed by Srinivasan and Chidambaram (2004), the value of order (N) of higher order harmonics is to be selected. Whereas in the asymmetric method of Srinivasan and Chidambaram (2003), no such value of N is required. In the asymmetric method, the model is identified first and then the controller settings are calculated. In the symmetrical method, the controller settings are calculated based on the ultimate values obtained from the relay test.

16.2.3 Simultaneous relay autotuning of cascade controllers

The modified analysis of relay autotuning proposed for a single feedback system by Srinivasan and Chidambaram (2004) is extended to tune series cascade controllers by simultaneous tuning method. The Srinivasan and Chidambaram method effectively takes care of higher order harmonics. The performances of the PI–PID controllers are compared with that of the principal harmonics analysis proposed by Saraf et al. (2003). The present methods give a better performance than that of the conventional analysis. The above work on series cascade control is also extended to parallel cascade controllers.

The modified analysis of relay autotuning proposed for a single feedback system by Srinivasan and Chidambaram (2004) is also extended to tune cascade controllers by simultaneous tuning method. The Srinivasan and Chidambaram method effectively takes care of higher order harmonics. The performances of the PI–PID controllers are compared with the Saraf et al. (2003) conventional relay analysis (principal harmonic analysis). The present methods give a better performance than that of the conventional analysis.

16.3 A Simple Method of Designing Cascade Controllers

A simple method is proposed to design PID controllers for a series cascade control system. FOPTD models are assumed for both the inner and outer loop sub-systems. The method is based on matching the coefficients of corresponding powers of s and s^2 in the numerator to α_1 and α_2 times that in the denominator of the closed loop transfer function model for a servo problem. The method is first applied to design a proportional (*P*) controller for the inner loop and then to design a proportional plus integral (PI) controller for the outer loop. The performance of the proposed controllers is evaluated for FOPTD models of the inner loop and outer loop process transfer functions. The response and robustness due to perturbations in the model parameters are evaluated and compared with the method proposed by Krishnaswamy et al. (1990) and that proposed by Lee et al. (1998b). The proposed method gives the best servo and regulatory performances. The above method is also applied to design controllers for the parallel cascade schemes.

16.4 Improved Saturation Relay Test for Systems with Large Dead Time

Yu (1999) introduced the saturation relay feedback test for determining the ultimate gain and ultimate frequency of the system that are required for the design of PI/PID controllers. In the present work, two modifications are proposed to improve the efficiency of the saturation relay feedback test, especially for systems with a large dead time. In the first method, the safety factor is corrected based on the ideal relay feedback test. In the second method, higher order harmonics are included in the analysis of ideal relay. Simulation results are given to show the improvement of the proposed method over that of the Yu method for an FOPTD system and a higher order system. The methods proposed by Srinivasan and Chidambaram (2004) and Luyben (2001) for identifying the FOPTD model parameters are compared. Two example systems are considered for the identification of an FOPTD model. For both the examples, after identification, the IMC tuning method is used for the controller design. The closed loop performances are compared with that of the actual system. The proposed method is shown to give an improved performance.

16.5 Model Parameters using Single Symmetrical Relay Test

16.5.1 FOPTD systems

A method is proposed to identify all the three parameters of a first order plus time delay (FOPTD) model using a single symmetrical relay test. On identifying the higher order system by an FOPTD model, the conventional method identifies a negative time constant (Li et al., 1991) due to the error in neglecting higher order dynamics in the system output. In the present work, all the parameters of an FOPTD model are estimated with an adequate accuracy. Three simulation examples are given. The estimated model parameters of an FOPTD model are compared with that obtained by Li et al. (1991) and also that with exact model parameters of the system. The proposed method gives performance close to that of the actual system. Simulation results for stable and unstable systems are given. Similar method is also proposed for stable SOPTD systems with equal time constants. The method is also extended to identify the three parameters of the integrating plus FOPTD systems.

16.5.2 Autotuning of unstable FOPTD system

Using a single symmetric relay feedback test, a method is proposed to identify all the three parameters of an FOPTD unstable model. It is found by simulation that the relay autotune method gives –23% error in the calculation of k_u, when $D/\tau = 0.6$. In the present work, a method is proposed by incorporating the higher order harmonics to explain the error in k_u calculation. A method is proposed to estimate all the parameters of an unstable FOPTD system. Two simulation results are given on unstable first and second order plus time delay (SOPTD) transfer function models. The estimated model parameters of the unstable FOPTD model are compared with the methods of Majhi and Atherton (2000) and Thyagarajan and Yu (2003). PID controllers are designed for the identified model and for the actual system. The proposed method gives the performance close to that of the actual system. For a second order plus time delay system, the latter two methods fail to identify the FOPTD model. The proposed method of identification of an unstable FOPTD system is applied to a non-linear continuous bioreactor that exhibits output multiplicity (Agarwal and Lim, 1984). The PID controller designed on the model identified by the proposed method gives a performance close to that of the controller designed on the locally linearized model.

16.5.3 Autotuning of PID controllers for critically damped stable SOPTD system

Using a single symmetric relay feedback test, a method is proposed to identify all the three parameters of a stable SOPTD model with equal time constants [$k_p \exp(-Ds)/(\tau s+1)^2$]. The conventional analysis of relay autotune method gives 27% error in the calculation of k_u, for a large D/τ. In the present work, a method is proposed to improve the accuracy in k_u calculation by incorporating the higher order harmonics. Three simulation examples are given on SOPTD and higher order time delay transfer function models. The open loop performance of the identified model is compared with that of the actual system. The proposed method gives performances close to that of the actual system. Simulation results are also given for a non-linear bioreactor system. The closed loop performance of the model identified by the proposed method gives a better performance than that proposed by Li et al. (1991).

16.6 Estimation of Model Parameters of SOPTD System

A method is proposed for estimating the parameters of an SOPTD transfer function model using an asymmetric relay test. The advantages of this method are; single relay feedback test, four parameters can be identified and the method can also be applied to open loop unstable processes without any change in the procedure. Simulation results show that, for three case studies, the proposed method gives improved parameter estimates, whereas for case study 4, it gives similar results as obtained from asymmetrical relay test with additional lead-lag element. The proposed method takes less time for system identification as it requires only a single test. The method is also robust for measurement noise.

16.7 Identification of Five Parameters of Unstable SOPTD System

Using an asymmetric relay, a method is proposed to identify all the five parameters of an SOPTD system with zero.

$$G(s)=k_p(\tau_1 s+1)\exp(-Ds)/\left[(\tau_2 s+1)(\tau_3 s-1)\right] \tag{16.12}$$

An asymmetric relay is placed in the feedback loop. The system input-output data are recorded. The open loop process gain (k_p) of the SOPTD model is obtained by integrating the system input-output response as

$$k_p=\frac{\int_{t_o}^{t_o+p_u} y(t)dt}{\int_{t_o}^{t_o+p_u} u(t)dt} \tag{16.13}$$

where y is output variable and u is the input (manipulated) variable. Here 't_o' is the starting time of the integration after initial dynamics has died out, p_u is the period of oscillations. Using an asymmetric relay, the dead time (D) is obtained by aligning output response (y) to the input response (u) by shifting to the left. The dead time is set as time required to the peak amplitude. By substituting $s = j\omega$ and expanding $\exp(-Ds) = 1-Ds$, in Eq.(11.12) we get

$$a+jb=k_p(\tau_1\ j\omega+1)\cos(-D\omega)-j\sin(-D\omega)/\left[(\tau_2 s\ j\omega+1)(\tau_3\ j\omega-1)\right] \tag{16.14}$$

From the simulation data, real part (a) and imaginary part (b) are calculated using Eq.(15.7) and Eq.(15.8). On cross-multiplying and equating the resulting real and imaginary parts to zero, two equations can be obtained. Substituting $s = s_1$ in Eq.(15.12) and from the response,

$y(s_1)$ and $u(s_1)$ are calculated using the definition of Laplace transform. Here, $s_1 = 8/t_s$ and t_s is the time taken after initial dynamics are died out. Hence, we get an equation

$$\left[y(s_1)/u(s_1)\right] = \left[k_p(\tau_1 s_1 + 1)\exp(-Ds_1)\right] / \left[(\tau_2 s_1 + 1)(\tau_3 s_1 - 1)\right] \quad (16.15)$$

From the three equations formulated, the remaining three parameters (τ_1, τ_2 and τ_3) can be calculated.

The open loop unstable process transfer function, considered for simulation, is given by Yu (1999):

$$G(s) = 1.0(3s+1)\exp(-s) / \left[(10s+1)(3s-1)\right] \quad (16.16)$$

Using the proposed method, the intermediate values are calculated as $y(s_1) = 0.0513$, $u(s_1) = 0.0048$, $a = -0.1216$ and $b = -0.0122$. The identified model parameters are $k_p = 1.0$, $D = 1.0$, $\tau_1 = 2.99$, $\tau_2 = 9.97$ and $\tau_3 = 3.05$. The method identifies all the five parameters of the model with adequate accuracy.

16.8 Identification of FOPTD Multivariable Systems

The asymmetric relay feedback method for a scalar system (Srinivasan and Chidambaram, 2003) is extended to identify multivariable FOPTD systems. This method identifies each element of transfer function matrix without any *a priori* knowledge of model parameters as time delay or steady-state gain. This identification method is also fully closed, which is desirable to cope with disturbance and measurement noise. Also this method can be used for unstable multivariable processes. The examples illustrated show that this method gives accurate process model. Also comparison of closed loop performance and ISE values show that identified model is close to the actual model.

16.9 Identification of SOPTD Multivariable Systems

Asymmetric relay feedback method for scalar system (Ramakrishnan and Chidambaram, 2003) is extended to a multivariable system. Decentralized relay feedback tests are carried out. The method identifies each element of the transfer function matrix as either FOPTD or SOPTD model without any *a priori* knowledge of model parameters such as time delay or steady-state gain. It is shown that for some higher order processes when identified as FOPTD model gives a negative time constant. Modeling the system as an SOPTD gives a better performance. Simulation results on two examples show that the present method gives identified model giving a similar closed performance as that of the actual system with the same controllers.

16.10 Comparison of Multivariable Controllers Tuning Methods

A simple tuning method proposed by Davison (1976) is applied to multivariable non-minimum phase systems. Performance of the controller designed is compared using ISE values with that of the decoupled IMC method proposed by Wang et al. (2002). Simulation examples have illustrated that a simple tuning method gives an improved performance than that of the proposed decoupled IMC method. Yet, the resulting controller obtained by simple tuning method is in the form of a simple PI and easy to implement. The controller obtained by decoupled IMC method is of second order and difficult to implement. Perturbation studies are also made separately for 10%, 15% and 20% change in steady-state gains. In all the cases, the controller designed by simple tuning method is giving improved performances with the decoupled IMC controller. The controller designed by decoupled IMC method becomes complicated for the systems other than 2 × 2 or systems having individual higher order transfer functions. By a simple tuning method, it is very easy to design the controllers and the method can be applied to any multivariable stable systems irrespective of their order.

16.11 Tuning of Multivariable Controllers by Genetic Algorithm

Genetic algorithm for the tuning of decentralized and centralized controllers is applied. The knowledge of the system is used to improve the search of genetic algorithms. All the loops are first tuned independently using the Ziegler–Nichols method or the synthesis method. The detuning factors of 1.5 and 2 are used to get the range of the controller parameters. The knowledge-based genetic algorithm is converged faster than the method proposed by Vlachos et al. (1999).

16.12 Conclusions

The following conclusions are obtained from the present study:

1. a. A method is proposed to consider the higher order harmonics in the analysis of relay tuning methods to get improved ultimate values of the controller gain.
   b. A simple method is proposed to estimate the three parameters of FOPTD system using a single asymmetrical relay test.
2. a. Consideration of higher order harmonics in the symmetric relay analysis gives an improved result in k_u calculations in the sequential tuning of series and parallel cascade control systems. The cascade controller designed by the conventional relay tuning method gives an oscillatory response. Similar results are also obtained for the simultaneous tuning method.
   b. The improved method of identification using an asymmetric relay (Srinivasan and Chidambaram, 2003) is extended to simultaneous tuning of series and parallel cascade control systems. The controllers designed on the identified model by the present method give improved closed loop performances.
3. A simple method to design a controller for a single loop (Padmasree and Chidambaram, 2003) is applied to series and parallel cascade control systems. The performance of the

controller is compared with the recent methods, giving 10–30% improvement in ITAE calculations. The controller designed by the proposed method is more robust than the earlier reported methods.

4. In the saturation relay methods for large D/τ values, two methods are proposed to improve the estimates of k_u. The proposed method gives an accurate value of k_u (with less than 1% error) as compared to the conventional method (with 15–19% error). Four simulation examples are considered. A PID controller is designed by IMC tuning rules proposed by Luyben (2001). The performance comparison of the proposed method and that of the non-relay methods shows that the ISE of the proposed method is very close to that of the actual system.

5. The improved symmetric relay identification method proposed by Srinivasan and Chidambaram (2004) is extended to (a) an unstable FOPTD system and (b) stable SOPTD system with equal time constants. A theoretical derivation for the occurrence of –23% and +27% errors involved in the principal harmonics analysis for unstable FOPTD systems and stable SOPTD systems equal time constants are given.

6. A symmetric relay identification method is proposed for estimating (a) the three parameters of an FOPTD system and SOPTD system with equal time constants and (b) the five parameters of an unstable SOPTDZ model. Simulation results on various transfer function models show the effectiveness of the proposed method.

7. The improved asymmetric relay tuning method for identification of FOPTD transfer function model for a single input single output is extended to identification of transfer function matrix of a multivariable system.

8. The improved asymmetric relay tuning method for identification of SOPTD transfer function model for a single input single output is extended to identification of transfer function matrix of a multivariable system.

9. A simple method of designing centralized PID controllers for multivariable system is reviewed for non-minimum phase multivariable systems.

10. Tuning of multivariable PI controllers using the settings by the Ziegler–Nichols method as initial guess values improves the speed of convergence of genetic algorithm.

Appendix A

A.1 Simulink Block Diagrams for Case Studies in Chapter 3

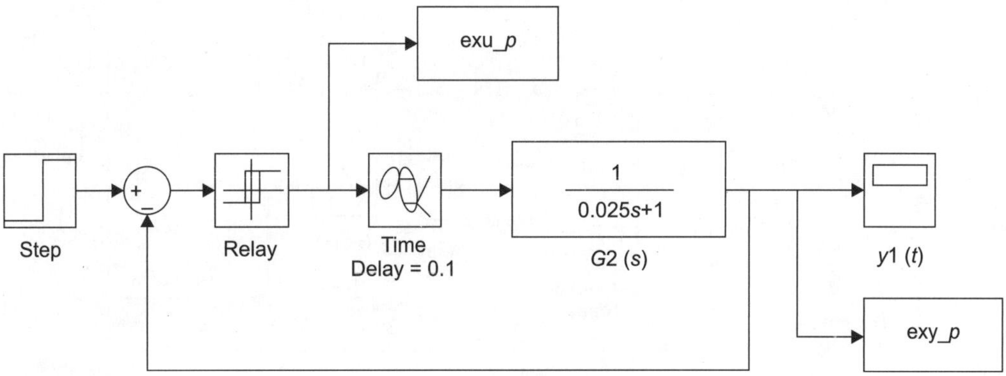

Fig. A.1 *Simulink block diagram for inner loop tuning for case study* 1

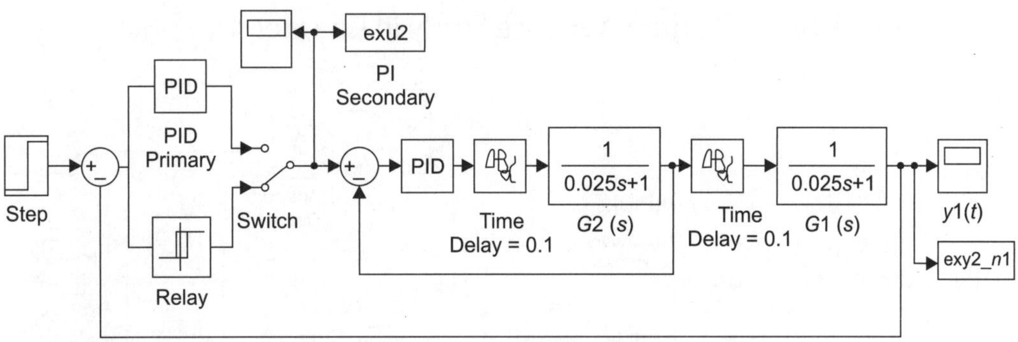

Fig. A.2 *Simulink block diagram for outer loop tuning for case study* 1

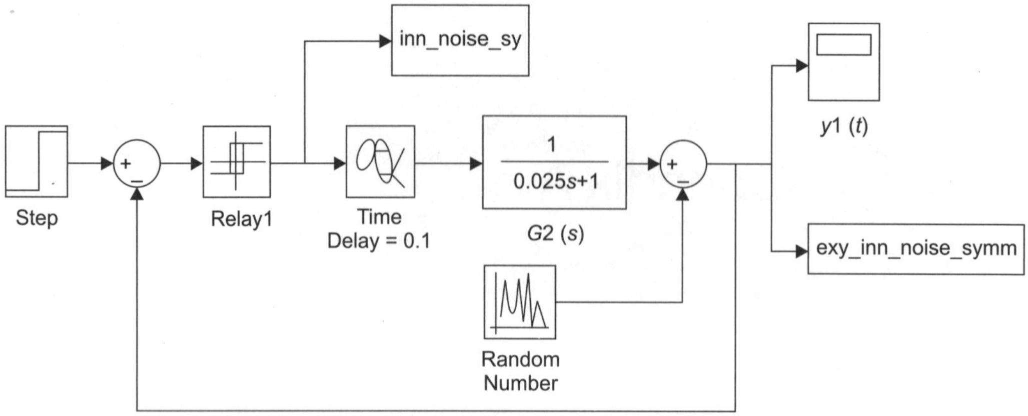

Fig. A.3 *Simulink block diagram for measurement noise in inner loop for case study 1*

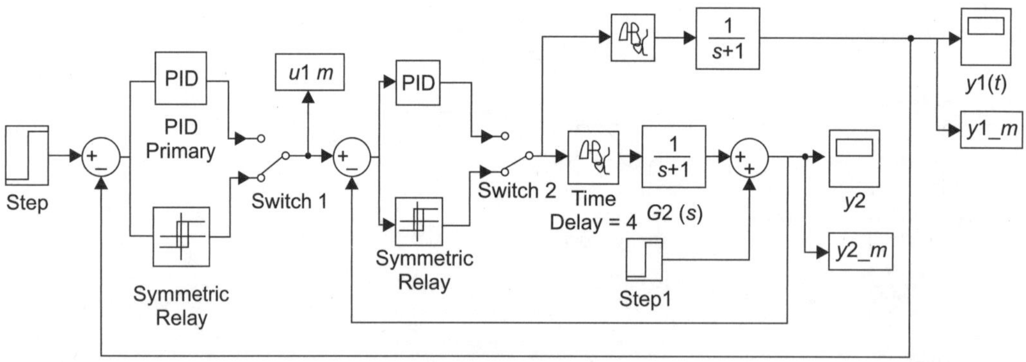

Fig. A.4 *Simulink block diagram for parallel cascade control system*

A.2 Simulink Block Diagram for Case Study in Chapter 4

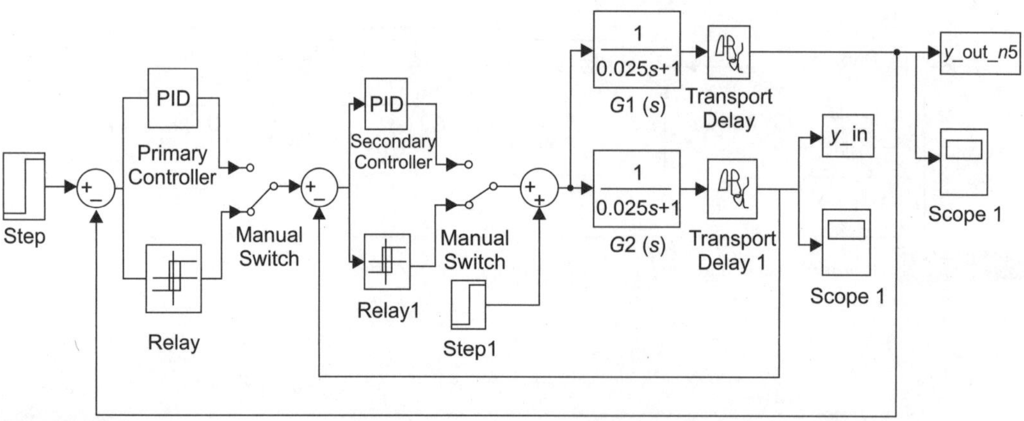

Fig. A.5 *Simulink block diagram for inner loop and outer loop tuning for case study 1*

A.3 Simulink Block Diagram for Case Study in Chapter 6

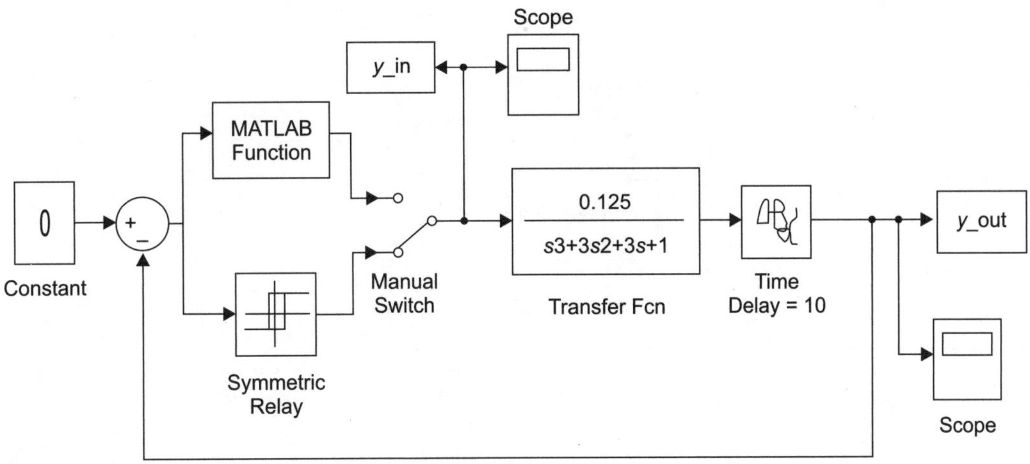

Fig. A.6 *Simulink block diagram for improved relay tuning for case study 1*

A.4 Simulink Block Diagrams for Case Studies in Chapter 8

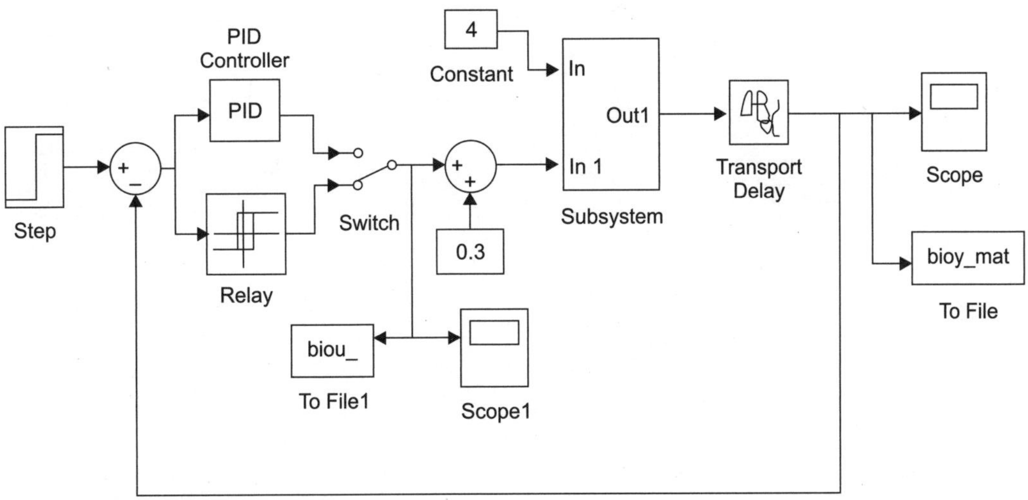

Fig. A.7 *Simulink block diagram for identification for the non-linear continuous bioreactor system*

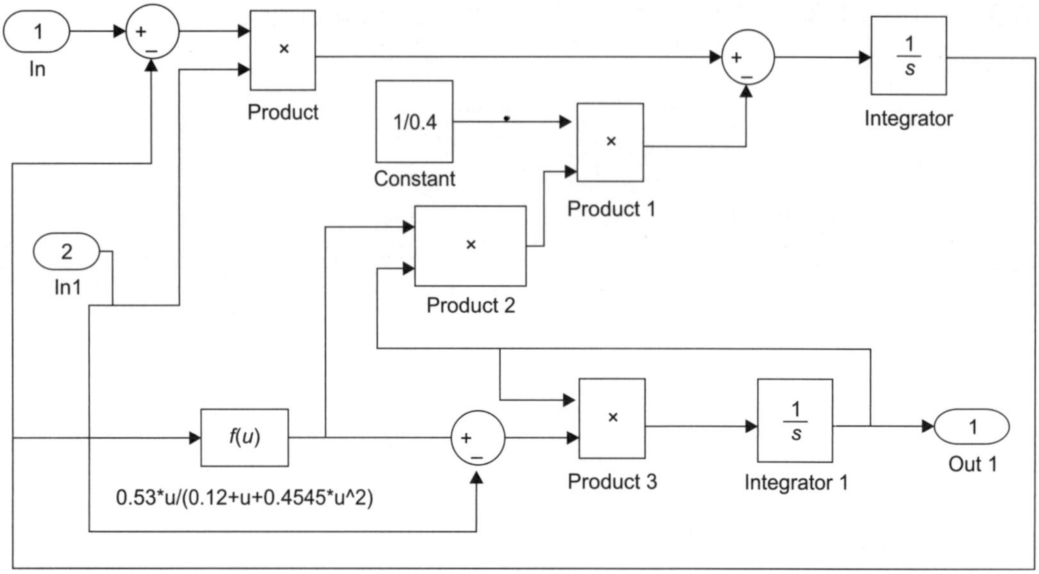

Fig. A.8 *Simulink block diagram for non-linear continuous bioreactor system*

A.5 Simulink Block Diagrams for Case Studies in Chapter 11

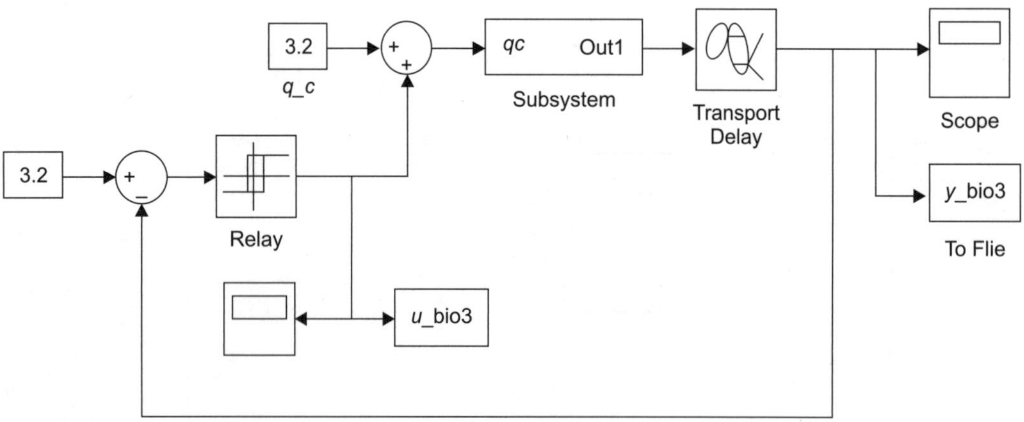

Fig. A.9 *Simulink block diagram for control of the non-linear CSTR process*

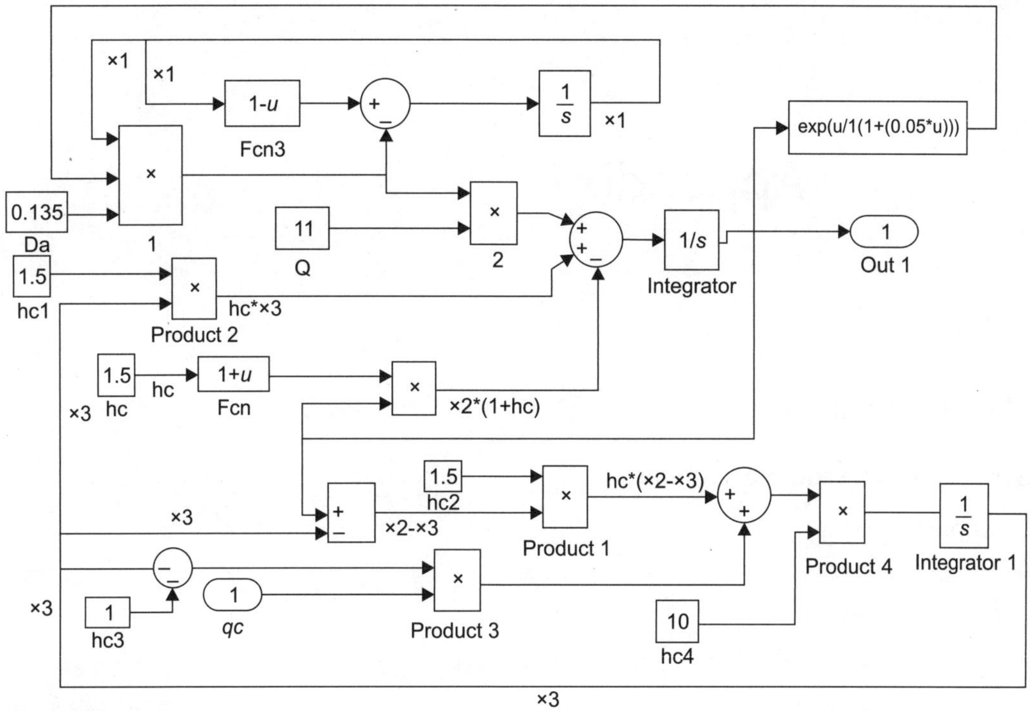

Fig. A.10 *Simulink block diagram for non-linear CSTR system*

Appendix B

MATLAB Program used for Case Studies in Chapter 3

Series cascade: Sequential tuning

Example 1

```
% Program for the intermediate values [a, b, kp] required for
identification
% uses input (exu_p.mat) and output (exy_p.mat) from relay tuning
of the simulink
% model (***)

clear all; clear all
load exy_p.mat; load exu_p.mat;
y = y_p(2,:); time = y_p(1,:); u = u_p(2,:);
% To Plot y
plot(time,y,'k')
grid on; xlabel('time'); ylabel('process output')
zoom

%Enter period of oscillation and starting time of integration
through keyboard
pu = input('Period of the oscillation: ')
ti = input('Starting time of integration: ')
st = 0.001;  w  = (2*pi)/(pu);
frequency = w
to = ti/st; p  = pu/st;
% To evaluate Kp
% Numerical integration - Simpson's 1/3 rule
```

```
i  = to+2;
c1 = 0;
while i<=((to+p)-2)
   c1 = c1+ 4*y(i);
   i  = i+2;
end
 i  = to+1;
 c2 = 0;
while i<=((to+p)-1)
   c2 = c2+ 2*y(i);
   i  = i+2;
 end
num =(st/3)*(y(to)+y(to+p)+c1+c2);
 % Numerical integration - Simpson's 1/3 rule
i  = to+2;
c1 = 0;
while i<=((to+p)-2)
   c1 = c1+ 4*u(i);
   i  = i+2;
end
i  = to+1;
c2 = 0;
while i<=((to+p)-1)
   c2 = c2+ 2*u(i);
   i  = i+2;
end
den =(st/3)*(u(to)+u(to+p)+c1+c2);
Kp  = num/den
% To evaluate G(jw)
for i = to:to+p
   yreal(i)= y(i)*cos(w*i*st);
end
 % Numerical integration - Simpson's 1 by 3 rule
i  = to+2;
c1 = 0;
```

```
while i<= ((to+p)-2)
  c1 = c1+ 4*yreal(i);
  i  = i+2;


end
i  = to+1;
c2 = 0;
while i<= ((to+p)-1)
   c2 = c2+ 2*yreal(i);
   i  = i+2;
end
yr = (st/3)*(yreal(to)+yreal(to+p)+c1+c2);
 % Evaluation of imaginary part of integrand (process output)
for j = to:to+p
   yimg(j)= y(j)*sin(w*j*st);
end
 % Numerical integration - Simpson's 1 by 3 rule
i  = to+2;
c1 = 0;
while i<= ((to+p)-2)
  c1 = c1+4*yimg(i);
  i  = i+2;
end
i  = to+1;
c2 = 0;
while i<= ((to+p)-1)
  c2 = c2+2*yimg(i);
  i  = i+2;
end
yi = (st/3)*(yimg(to)+yimg(to+p)+c1+c2);
 % Evaluation of real part of integrand (relay output)
for j = to:to+p
   ureal(j) = u(j)*cos(w*j*st);
end
```

```
% Numerical integration - Simpson's 1 by 3 rule
i  = to+2;
c1 = 0;
while i<=((to+p)-2)
   c1 = c1+ 4*ureal(i);
   i  = i+2;
end
i  = to+1;
c2 = 0;
while i<= ((to+p)-1)
   c2 = c2+ 2*ureal(i);
   i  = i+2;
end
ur = (st/3)*(ureal(to)+ureal(to+p)+c1+c2);
 % Evaluation of imaginary part of integrand (relay output)
for j = to:to+p
   uimg(j) = u(j)*sin(w*j*st);
end
 % Numerical integration - Simpson's 1 by 3 rule
 i  = to+2;
 c1 = 0;
while i<= ((to+p)-2)
   c1 = c1+ 4*uimg(i);
   i  = i+2;
end
i  = to+1;
c2 = 0;
while i<= ((to+p)-1)
   c2 = c2 + 2*uimg(i);
   i  = i+2;
end
ui    = (st/3)*(uimg(to)+ uimg(to+p)+c1+c2);
Greal = ((yr*ur)+(yi*ui))/(ur^2+ui^2)
Gimg  = ((yr*ui)-(yi*ur))/(ur^2+ui^2)
```

```
frequency = w
% To evaluate y(s1) and u(s1)
s1 = 8/ti
  s1om = s1/frequency
  c = 0;
for i = 1:to
       c       = c+(y(i)*exp(-s1*i*st));
       ys(i)   = c;
       t1(i)   = i;
end
ys1 = c*st
c1  = 0;
for k = 1:to
       c1      = c1+(u(k)*exp(-s1*k*st));
       us(k)   = c1;
       t2(k)   = k;
end
us1 = c1*st
```

```
% Program to identify inner loop  FOPTD model parameters
% Input: Intermediate values [a, b, kp]

function f = foptdsolve(x)
a = -0.8478;b = 0.0223; kp = 0.8722;
f = [a-b*x(1)-kp*cos(x(2));a*x(1)+b+kp*sin(x(2))];
```

MATLAB Program used for Case Studies in Chapter 6

```
% Program for saturation relay used as MATLAB function in Simulink(*)
% Input: ultimate gain ku from ideal relay simulation, relay height = h

function [u] = vivek_high(e)
k = 8.9354;kp = 0.125;h = 1;
u = k*e;
if kp>0
```

```
    if e>0
        if u>h
            u = h;
        end
    end
    if u<-h
        u = -h;
    end
end
if e>=0
    if u<-h
        u = h;
    end
end
if e = 0
    u = h;
end
```

Example 3

```
% Program for the intermediate values [s1, y(s1), u(s1)] required
for identification
% uses input (u_in.mat) and  output (y_in.mat) from relay tuning
of the simulink
% model (***)

clear all
close all
omega = 0.2431;
load y_out.mat;load y_in.mat;
x10 = y_out(1,:);y10 = y_out(2,:);
x20 = y_in(1,:);y20 = y_in(2,:);
t  = 121.52;
s1 = 8/t
s1byomega = s1/omega
```

```
for i = 1:1:20001;
    in1(i) = y10(i)*(exp(x10(i)*(-0.0658)));
end
h1 = 0.01;
f1 = in1(1)+in1(20001);
m1 = 0;
for i = 2:2:20000;
    m1 = m1+in1(i);
end
l1 = 0;
for i  = 3:2:20001;
    l1 = l1+in1(i);
end
area1 = (h1/3)*(f1+(4*m1)+(2*l1));

for i = 1:1:20001;
    in2(i) = y20(i)*(exp(x20(i)*(-0.0658)));
end
h2 = 0.01;
f2 = in2(1)+in2(20001);
m2 = 0;
for i = 2:2:20000;
    m2 = m2+in2(i);
end
l2 = 0;
for i  = 3:2:20001;
    l2 = l2+in2(i);
end
area2 = (h2/3)*(f2+(4*m2)+(2*l2));
ratio = area1/area2
```

```
% This program identifies parameters using higher order harmonics.
% Input: Intermediate values [s1, y(s1),u(s1)]
```

```
function f = tri_10(x)
ku = 8.7205;omega = 0.2431;
```

```
s1byomega = 0.2707;ratio = 0.0535;
v = ((ku*x)^2-1)^0.5
n = -s1byomega*(pi-atan(v))
der = (pi-atan(v))/omega
f = (ratio*((s1byomega*v)+1))-(x*exp(n));
```

```
% PID controller settings using IMC method for FOPTD system
% input: FOPTD model parameters

clear all; close all
D = 11.18; tao = 1.85; kp = 0.125;
lam = 1.7*D
kc = ((2*tao+D)/(2*lam))*(1/kp)
taoi = tao+(D/2)
```

MATLAB Program used for Case Studies in Chapter 8

Case study 1

```
% Program for the intermediate values [s1, y(s1), u(s1)] required
for identification
% uses input (exu2.mat) and  output (exy2.mat) from relay tuning
of the simulink
 % model (***)

clear all; close all
load exy2.mat; load exu2.mat
x10 = y2(1,:); y10 = y2(2,:);
x20 = u2(1,:); y20 = u2(2,:);
s = 4.9; s1 = 8/s
for i = 1:1:5001;
    in1(i) = y10(i)*(exp(x10(i)*(-s1)));
end
h1 = 0.01;
f1 = in1(1)+in1(5001);
m1 = 0;
```

```
for i  = 2:2:5000;
    m1 = m1+in1(i);
end
l1 = 0;
for i  = 3:2:5001;
    l1 = l1+in1(i);
end
y_s1 = (h1/3)*(f1+(4*m1)+(2*l1))

for i = 1:1:5001;
    in2(i) = y20(i)*(exp(x20(i)*(-s1)));
end
h2 = 0.01;
f2 = in2(1)+in2(5001);
m2 = 0;
for i  = 2:2:5000;
    m2 = m2+in2(i);
end
l2 = 0;
for i  = 3:2:5001;
    l2 = l2+in2(i);
end
u_s1 = (h2/3)*(f2+(4*m2)+(2*l2))

% Program to identify UFOPTD model parameters using principal
harmonics
% Input: Intermediate values [s1, y(s1),u(s1)]
function f = tri2(x);
ku = 5.8621; omega = 6.9813; s1 = 1.6327; ys1 = 0.0173; us1 =
0.0155; ratio = ys1/us1;
v = ((ku*x)^2-1)^0.5
der = atan(v)/omega
f = (ratio*((v*s1/omega)-1))-(x*exp(-s1*der))
kp = x
```

```
tao = v/omega
taod = der
```

```
% Program to identify UFOPTD model parameters using higher order
harmonics
% Input: Intermediate values [s1, y(s1),u(s1)], value of N=3 (no
of terms in series)
function f = ntri2(x);
ku = 7.2664; omega = 6.9813; s1 = 1.6327; ys1 = 0.0173; us1 =
0.0155; ratio = ys1/us1;
v = ((ku*x)^2-1)^0.5
der = atan(v)/omega
f = (ratio*((v*s1/omega)-1))-(x*exp(-s1*der))
kp = x
tao = v/omega
taod = der
```

Appendix C

C.1 Relay Tuning of Integrating Plus FOPTD Systems

The most used dynamic model for chemical engineering process is the FOPTD model. In many cases, time constant is very large; in some cases, dead time is the dominant one. In the literature, industrial examples of large time constants are reported by McNeill and Sacks (1969) for distillation and by Westerlund et al. (1980) for cement production. Some of the processes such as heating boilers, liquid storage tanks and batch chemical reactors are examples of integrating processes in industrial and chemical plants (Liu et al., 2005)

Zhang et al. (1999) proposed PID controller design method for integrating process with dead time and time constant. Kwak et al. (1997) proposed an online identification and autotuning method for integrating process. Kookos et al. (1999) proposed online PI controller tuning for integrating/dead time processes. Tian and Gao (1999) proposed a control scheme for integrating process with dominant time delay. The proposed scheme consist of a local proportional feedback to pre-stablize the process, a proportional controller for set-point tracking and a PD controller for load disturbance rejection. Huzmezan et al. (2002) designed a PID controller based on adaptive predictive control strategy to handle integrating type processes with large time constant. An industrial example of temperature control of a process that involves the heating and cooling of a batch reactor is studied.

C.2 Improved Analysis of Relay Tuning

The transfer function model for the integrating plus FOPTD system is given by K_p exp($-Ds$)/ [$s(\tau s+1)$]. Using Fourier series analysis, the output response can be represented as a combination of many sinusoidal waves:

$$y(t)=\left[a_1\sin(\omega_u t+\phi_1)+(1/3)a_3\sin(3\omega_u t+\phi_3)+(1/5)a_5\sin(5\omega_u t+\phi_5)+\ldots\right] \tag{C1}$$

where,

$$a_1=1\Big/\left\{\omega_u\left[1+(\tau\omega_u)^2\right]^{0.5}\right\};\quad a_3=1\Big/\left\{3\,\omega_u\left[1+(3\tau\omega_u)^2\right]^{0.5}\right\};\ \ldots\text{ etc.} \tag{C2}$$

$$\phi_1=-(D\omega_u)-(\pi/2)-\tan^{-1}(\tau\omega_u)=-\pi \tag{C3}$$

$$\phi_3 = -3D\omega_u - (\pi/2) - \tan^{-1}(3\tau\omega_u) = 3(-(D\omega_u) - (\pi/2) - \tan^{-1}(\tau\omega_u)) + \pi = -2\pi \quad \text{(C4)}$$

$$\phi_5 = -3\,\pi;\ \phi_7 = -4\,\pi, \text{ etc.} \quad \text{(C5)}$$

Eq.(C1) can be written as

$$y(t) = [a_1 \sin(\omega_u t - \pi) + (a_3/3)\sin(3\omega_u t - 2\pi) + (a_5/5)\sin(5\omega_u t - 3\pi) + \ldots] \quad \text{(C6)}$$

$$y(t) = [-a_1 \sin(\omega_u t) + (a_3/3)\sin(3\omega_u t) - (a_5/5)\sin(5\omega_u t) + \ldots] \quad \text{(C7)}$$

Let at $t^* = 0.5\,\pi/\omega_u$. Then from Eq.(C7) we get

$$y(t^*) = [-a_1 - (a_3/3) - (a_5/5) - (a_7/7) - (a_9/9) - \ldots] \quad \text{(C8)}$$

$$y(t^*) = -a_1[1 + (b_3/3) + (b_5/5) + (b_7/7) + (b_9/9) + \ldots] \quad \text{(C9)}$$

where,

$$b_3 = (a_3/a_1) = \left\{\omega_u\left[1 + (\tau\omega_u)^2\right]^{0.5}\right\} \Big/ \left\{(3\omega_u)\left[1 + (\tau\omega_u)^2\right]^{0.5}\right\}$$

Case 1: For smaller values of $\tau\omega_u$

For smaller values of $\tau\omega_u$, the value of $\tau\omega_u$ can be neglected when compared to 1 and hence the values of $b_3 = 1/3$, $b_5 = 1/5\ldots$

$$y(t^*) = A_1[1 + (1/9) + (1/25) + \cdots + (1/j2)] \quad \text{(C10)}$$

Here, 'A_1' is the modulus of a_1.

$$A_1 = 1.273\, y(t^*) \quad \text{(C11)}$$

In the conventional method, we use only the first term in Eq.(C10) [i.e., $A_1 = y(t^*)$]. Whereas the consideration of all the higher order harmonics gives $A_1 = 1.273\, y(t^*)$. This limiting value shows that a maximum error of 27% in k_u is obtained by using the conventional analysis (principal harmonics method).

Case 2: Limiting case of larger $\tau\omega_u$

Let us derive approximate evaluation of $y(t)$ for the limiting cases of larger $\tau\omega_u$. The phase angle criterion is given by

$$\phi_1 = -D\omega_u - (\pi/2) - \tan^{-1}(\tau\omega_u) = -\pi \quad \text{(C12)}$$

For larger $\tau\omega_u$ value, the $\tan^{-1}(\tau\omega_u) \cong \pi/2$. After substituting in Eq.(C12) gives, $D\omega_u = 0$ (since D and ω_u will have some value, the product cannot be equal to zero).

Let us consider $D\omega_u = q$ (C13)

$$\phi_3 = -3D\omega_u - (\pi/2) - \tan^{-1}(3\tau\omega_u) = -3\,q - \pi$$

$$\phi_5 = 5\,q - \pi;\ \phi_7 = 7\,q - \pi;\ \text{etc} \tag{C14}$$

Using values of ϕ_1, ϕ_3, ϕ_5 … in Eq.(C 1) we get

$$y(t) = \left[a_1 \sin(\omega_u t - \pi) + (a_3/3)\sin(3\omega_u t - 3q - \pi) + (a_3/5)\sin(5\omega_u t - 5q - \pi) + \cdots\right] \tag{C15}$$

$$y(t) = \left[-a_1 \sin(\omega_u t) - (a_3/3)\sin(3\omega_u t - 3q) - (a_5/5)\sin(5\omega_u t - 5q) - \cdots\right] \tag{C16}$$

$$y(t) = a_1\left[-\sin(\omega_u t) - (b_3/3)\sin(3(\omega_u t - 3q)) - (b_3/5)\sin(5(\omega_u t - 5q)) - \cdots\right] \tag{C17}$$

$$\text{where } b_3 = (a_3/a_1) = \left\{\omega_u\left[1 + (\tau\omega_u)^2\right]^{0.5}\right\} \Big/ \left\{(3\omega_u)\left[1 + (\tau\omega_u)^2\right]^{0.5}\right\} \tag{C18}$$

$$\text{let } a_{1m} = a_1\omega_u = 1 \Big/ \left\{\omega_u\left[1 + (\tau\omega_u)^2\right]^{0.5}\right\} \tag{C19}$$

$$\tau\omega_u = \left[1 - (a_{1m})^2\right]^{0.5} \Big/ a_{1m} \tag{C20}$$

For larger values of $\tau\omega_u$, the value of 1 can be neglected when compared to the value of $\tau\omega_u$ in Eq.(C 18) and hence the values of b_3, can be approximated as

$$b_3 = (1/9)\{[1/(\tau\omega_u)] + 1\} = (1/9)\left\{\left[a_{1m}\Big/\left(1 - a_{1m}^{\ 2}\right)^{0.5}\right] + 1\right\};\quad b_5 = (1/25)\left\{\left[a_{1m}\Big/\left(1 - a_{1m}^{\ 2}\right)^{0.5}\right] + 1\right\};$$

$$b_j = (1/j^2)\left\{\left[a_{1m}\Big/\left(1 - a_{1m}^{\ 2}\right)^{0.5}\right] + 1\right\};,\quad \text{etc.} \tag{C21}$$

Hence, Eq.(C17) becomes

$$y(t) = (a_{1m}/\omega_u)\left[-\sin(\omega_u t) - \left[(1/27)\left\{\left[a_{1m}\Big/\left(1 - a_{1m}^{\ 2}\right)^{0.5}\right] + 1\right\}\sin(3(\omega_u t - q))\right] - \cdots\right] \tag{C22}$$

$$y(t) = a_{1m}\left[-\sin(\omega_u t) - \left[(1/27)\left\{\left[a_{1m}\Big/\left(1 - a_{1m}^{\ 2}\right)^{0.5}\right] + 1\right\}\sin(3(\omega_u t - q))\right] - \cdots\right] \tag{C23}$$

The value of 'a_{1m}' is not the amplitude what we observe from the output oscillation.

Let us consider the time (t^*) at which

$$\omega_u\, t^* = 0.5\pi. \tag{C24}$$

where $\varepsilon = D/\tau$. For the various values of ε, ranging from 0.001 to 1.0, q value is calculated. It is found that the smallest value of q is 0.0351 (i.e., 2.01°). Using this q value in Eq.(C22), we get

$$y^*(t^*) = \left[a_{1m}(0.9682)\right] - \left\{\left[a_{1m}^{\ 2}\Big/\left(1 - a_{1m}^{\ 2}\right)^{0.5}\right](0.0318)\right\} \tag{C25}$$

using a_{1m} ranging from 0 to 1.0, a linear model is used to relate $y^*(t^*)$ and a_{1m} gives

$$y^*(t^*) = 0.91^* a_{1m} + 0.011 \tag{C26}$$

Rearranging, we get

$$a = 1.0989(y(t) - 0.012) \tag{C27}$$

The consideration of all higher order harmonics gives $a \cong 1.1y(t^*)$. This limiting value shows that a maximum error −10% in k_u is obtained by using the conventional analysis (principal harmonics method).

C.3 Parameters Estimation by Proposed Method 1

Let us consider an integrating plus first order plus time delay (IFOPTD) system:

$$y(s)/u(s) = k_p \exp(-Ds)/[s(\tau s - 1)] \tag{C28}$$

Where k_p = process gain, D = time delay and τ = time constant, are the parameters to be estimated. Using symmetric relay, with relay height $\pm h$, the output $[y(t)]$ and input $[u(t)]$ data are recorded. From the definition of Laplace transform (Kreyszig, 1996) we get

$$y(s) = \int_0^\infty y(t) \exp(-st)dt \tag{C29}$$

The above integral can be evaluated for a particular value of s (say s_1). It is suggested to use the value $s_1 = 8/t_s$, where 't_s' is the time at which three repeated cycles of oscillations appear in the output. The reason for taking $s_1 = 8/t_s$ is that, for $t > t_s$, because of a very small value of the term $\exp(-s_1 t)$, the contributions by subsequent terms is negligible while evaluating the integral. Let the above resulting integral value be denoted as $y(s_1)$. Similarly,

$$u(s) = \int_0^\infty u(t)\exp(-st)dt \tag{C30}$$

The above equation is also evaluated using $s_1 = 8/t_s$. With the numerical values of $y(s_1)$ and $u(s_1)$, using the assumed model [Eq.(C27)], an equation is formulated here

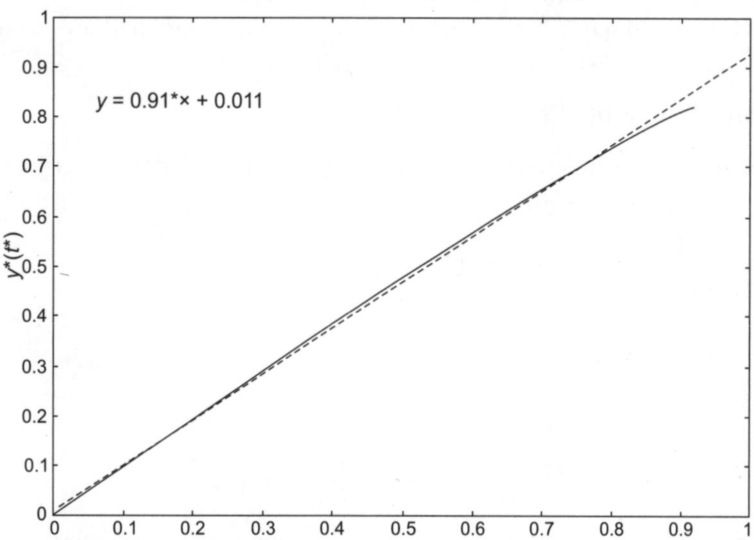

Fig. C.1 *$y^*(t^*)$ versus a_{1m} corresponding to Eq.(C24)*

$$y(s_1)/u(s_1) = k_p \exp(-Ds_1)/[s(\tau s_1 + 1)] \tag{C31}$$

On cross-multiplying, we get

$$\{[y(s_1)/u(s_1)][s(\tau s_1 + 1)]\} - k_p \exp[-Ds_1] = 0 \tag{C32}$$

From the amplitude criterion, we get

$$k_p k_u / \{\omega_u [(\tau\omega_u)^2 + 1]\} = 1 \tag{C33}$$

k_u used in the above equation is the corrected k_u giving

$$v = \tau\omega_u = \left[(k_p k_u/\omega_u)^2 - 1\right]^{0.5} \tag{C34}$$

From the phase angle criterion, we get

$$D = \{[(\pi/2) - \tan^{-1}(v)]/\omega_u\} \tag{C35}$$

Eq.(C30) becomes

$$([y(s_1)/u(s_1)]\{s_1[(v/\omega_u)s_1 + 1]\}) - (k_p \exp\{-(s_1/\omega_u)[(\pi/2) - \tan^{-1}(v)]\}) = 0 \tag{C36}$$

Solving numerically, Eq.(C34), we get the value for k_p. Since

$$v = \tau\omega_u = \left[(k_p k_u/\omega_u)^2 - 1\right]^{0.5} \tag{C37}$$

using $v = \tau\omega_u$ and Eq.(C35), we get τ and D.

C.4 Parameter Estimation by Proposed Method 2

In this method, the model parameters are identified without using k_u. Using the input data $u(t)$ and output data $y(t)$, the parameters are identified.

Using Eq.(C27) and Eq.(C28), the numerical values of $y(s_1)$ and $u(s_1)$ are calculated. Using the assumed model, an equation is formulated as given below:

$$y(s_1)/u(s_1) = k_p \exp(-Ds_1)/[s(\tau s_1 + 1)] \tag{C38}$$

Since $y(t)$ and $u(t)$ are piecewise continuous and periodic, the Laplace transform of $y(t)$ and $u(t)$ can be written as (Kreyszig, 1996)

$$y(s) = \frac{1}{1 - \exp(-Ps)} \int_0^P y(t) \exp(-st)\, dt \tag{C39}$$

$$u(s) = \frac{1}{1 - \exp(-Ps)} \int_0^P u(t) \exp(-st)\, dt \tag{C40}$$

From Eq.(C38) and Eq.(C39), we get

$$G(s) = y(s)/u(s) = \left[\int_0^P y(t)\exp(-st)\,dt\right] \Big/ \left[\int_0^P u(t)\exp(-st)\,dt\right] \tag{C41}$$

Where $P = 2\pi/\omega$ and ω is the frequency of oscillations observed in the output.

By substituting $s = j\omega$ and using Euler's expansion for $\exp(-j\omega t)$ in Eq.(C27) we get

$$G(j\omega) = y(j\omega)/u(j\omega) = (c_1 - jd_1)/(c_2 - jd_2) \tag{C42}$$

Where

$$c_1 = \int_0^P y(t)\cos(\omega t)\,dt \tag{C43a}$$

$$d_1 = \int_0^P y(t)\sin(\omega t)\,dt \tag{C43b}$$

$$c_2 = \int_0^P u(t)\cos(\omega t)\,dt \tag{C43c}$$

$$d_2 = \int_0^P u(t)\sin(\omega t)\,dt \tag{C43d}$$

Equations (C42a) to (C42d) can be evaluated numerically using $y(t)$ and $u(t)$ data of relay test. Hence, Eq.(C41) written as

$$G(j\omega) = p + jq \tag{C44}$$

Where

$$p = (c_1 c_2 + d_1 d_2)/(c_2^2 + d_2^2) \tag{C45a}$$

$$q = (d_2 c_1 - d_1 c_2)/(c_2^2 + d_2^2) \tag{C45b}$$

By substituting $s = j\omega$ in Eq.(C27), we get

$$G(j\omega) = k_P \exp(-\theta j\omega)/\left[(j\omega)(\tau j\omega + 1)\right] \tag{C46}$$

On substituting $\exp(-\theta j\omega) = \cos(\theta\omega) - j\sin(\theta\omega)$ in Eq.(C45), we get

$$a + bj = k_p\left[\cos(D\omega) - j\sin(D\omega)\right]/\left[(j\omega)(\tau j\omega + 1)\right] \tag{C47}$$

On cross-multiplying and separating the real and imaginary terms, we get

$$-(a\,\tau\,\omega^2) - (\omega b) - \left[k_p \cos(D\omega)\right] = 0 \tag{C48}$$

$$\omega a - (b\,\tau\,\omega^2) + \left[k_p \sin(D\omega)\right] = 0 \tag{C49}$$

From the relay test, the three linear algebraic equations Eqs (C37), (C47) and (C48) can be solved to estimate the three unknown parameters: K_P, θ and τ. In Eq.(C47) and Eq.(C48) the frequency of oscillation of the system output is used.

In proposed method 2, we do not use the equation $K_u = 4h/(\pi a)$ and $\omega_u = 2\pi/P_u$. Thus, using single symmetrical relay test, all the three parameters of an FOPTD system are calculated.

Simulation Results

To verify effectiveness of proposed method 1 and proposed method 2, a simulation example is considered as

$$y(s)/u(s) = 1.0\ \exp(-2s)/[s(s+1)] \tag{C.50}$$

The intermediate values calculated are $s_1 = 2045$, $y(s_1) = 0.6053$, $u(s_1) = 0.2295$, $a = -1.6638$ and $b = 0.0336$. Table C.1 gives the model parameters identified by the proposed method 1 using the principal harmonics analysis ($N = 1$), higher order harmonics ($N = 3$) analysis and that identified by the proposed method 2. Based on the identified model, a PID controller is designed using the IMC method proposed by Skogestad (2003). The servo performance comparison shows that the proposed method 1 with $N = 3$ and the proposed method 2 give ISE and ITAE values closer to that of the actual system.

Table C.1 *Estimated parameters of the integrating FOPTD model by the proposed methods*

Parameter	Actual	Method 1		Method 2
		k_u		
		$N = 1$	$N = 3$	
k_p	0.98	0.99	0.99	0.98
τ	0.6	1.2	1.2	0.904
D	2.43	1.83	1.83	2.07
k_c	0.25	0.21	0.275	0.25
τ_I	16	19.44	14.64	16.56
τ_D	1	0.6	1.2	0.904
IAE	9.58	10.28	9.2	9.61
ISE	5.06	5.33	4.9	5.07
ITAE	100.58	120.4	91.6	101.9

$P_u = 11.84$

Table C.2 *Effect of including higher order harmonics on* k_u

D/τ	k_u					k_u exact
	$N = 1$	$N = 3$	$N = 5$	$N = 7$	$N = 9$	
4	0.599	0.688	0.707	0.716	0.72	0.67
8	0.61	0.71	0.73	0.74	0.743	0.71
10	0.617	0.714	0.734	0.743	0.748	0.721
20	0.627	0.723	0.743	0.752	0.76	0.75

Table C.3 *Details of calculations for Table C.2*

D / τ	a_0 observed amplitude	ω_u frequency	t^* (=0.5*π/ω_u)	$y(t^*)$	a calculated amplitude
4	2.151	0.629	2.498	2.13	1.85 ($N = 3$)
8	2.077	0.698	2.25	2.07	1.795 ($N = 3$)
10	2.061	0.714	2.20	2.053	1.783 ($N = 3$)
20	2.031	0.748	2.10	2.026	1.712 ($N = 5$)

Nomenclature

a	amplitude of oscillation corresponds to the principal harmonics
a_C	corrected amplitude as per Eq.(6.6) for saturation relay tuning
a_0	amplitude of oscillation observed from the process output
a_1	corrected amplitude of output oscillation
A_1 to A_5	defined by Eq.(5.14) for series cascade control tuning
A_1 to A_6	defined by Eq.(5.35) for parallel cascade control tuning
A_m	gain margin
b_3, b_5	defined by Eq.(2.29) or Eq.(8.15)
c_1, c_2	defined by Eq.(2.43) or Eq.(7.11)
c_1, c_2	defined by Eq.(10.10)
d_1, d_2	defined by Eq.(10.10)
d_1, d_2	defined by Eq.(7.11)
d_1	disturbance entering outer loop
d_2	disturbance entering inner loop
D	time delay
D_1	dilution rate, h^{-1}
e	$y_r - y$
G	process transfer function
h	relay height
k_c	controller gain
K_c	PID controller parameter
K_p	steady state process gain
K_u	ultimate gain
k_{c1}, τ_I	outer loop PI controller settings
k_{c2}	inner loop controller gain
$k_{c,\max}$, k_u	controller ultimate gain

k_p	process gain
$(k_pG_p)_1$	transfer function of the outer loop process
$(k_pG_p)_2$	transfer function of the inner loop process
$(k_LG_L)_1$	transfer function for load disturbance in the outer loop
$(k_LG_L)_2$	transfer function for load disturbance in the inner loop
K_1	$k_{p1}k_{c1}$
K_2	$k_{p2}k_{c2}$
K_m	constant in the expression for μ
L	dead time (in saturation relay tuning)
N	number of terms considered for higher order harmonics calculations
P	period of process output at constant cycling
P_u	ultimate period of oscillations
p	defined by Eq.(7.13) for chapter 7
p	defined by Eq.(9.69) for chapter 9
q	defined by Eq.(7.13) for chapter 7
q	defined by Eq.(9.69) for chapter 9
q	$D\omega_u$ in Chapter 8
P_1	K_1K_2
P_2	P_1/τ_I^*
P_u	period of output oscillation
s	Laplace variable
s_1	$8/t_s$
S_f	safety factor as in Eq.(6.2) for saturation relay tuning
t_s	time taken to reach 3 invariant cycles of oscillation in the output
t	time
t_o	starting time for integration
t_s	time for repeated cycles of oscillations
t^*	$0.5\ \pi/\omega_u$
u	input variable
ν	$[(k_uk_p)^2-1]^{0.5}$
X_1	cell mass concentration, g/l
X_2	substrate concentration in the bio reactor, g/l
$X_{2,f}$	feed substrate concentration, g/l
y	process output
y_1	outer loop output variable

y_2	inner loop output variable
y_r	reference value for y

Greek letters

α, β	lead and lag time constants (Chapter 9)
α_1, α_2	tuning parameters for the equating coefficient method (Chapter 5)
τ	process time constant
τ_1, τ_2, τ_3	time constants of SOPTD model
τ_d	process time delay
τ_I	integral time
τ_D	derivative time
ω	frequency of oscillation
ω_u	ultimate frequency of oscillation
α_1, α_2	tuning parameters for cascade controller tuning
ε	D/τ (Chapter 8)
ε	L_1/τ_2
λ	L_2/τ_2
γ	displacement in the relay height (Chapter 7)
ϕ	τ_1/τ_2
μ	specific growth rate, h^{-1}
μ_m	maximum specific growth rate
τ_I^*	τ_I/τ_2
η	$(\varepsilon+\lambda)$
ψ	$\eta+\varepsilon$
ω_p	phase cross-over frequency
ω_g	gain cross-over frequency
ϕ_m	phase margin

Subscript

0	outer loop
1	inner loop

Problems

1. Consider a system given by $y(s)/u(s) = 2/(I + 1)^3$. With a proportional controller (withgain $k=2$), get the response in y for a set point change of 0.1 of the closed loop system. Similarly, with each value of $k_c= 2.5, 3.0, 3.5$ and 4, get the response for a step change in the set point. Can you get a sustained oscillation in the output when $k_c=4$? From the period of oscillation, and $k_{c,max}$ value, design a suitable PID controller using the Ziegler–Nichols tuning formulae. With the designed controller, obtain the response in the output for a step change in the set point.

2. For the system given in problem 1, replace the proportional gain by an on–off symmetric relay ($h = +1$ and -1 height). No change in the set point is considered. Check whether you can get a sustained oscillation in the output. Note down the period of oscillation. Using Eq.(1.8) calculate the controller ultimate gain. Hence, calculate the PID settings using the Ziegler–Nichols method based on continuous cycling oscillation method. With the designed controller, obtain the response in the output for a step change in the set point.

3. Consider the system $y(s)/u(s) = 2 \exp(-2s)/[(8s+1)(s+1)]$. Using a symmetric relay with $h = +1$ and -1, obtain the oscillation in the output and identify the model parameters of FOPTD as given in section 2.2. Use the method of Li et al. (1991) to get the model parameters after noting the delay time from the response. Calculate the values of PID settings by the Ziegler–Nichols tuning method. With the designed controller, obtain the closed loop servo response of the original system.

4. From the relay oscillations of problem 3, consider the higher order harmonics to get an improved estimate of k_u. Calculate the values of PID settings by the Ziegler–Nichols tuning method. With the designed controller, obtain the closed loop servo response of the original system.

5. Consider $y(s)/u(s) =3 \exp(-0.2s)/[3s+1)(s+1)]$ with an asymmetric relay $h = 1$ and $\gamma = 2$, and get the closed loop oscillation in the output y. Using the method given in section 2.6, obtain the values of FOPTD model parameters. Apply the method of Huang et al. (1996a, b) in order to get the model parameters of an FOPTD system. Design a PID controller by IMC method (Skogestad, 2003). Compare the servo responses of the original system for a unit step change in the set point.

6. Consider a series cascade control scheme with inner loop process $(k_pG_p)_2 = 2 \exp(-0.2s)/(0.05s + 1)$ and outer loop process $(kpGp)_1 = 1.5 \exp(-0.2s)/(0.05s + 1)$. Consider a symmetric

square relay with h=1 and -1. Carry out the sequential loop relay tuning method 1as given in Chapter 3 to get the ultimate controller gain and the period of oscillation. Use the principal harmonics and higher order harmonics methods to get the value of k_u. Based on the Ziegler–Nichols method, design P–PI controllers. Compare the closed loop servo responses for both these approaches.

7. Consider a series cascade control scheme with inner loop process $(k_pG_p)_2$ = 2 exp (−0.2*s*)/(0.05*s* + 1) and outer loop process $(k_pG_p)_1$ = 1.5 exp(−0.2*s*)/(0.05*s* + 1). Consider an asymmetric square relay with h = 1 and γ = 2. Carry out the method 2 of Chapter 3 (sequential loop relay tuning method) to get the FOPTD model parameters. Design P–PI controllers by IMC method and get the closed loop servo and regulatory responses.

8. Consider a parallel cascade control scheme with inner loop process $(k_pG_p)_2$ = 2 exp(−0.2*s*)/(0.05*s* + 1) and outer loop process $(k_pG_p)_1$ =1.5 exp(−0.2*s*)/(0.05*s* + 1). Consider a symmetric square relay with h = 1 and −1. Carry out the sequential relay tuning method as given in the section 3.5 to get the ultimate controller gain and oscillations. Use the principal harmonics and higher order harmonics methods. Design P–PI controllers based on the Ziegler–Nichols continuous cycling method. Compare the closed loop servo responses for both these approaches.

9. Consider a parallel cascade control scheme with inner loop process $(k_pG_p)_2$ = 2 exp (−0.2s)/(0.05*s* + 1) and outer loop process $(k_pG_p)_1$ = 1.5 exp(−0.2*s*)/(0.05*s* + 1). Consider an asymmetric square relay with h = 1 and gamma = 2. Carry out the method 2 of Chapter 3 (sequential loop relay tuning method as given in Chapter 3, section 3.5) to get the FOPTD model parameters. Design P–PI controllers by IMC method and get the closed loop servo and regulatory responses.

10. Consider a series cascade control scheme with inner loop process $(k_pG_p)_2$ = 2 exp(−0.2*s*)/(0.05*s*+1) and outer loop process $(k_pG_p)_1$ = 1.5 exp(−0.2*s*)/(0.05*s*+1). Consider a symmetric square relay with h = 1 and −1. Carry out the simultaneous relay tuning method as given in Chapter 3 to get the ultimate controller gain and oscillations. Use the principal harmonics and higher order harmonics methods. Design P–PI controllers by the Ziegler–Nichols tuning method. Compare the closed loop servo responses for both these approaches.

11. Consider a series cascade control scheme with inner loop process $(k_pG_p)_2$ = 2 exp(−0.2s)/(0.05*s* + 1) and outer loop process $(k_pG_p)_1$ = 1.5 exp(−0.2s)/(0.05*s* + 1). Consider an asymmetric square relay with h = 1 and (symbol see problem 7) = 2. Carry out the simultaneous relay tuning method as given in Chapter 3 to get the FOPTD model parameters. Design P–PI controllers by IMC method and get the closed loop servo responses.

12. Consider a parallel cascade control scheme with inner loop process $(k_pG_p)_2$ = 2 exp(−0.2s)/(0.05*s* + 1) and outer loop process $(k_pG_p)_1$ = 1.5 exp(−0.2*s*)/(0.05*s* + 1). Consider a symmetric square relay with h = 1 and −1. Carry out the simultaneous relay tuning method as given in Chapter 3 to get the ultimate controller gain and oscillations. Use the principal harmonics and higher order harmonics methods. Design P–PI controllers by the Ziegler–Nichols continuous cycling method. Compare the closed loop servo responses for both these approaches.

13. Consider a parallel cascade control scheme with inner loop process $(k_pG_p)_2$ = 2 exp $(-0.2s)/(0.05s+1)$ and outer loop process $(k_pG_p)_1 = 1.5 \exp(-0.2s)/(0.05s+1)$. Consider an asymmetric square relay with $h = 1$ and $\gamma = 2$. Carry out the method 2 of Chapter 3 (simultaneous relay tuning method as given in Chapter 3, section 3.5) to get the FOPTD model parameters. Design P–PI controllers by IMC method and get the closed loop servo responses.

14. Consider a series cascade control scheme with inner loop process $(k_pG_p)_2 = 2 \exp(-0.2s)/(0.05s+1)$ and outer loop process $(k_pG_p)_1 = 1.5 \exp(-0.2s)/(0.05s+1)$. Design P controller for the inner loop and PI controller for the outer loop by equating coefficient (EC) method of Chapter 5. Compare the closed loop servo responses and regulatory responses of the present (EC) method with that of the Ziegler–Nichols tuning method.

15. Consider a parallel cascade control scheme with inner loop process $(k_pG_p)_2 = 2 \exp(-0.2s)/(0.05s+1)$ and outer loop process $(k_pG_p)_1 = 1.5 \exp(-0.2s)/(0.05s+1)$. Design P controller for the inner loop and PI controller for the outer loop by equating coefficient (EC) method of Chapter 5. Compare the closed loop servo responses and regulatory responses of the (EC) method with that of the Ziegler–Nichols tuning method.

16. Consider an FOPTD system with $k_p = 1$, $\tau = 1$ and $D = 5$. Carry out the saturation relay test given in section 6.4 to get the model parameters and hence design PI controllers for the identified model and the original systems. Compare the servo and regulatory responses.

17. Consider the system with $y(s)/u(s) = 1.5 \exp(-2s)/[(8s+1)(s+1)]$ and a symmetric relay h = 1 and –1. Using the oscillatory response, identify all the three parameters of FOPTD system using the method given in Chapter 7.

18. Consider the system $y(s)/u(s) = 1.5 \exp(-2s)/[(4s-1)]$. Using a symmetric relay with h = +1 and –1, get the oscillation. Calculate the improved value of k_u by considering the higher order harmonics by the method given in section 8.2. Estimate the FOPTD model parameters by the method given in section 8.4. For the original model and the identified model, design PID controllers by the method of Padmasree et al. and compare the closed loop servo responses.

19. Consider the system $I(s)/u(s) = 1.5 \exp(-2s)/[(4s+1)^2]$. Using a symmetric relay with h = +1 and –1, get the oscillation. Calculate the improved value of k_u by considering the higher order harmonics by the method given in section 9.2. Estimate the SOPTD model (with equal time constants) parameters by the method given in section 9.4. For the original and the identified models, design PID controllers by IMC method and compare the closed loop servo responses.

20. Consider the system $y(s)/u(s) = 1.5 \exp(-2s)/[(4s+1)^3]$. Using an asymmetric relay with $h = +1$ and $\gamma = 2$, get the sustained oscillation. Estimate the SOPTD model with equal time constants parameters by the method given in section 10.5. For the original model and the identified model, design PID controllers by the method of Huang and Chen (1999) and get the closed loop servo responses

21. Consider the system $y(s)/u(s) = 1.5 \exp(-2s)(2s + 1)/[(4s+)(8s + 1)]$. Using an asymmetric relay with $h = +1$ and $\gamma = 2$, get the sustained oscillation. Estimate all the 5 parameters of a general SOPTD model by the method given in section 11.5. For the original model and the identified model, design PID controllers by the method of Rao and Chidambaram (2007) and compare the closed loop servo responses

22. For the multivariable system

$$G(s) = \begin{bmatrix} \dfrac{3e^{-2s}}{1+8s} & \dfrac{e^{-s}}{1+4s} \\ \dfrac{(s+1)e^{-s}}{(s^2+5s+6)} & \dfrac{e^{-6s}}{1+2s} \end{bmatrix}$$

conduct the asymmetric relay test to calculate the estimation of the FOPTD model by decentralized relay feedback method. Design IMC controllers by the Tanttu and Lieslehto method (1991). Obtain the closed loop servo responses.

23. For the multivariable system given in problem 22, conduct the asymmetric relay test to calculate the estimation of the general SOPTD model by decentralized relay feedback method. Design IMC controllers by the Tanttu and Lieslehto method(1991). Obtain the closed loop servo responses.

24. Balachandran and Chidambaram (1996) presented a method of designing robust multivariable PID controllers for crude distillation columns, which will give good performances on three different crudes. Give a summary of the work reported by them.

25. Govindhakannan and Chidambaram (1997) and Govindhakannan and Chidambaram (2000) reported simple methods of designing multivariable PID controllers for unstable systems. Give a summary of the work reported by them.

Suggestive Reading

Ganesh, D.S. and M. Chidambaram. 2005. Identification of multivariable systems using relay tuning, *Chem. BioChem. Engg. Q.*, 19(2), 153–58. [Figs 13.1 to 13.2]

Ramakrishnan, V. and M. Chidambaram. 2003. Estimation of a SOPTD transfer function model using a single asymmetrical relay feedback test, *Comp and Chem. Engg.*, 27, 1779–784. [Figs 10.1 to 10.5]

Reddy, P.D.S., M. Pandit and M. Chidambaram. 2006. Comparison of multivariable controllers for non-minimum phase systems, *Int. J. Modeling and Simulation,* 26, 237–43. [Figs 14.1 to 14.4]

Srinivasan, K. and M. Chidambaram. 2003. Modified relay feedback method for improved system identification, *Comp and Chem. Eng.*, 27, 727–32. [Figs 2.8 to 2.12]

Srinivasan, K. and M. Chidambaram. 2004. An improved auto tune identification method, *Chem. BioChem. J.*, 18, 249–56. [Figs 2.1 to 2.7]

Sivakumar, E.,Vivek Sathe and M. Chidambaram. 2005. Improved saturation relay tuning test for systems with larger dead time, *Ind. and Eng. Chem. Res.*, 44, 2183–190. [Figs 6.1 to 6.10]

Vivek Sathe, S. and M. Chidambaram. 2005a. Identification using single symmetrical relay feedback test, *Comp and Chem. Engg.*, 29, 1625–630. [Figs 7.1 to 7.5]

Vivek Sathe, S. and M. Chidambaram. 2005b. An improved auto tuning of PID controllers for unstable FOPTD system, *Comp and Chem. Engg.*, 29, 2060–68. [Figs 8.1 to 8.6]

Vivek Sathe, S. and M. Chidambaram. 2006. A simple method for tuning parallel cascade controllers, *Indian Chemical Engineer,* 48, 262–69. [Figs 3.7 to Fig 3.11]

Vivek Sathe, S. and M. Chidambaram. 2004. Cascade controller tuning by relay auto tune method, *J. Indian Institute of Sci.*, 84, 89–97. [Figs 3.1 to 3.6]

Vivek Sathe, S. and M. Chidambaram. 2008. Simultaneous relay tuning of series cascade controllers, *Indian Chemical Engineer,* 49(4), 324–29. [Figs 4.1 to 4.9]

Vivek Sathe, S. and M. Chidambaram. 2004. A simple method of tuning cascade controllers, *J. Indian Institute of Sci.*, 84, 233–42. [Figs 5.1 to 5.11]

Vivek Sathe, S. and M. Chidambaram. 2012. Identification of five parameters of a SOPTD model by relay tuning, *Indian Chemical Engineer,* 54(2), 79–96. [Figs 11.1 to 11.7]

Vivek Sathe, S. and M. Chidambaram. 2012. An improved relay auto tuning of PID Controllers for critically damped SOPTD systems, *Chem. Eng. Commun.*, 119(11), 1437–462. [Figs 9.1. to 9.14]

Conferences

Ganesh, D.S. and M. Chidambaram. January 3–4, 2003. Multivariable system identification using relay feedback, *Proc. Int. Symp. Process System Engg. and Control,* 265–69. [Fig 12.1 to 12.2]

Sadasivarao, M.V and M. Chidambaram. December 4–6, 2003. Multivariable controller tuning by genetic algorithms, *Proc. ICECON'03, Thiruchirappalli.* [Fig 15.1 to 15.5]

References

Abbas, A. (1997) A new set of controller tuning regulations, *ISA Transactions*, 36, 183–87.

Agarwal, P. and H.C. Lim (1984) Analysis of various control schemes for continuous bioreactors, *Advances in Biochemical Eng./Biotechnology*, 30, 61–90.

Ang, K.H., G. Chong and Y. Li (2005) PID control system analysis, design and technology, *IEEE Trans Cont. Sys. Tech.*, 13, 559–76.

Åström, K.J. and T. Hägglund (1984) Automatic tuning of simple regulators with specification on phase angle and amplitude margins, *Automatica*, 20, 645–51.

Åström, K.J. and T. Hägglund (1995) PID controllers: Theory, design and tuning, *Instru. System. Control Soc.*, North Carolina.

Åström, K.J., P. Albertos and J. Queveda (2001) PID Control, *Control Engineering Practice*, 9, 1159–161.

Åström, K.J. and T. Hägglund (2001) The future of PID control, *Control Engineering Practice*, 9, 1163–175.

Åström, K.J. and T. Hägglund (2004) Revisiting the Ziegler-Nichols step response method for PID control, *J. Process Control*, 14(6), 635–50.

Atherton, D.P. (1975) Non-linear control engineering, *Van Nostrand Reinhold*, London.

Atherton, D.P. (2000) Relay auto tuning, use of old ideas in a new setting, *Trans. Inst. Measu. Control*, 22, 103–22.

Balachandran, R. and M. Chidambaram (1996) Comparison of performances of decentralized controllers, *Process Control and Quality*, 8, 123–31.

Bequette, B.W. (2003) Process control: Modeling, design and simulation, *Prentice Hall India*, New Delhi.

Brambilla, A. and D. Semino (1992) Nonlinear filter in cascade control schemes, *Ind. Eng. Chem. Res.*, 31(12), 2694–699.

Chandrasekhar, R., R. Padmasree and M. Chidambaram (2002) Design of PI/PID controllers for unstable systems with time delay by synthesis method, *Indian Chem. Engr.*, 44, 82–88.

Chen, C.L. (1989) A simple method of on-line identification and controller tuning, *AIChE J.*, 35, 2037–203.

Chen, D. and D.E. Seborg (2002) PI/PID controller design based on direct synthesis and disturbance rejection, *AIChE J.*, 41, 4807–822.

Chiang, R.C., S.H. Shen and C.-C, Yu (1992) Derivation of transfer function from relay feedback systems, *Ind. Eng. Chem. Res.*, 31, 855–60.

Chidambaram, M. (1997) Control of unstable systems: A review, *Journal of Energy, Heat and Mass transfer*, 19, 49–57.

Chidambaram, M. (1998) Applied process control, *Allied Publishers Ltd.*, New Delhi.

Chidambaram, M. and R. Padmasree (2003) A simple method of tuning PI controllers for integrating plus time delay systems, *Comp. & Chem. Eng.*, 27, 211–15.

Chien, I-L. and P.S. Fruehauf (1991) Consider IMC tuning to improve controller performance, *Chemical Engineering Progress*, Oct., 33–41.

Chipperfield, A., P. Fleming, H. Pohlheim and C. Fonseca (1986) Genetic algorithm MATLAB tool box, *Department of Automatic Control and Systems Engineering, University of Sheffield*, Sheffield.

Clement, V. and M. Chidambaram (1997a) PID control of unstable FOPTD systems, *Chem. Eng. Comm.*, 162, 63–74.

Clement, V. and M. Chidambaram (1997b) PI control of unstable FOPTD systems, *Indian Chemical Engineer*, 39, 9–14.

Cohen, G.H. and G.A. Coon (1953) Theoretical considerations of retarded control, *Trans. ASME*, 75, 827–34.

Cominos, P. and N. Munro (2002) PID controllers: Recent tuning methods and design to specification, *IEE Proc. - CTA*, 149(1), 46–53.

Datta, A., M. Ho and S.P. Bhattacharya (2000) Structure and synthesis of PID controllers, *Springer*, Berlin.

Davison, E.J. (1976) Multivariable tuning regulators: The feed forward and robustness of general servo-mechanism problem, *IEEE Trans. Aut. Contr.*, AC-21, 35–41.

DePaor, A.M. and M. O'Malley (1989) Controllers of Ziegler-Nichols type for unstable system with time delay, *Int. J. Control*, 41, 1025–036.

Edgar, T.F., R.C. Heeb and O.J. Hougen (1982) Computer aided process control system design using iterative graphics, *Comp. & Chem. Eng.*, 5(4), 225.

Frank, P.M. (1974) Entwurf von regelkreisen mit vorgeschrieben verhalten, *G. Braun, Karlsruhe.*

Foley, M.W., N.R. Ramharack and B.R. Copeland (2005) Comparison of PI controller tuning methods, *Ind. Eng. Chem. Res.*, 44, 6741–750.

Friman, M., K.V. Waller (1994) Auto tuning of multi loop control system, *Ind. Eng. Chem. Res.*, 33, 1708–717.

Friman, M. and K.V. Waller (1997) A two channel relay for auto tuning, *Ind. Eng. Chem. Res.*, 37, 2662–671.

Ganesh, D. and M. Chidambaram (2003) Multivariable system identification using relay feedback, *Proc. Int. symp. Process syst. Engg. Control*, Mumbai, Jan 4–5.

Ganesh, D. and M. Chidambaram (2005) Identification of multivariable system using relay tuning, Chem. Biochem. Eng., Q-19(2), 153–58.

Goldberg, D.E. (1989) Genetic algorithms in search, optimization and machine learning, *Addison-Wesley Publications,* New York.

Govindhakannan, J. and M. Chidambaram (1997) Multivariable PI control of unstable systems, *Process Control and Quality*, 10, 319–29.

Govindhakannan, J. and M. Chidambaram (2000) Two-stage multivariable PID controllers for unstable plus time delay systems, *Indian Chemical Engineer*, 42, 89–93.

Grefenstette, J.J. (1986) Optimization of control parameters for genetic algorithms, *IEEE Trans.,* SMC-16(1), 122–28.

Hang, C.C., K.J. Astrom and W.K. Ho (1993) Relay auto tuning in the presence of static load disturbance, *Automatica*, 29, 563–64.

Hang, C.C., A.P. Loh and V.U. Vasnani (1994) Relay feedback auto tuning of cascade controllers, *IEEE Control sys. Tech*, CST-2, 42–45.

Hang, C.C., K.J. Astrom and Q.G. Wang (2002) Relay feedback auto tuning of process controllers: A tutorial review, *J. Process Control*, 12, 143–62.

He, J.B., Q.G. Wang and T.H. Lee (2000) PI/PID controller tuning via LQR approach, *Automatica*, 55, 2429–439.

Ho, W.K., C.C. Hang and L.S. Cao (1995) Tuning of PID controllers based on gain and phase margin specifications, *Automatica,* 31(3), 497–502.

Ho, W.L., O.P. Gan, E.B. Tag and E.L. Ang (1996) Performance and gain and phase margins of well-known PID tuning formulas, *IEEE Trans. Contr. Syst. Technol.,* 4, 475–77.

Ho, W.K. and W. Xu (1998) PID tuning for unstable processes based on gain and phase margin specification, *IEE Proc.- CTA,* 145, 393–96.

Holt, B.R. and M. Morari (1985) Design of resilient processing plants – VI. The effect of right – half plane zeros on dynamic resilience, *Chemical Engg. Sci.,* 40, 59–74.

Huang, H.P., C.I. Chen, C.W. Lai and G.B. Wang (1996a) Autotuning for model based PID controllers, *AIChE J.*, 42, 2687–691.

Huang, H.P., C.L. Chen, G.B. Guo and C.W. Lai (1996b) Estimation of SOPTD transfer function modles using an auto-tuning test, *J Chinease I. Ch. E.*, 27(3), 153.

Huang, H.P. and C.C. Chen (1997) Control system synthesis for open loop unstable processes with time delay, *IEE Proc.* – CTA, 144, 334–46.

Huang, H.P. and C.C. Chen (1999) Auto tuning of PID controllers for second order unstable process having dead time, *J. Chem. Eng. Japan*, 32, 486–97.

Huzmezan, M., W.A. Gough, G.A. Dumont and S. Kovac (2002) Time delay integrating systems: A challenge for process control industries, a practical solution, *Control Engineering Practice*, 10, 1153–161.

Isaksson, A. and T. Hagglund (2002) Special section on PID Control, *IEE Proceedings*-CTA, 149, 1.

Jacob, E.F. and M. Chidambaram (1996) Robust control of an unstable bioreactor, *Process Contr. & Quali.*, 8, 167–75.

Johansson, K.H. (2000) The quadruple tank process: A multivariable laboratory process with an adjustable zero. *IEEE Transactions on Control Systems Technology*, 8, 456–65.

Johnson, M.A. and M.H. Moradi (2005) PID control: New identification and design methods, *Springer*, London.

Jhunjhunwala, M. and M. Chidambaram (2001) PID controller tuning for unstable system by optimization method, *Chem. Eng. Commun.*, 185, 91–113.

Jutan, A. and E.S. Rodriguez (1984) Extensions of a new method of on-line controller tuning, The *Canadian. J. Chem. Eng.*, 62, 802–07.

Kausthubh, S.M. and M. Chidambaram (2000) Control of isothermal crystallizer, *Proc. of CHEMCON*, Calcutta, Dec 18–21.

Kaya, I. and D.P. Atherton (1999a) Using limit cycle data for parameter estimation, *Trans. Inst. Meas. Control*, 21, 21–29.

Kaya, I. and D.P. Atherton (1999b) Parameter estimation from relay auto tuning with asymmetric limit cycle data, *J. Process Control*, 11, 429–39.

Kavdia, M. and M. Chidambaram (1996) On-line tuning of unstable systems, *Comp & Chem. Eng.*, 20, 301–05.

Kookos, I.K., A.I. Lygeros and K.G. Arvanitis (1999) Online PI controller tuning for integrator dead time processes, *European J. of Control*, 5, 19–31.

Kreyszig, E. (1996) Advanced engineering mathematics, 5th ed., *Wiley*, New York.

Krishnaswamy, P.R., G.P. Rangaiah, R.K. Jha and P.B. Deshpande (1990) When to use cascade control, *Ind. Eng. Chem. Res.*, 29, 2163–166.

Kwak, H.J., S.W. Sung and I-B. Lee (1997) On-line process identification and autotuning for integrating processes, *Ind. Eng. Chem. Res.*, 36(12), 5329–338.

Lee, J. (1989) On-line controller tuning from a single closed loop test, *AIChE J.*, 35, 329–31.

Lee, T.H., Q.G. Wang and K.K. Tan (1996) Robust Smith predictor controller for uncertain delay systems, *AIChE J.*, 42(4), 1033–040.

Lee, Y., S.J. Park and M. Lee (1998a) PID controller tuning to obtain closed loop response for cascade control scheme, *Ind. Eng. Chem. Res.*, 37, 1859–865.

Lee, J., S. Park, M. Lee and C. Brosilow (1998b) PID controller tuning for desired closed loop responses for SI/SO systems, *AIChE J.*, 44, 106–15.

Lee, J., S.W. Sung and T.F. Edgar (2007) Integrals of relay feedback responses for extracting process information, *AIChE J.*, 53, 2329–338.

Leva, A. (2005) Auto tuning process controller with improved load disturbance rejection, *J. Process Control*, 15, 223–34.

Li, W., E. Eskinat and W.L. Luyben, (1991) An improved auto tune identification method, *Ind. Eng. Chem. Res.*, 30, 1530–541.

Lin, J.S., S.S. Jang, S.S. Shieh and M. Subramaniam (1999) Generalized multivariable dynamic artificial neural network modeling for chemical processes, *Ind. Eng. Chem.*, 38, 4700–711.

Liu, T., W. Zhang and D. Gu (2005) IMC based control strategy for open-loop unstable cascade processes, *Ind. Eng. Chem. Res.*, 44(4), 900–09.

Liu, T., F. Gao and Y. Wang (2008) A systematic approach for on-line identification of second order process model from relay feedback test, *AIChE J.*, 54, 1560–578.

Liu, T. and F. Gao (2009) A generalized relay identification method for time delay and non-minimum phase processes, *Automatica*, 45 ,1072–079.

Luyben, W.L. (1987) Derivation of transfer function for highly nonlinear distillation columns, *Ind. Eng. Chem. Res.*, 26, 2490–495.

Luyben, W.L. and M.L. Luyben (1997) Essentials of process control. *McGraw-Hill, International Edition*, New York.

Luyben, W.L. (2001) Getting more information from relay feedback tests, *Ind. Eng. Chem. Res.*, 40, 4391–402.

Majhi, S. and D.P. Atherton (1999) Auto tuning and controller design for processes with small time delay, *IEE Proc.-CTA*, 146, 415–25.

Majhi, S. and D. Atherton (2000) Online tuning of controllers for an unstable FOPTD processes, *IEE Proc.-CTA*, 147, 421–27.

Majhi, S. (2007) Relay based identification of a class of non-minimum phase SISO processes, *IEEE Trans Automatic Control*, 52(1), 134–39.

Manoj, K.J. and M. Chidambaram (2001) PID controller tuning for unstable systems by optimization method, *Chem. Eng. Comm.*, 185, 91–113.

Marchetti, G., C. Scali and D.R. Lewin (2001) Identification and control of open loop unstable processes by relay methods, *Automatica*, 37, 2049–055.

Marlin, T.E. (1995) Process control: Designing process and control systems for dynamic performance, *McGraw-Hill*, New York.

McNeill, G.A. and J.D. Sacks (1969) High performance column control, *Chem. Eng. Prog.*, 65(3), 33–39.

Morari, M., Zafiriou, E. and B.R. Holt (1987) Design of resilient processing plants: New characterization of the effect of RHP zeros. *Chemical Engineering Sci.*, 42, 2425–428.

Niederlinski, A. (1971) A heuristic approach to the design of linear multivariable interacting control systems, *Automatica*, 7, 691–701.

O'Dwyer, A. (2003) Handbook of PI and PID tuning rules, *Imperial College Press*, London.

Padmasree, R. and M. Chidambaram (2001) Identification of transfer function model for unstable system with a zero, Proc. *International Conference on Control, Instrumentation and Information Communication (Ed.)* A.K. Datta, A. Deb and S. Sengupta, *Kolkata, Dec* 13–15, 37–40.

Padmasree, R. and M. Chidambaram (2003) A simple method of designing PI controllers for stable inverse response systems, *J. Indian. Inst. Science*, 83, 73–85.

Padmasree, R., M.N. Srinivas and M. Chidambaram (2004) A simple method of designing PID controllers for stable and unstable systems, *Comp. & Chem. Eng.*, 28, 2201–218.

Padmasree, R. and M. Chidambaram (2005) Control of unstable systems, *Narosa Publishers*, New Delhi.

Panda, R.C. and C.C. Yu (2005) Shape factor of relay response curves and its use in autotuning, *J. Process Control*, 15, 893–906.

Park, J.H., S.W. Sung and I-B. Lee (1997) Improved relay auto-tuning with static load disturbance, *Automatica*, 33, 711–15.

Pedret, C., R. Vilanova, R. Moreno and I. Serra (2002) A refinement procedure for PID controller tuning, *Comp. & Chem. Eng.*, 26(6), 903–08.

Pomerleau, A., D. Hodouin, A. Desbiens and E. Gagnon (2000) A survey of grinding circuit control methods: From decentralized PID controllers to multivariable predictive controllers, *Powder Technology*, 108, 103–15.

Pomerleau, D. and A. Pomerleau (2001) Guidelines for the tuning and the evaluation of decentralized and decoupling controllers for processes with recirculation, *ISA Trans.*, 40, 341–51.

Pottmann, M., M.A. Henson, B.A. Ogunnaike and J.S. Schwaber (1996) A parallel control strategy abstracted from the baroreceptor reflex, *Chem. Eng. Sci.*, 51(6), 931–45.

Ramakrishnan, V. and M. Chidambaram (2003) Estimation of a SOPTD transfer function model using a single asymmetric relay feedback test, *Comp. & Chem. Eng.*, 27, 1779–764.

Rangaiah, G.P. and P.R. Krishnaswamy (1994) Estimating second order plus dead time model parameters, *Ind & Engg. Chem. Res.*, 33, 1867–871.

Rao. A.S. and M. Chidambaram (2007) Control of unstable processes with two RHP poles, a zero and time delay, *Asia Pacific Jl. Chem. Engg.*, 1, 63–69.

Reddy, P.D.S, M. Pandit and M. Chidambaram (2006) Comparison of multivariable controllers for non-minimum phase systems, *Int. Jl. Modeling and Simulation*, 26, 237–43.

Rosenbrock, K.H. (1962) The control of distillation columns, *Trans. Inst. Chem. Engrs.*, 40, 35.

Rotstein, G.E. and D.R. Lewin (1991) Simple PI and PID tuning of open loop unstable systems, *Ind. Eng. Chem. Res.*, 30, 1864–869.

Russo, L.P. and B.W. Bequette (1995) Impact of process design on the multiplicity behaviour of a jacketed exothermic CSTR, *AIChE J.*, 41, 135.

Rivera, D.E., M. Morari and S. Skogestad (1986) Internal model control, PID controller design, *Ind. Eng. Chem. Proc. Res. Dev.*, 25, 252–65.

Sadasivarao, M.V and M. Chidambaram (2003) Multivariable controller tuning by genetic algorithms, *Proc. ICECON'03*, Truchirappalli, 4–6th Dec.

Saraf, V., F. Zhao and B.W. Bequette (2003) Relay autotuning of cascade controlled open loop unstable reactors, *Ind. Eng. Chem. Re.*, 42, 4488–494.

Scali, C. and A. Rachid (1998) Analytical design of proportional–integral-derivation controllers for inverse response processes, *Ind. Eng. Chem. Res.*, 37, 1372–379.

Scali, C., G. Marchetti and D. Semino (1999) Relay with additional delay for identification and autotuning of completely unknown process, *Ind. Eng. Chem. Res.*, 38, 1987–997.

Semino, D. and C. Scali (1996) Multiloop autotuning using relay feedback: Limits and extensions, *Comp. & Chem. Eng.*, 20, 907–12.

Semino, D. and C. Scali (1998) Improved identification and auto tuning of PI controllers for MIMO processes by relay techniques, *J. Process Control*, 8, 219–27.

Shen, S-H. and C.C. Yu (1990) Selection of secondary measurement for parallel cascade control, *AIChE J.*, 36, 1267–271.

Shen, S.H. and C.C. Yu (1994) Use of relay feedback test for automatic tuning of multivariable systems, *AIChE J.*, 40, 627–45.

Shen, S.H., J.S. Wu and C.C. Yu (1996a) Use of biased relay feedback for system identification, *AIChE J.*, 42, 1174–180.

Shen, S.H., H.D. Yu and C.C. Yu (1996b) Use of saturation relay feedback for autotune identification, *Chem. Eng. Sci.*, 51, 1187–198.

Shen, S.H., J.S. Wu and C.C. Yu (1996c) Autotune identification under load disturbance, *Ind. Eng. Chem. Res.*, 35, 1642–651.

Skogestad, S. and I. Postlethwaite (1996) Multivariable feedback control, *John Wiley & Sons*, New York.

Skogestad, S. (2003) Simple analytic rules for model reduction and PID controller tuning, *J. Process control*, 13, 291–309.

Smith, C.A. and A.B. Corripio (2001) Principles and practice of automatic process control, *McGraw Hill*, New York.

Srinivasan, K. and M. Chidambaram (2003) Modified relay feedback method for improved system identification, *Comp. & Chem. Eng.*, 27, 727–32.

Srinivasan, K. and M. Chidambaram (2004) An improved auto-tune identification method, *Chem. Biochem. Eng.*, Q18, 249–56.

Stein, G. (2003) Respect the unstable, *IEEE Control Sys. Mag.*, 23, 12–25.

Stephanopoulos, G. (1984) Chemical process control. *Prentice-Hall, Englewood Cliffs*, New Jersey.

Sung, S.W. and I.B. Lee (1996) Limitations and countermeasures of PID controllers, *Ind. Eng. Chem. Res.*, 35, 2596–610.

Syrcos, G. and I.K. Kookos (2005) PID controller tuning using mathematical programming, *Chemical Eng. and Proc.*, 44, 41–49.

Szita, G. and C.K. Sanathanan (2000) A model matching approach for designing decentralized MIMO controllers, *J. Franklin. Inst.*, 337, 641–60.

Tan, K.K., Q-G. Wang, C.C. Hang and T. Hagglund (1999) Advances in PID Control, *Springer*, Berlin.

Tan, K.K., T.H. Lee and R. Ferdous (2000) Simultaneous online automatic tuning of cascade control for open loop stable processes, *ISA Transactions*, 39, 233–42.

Tanttu, J.T. and J. Lieslehto (1991) A comparative study of some multivariable PI controller tuning methods, in *Intelligent Tuning and Adaptive Control Methods,* edited by R. Devanathan, *Pergamon Press*, Oxford, 357–62.

Thyagarajan, T. and C.C. Yu (2003) Improved auto tuning using shape factor from relay feedback, *Ind. Eng. Chem. Res.*, 42, 4425–440.

Tian, Y-C. and F. Gao (1999) Control of integrator processes with dominant time delay, *Ind. Eng. Chem. Res.*, 38, 2979–983.

Toscano, R. (2005) A simple robust PI/PID controller design via numerical optimization approach, *J. Process control*, 15, 81–88.

Venkatashankar, V. and M. Chidambaram (1994) Design of P and PI controllers for unstable first order plus time delay systems, *Int. J. Control*, 49, 137–44.

Vlachos, C., D. William and J.B. Gomm (1999) Genetic approach to decentralized PI controller tuning for multi variable processes, *IEEE Proc.-Control Theory Appl.*, 146, 58–64.

Visioli, A. (2001) Optimal tuning of PID controllers for integrator and unstable processes, *IEE Proc.-CTA, 148,* 180–84.

Visioli, A. (2006) Practical PID control, *Springer-Verlag*, London.

Wang, Q-G., B. Zou, T-H. Lee and Q. Bi (1997) Auto-tuning of multivariable PID controller from decentralized relay feedback, *Automatica,* 33, 319–30.

Wang, Q-G., H.W. Fung and V. Zhang (1999a) PID tuning with exact gain and phase margins, *ISA transactions*, 38, 243–49.

Wang, Q-G., H.W. Fung, Q. Bi and Y. Zhang (1999b) PID tuning for improved performance, *IEEE Trans. Control Sys. Tech.*, 7, 457–65.

Wang, Y.G. and H.H. Shao (1999) PID auto tuner based on gain and phase margin specifications, *Ind. Eng. Chem. Res.,* 38, 3007–12.

Wang, Y.G. and H.H. Shao (2000) Optimal tuning for PI controller, *Automatica*, 36, 147–52.

Wang, Q-G., C.C. Hang and Y. Zhang (2001) Single loop controller design via IMC principles, *Automatica*, 37, 2041–048.

Wang, Q-G., Y. Zhang, M-S. Chiu (2002) Decoupling internal model control for multivariable systems with multiple time delays, *Chem. Engg. Sci.*, 57, 115–24.

Wang, Y.G. and W-J. Cai (2002) Advanced proportional–integral-derivative tuning for integrating and unstable processes with gain and phase margin specifications, *Ind. Eng. Chem. Res.*, 41, 2910–914.

Wang Q-G., T.H. Lee and C. Lin (2003) Relay feedback analysis: Identification and control, *Springer*, Singapore.

Wang, L.W. and S.H. Hwang (2003) On the identification and model reduction of processes under unknown load disturbances, *Ind. Eng. Chem. Res.*, 42, 1905–913.

Wang, L-W. and S-H. Hwang (2005) Identification and control for unstable processes of three dynamic types, *Chem. Eng. Comm.*, 192, 34–61.

Westerlund, T., H. Toivonen and K.E. Nyman (1980) Stochastic modeling and self tuning control of a continuous cement raw material mixing system, *Model Identification & Control*, 1, 17–37.

Wood, R.K. and M.W. Berry (1973) Terminal composition control of binary distillation column, *Chem. Eng. Sci.*, 28, 1707–717.

Wu, W-T., C.G. Tseng and Y.T. Chiu (1994) System Identification and on-line robust control of a multivariable system, *Int. J. Systems Sci.,* 25, 423–39.

Yu, C-C. (1988) Design of parallel cascade control for disturbance rejection, *AIChE J.*, 34, 1833–838.

Yu, C-C. (1999) Auto tuning of PID controllers: Relay feedback approach, *Springer*, London. [2nd Ed., 2006]

Yuwana, M. and D.E. Seborg (1982) A new method for on-line controller tuning, *AIChE J.*, 28, 434–40

Zhang, W., X. Xu and Y. Sun (1999) Quantitative performance design for integrating processes with time delay, *Automatica*, 35, 719–23.

Ziegler, J.G. and N.B. Nichols (1942) Optimum settings for automatic controller, *Trans ASME*, 64, 759–65.

Index